POPULAR LECTURES IN MATHEMATICS

Editors: I. N. SNEDDON AND M. STARK

VOLUME 12

Mathematical Problems and Puzzles from the Polish Mathematical Olympiads

TITLES IN THE POPULAR LECTURES IN MATHEMATICS SERIES

Vol. 1 *The Method of Mathematical Induction*
By I. S. Sominskii

Vol. 2 *Fibonacci Numbers*
By N. N. Vorob'ev

Vol. 3 *Some Applications of Mechanics to Mathematics*
By V. A. Uspenskii

Vol. 4 *Geometrical Construcions Using Compasses Only*
By A. N. Kostovskii

Vol. 5 *The Ruler in Geometrical Constructions*
By A. S. Smogorzhevskii

Vol. 6 *Inequalities*
By P. P. Korovkin

Vol. 7 *One Hundred Problems in Elementary Mathematics*
By H. Steinhaus

Vol. 8 *Complex Numbers and Conformal Mappings*
By A. Markushevich

Vol. 9 *The Cube Made Interesting*
By A. Ehrenfeucht

Vol. 10 *Mathematical Games and Pastimes*
By A. P. Domoryad

Vol. 11 *A Selection of Problems in the Teory of Numbers*
By W. Sierpiński

Vol. 12 *Mathematical Problems and Puzzles from the Polish Mathematical Olympiads*
By S. Straszewicz

Mathematical Problems and Puzzles

from the Polish Mathematical Olympiads

BY

S. STRASZEWICZ

TRANSLATED FROM POLISH BY

J. SMÓLSKA

PERGAMON PRESS

OXFORD · LONDON · EDINBURGH · NEW YORK
PARIS · FRANKFURT

PWN—POLISH SCIENTIFIC PUBLISHERS
WARSZAWA

Pergamon Press Ltd., Headington Hill Hall, Oxford
4 & 5 Fitzroy Square, London W.1

Pergamon Press (Scotland) Ltd., 2 & 3 Teviot Place, Edinburgh 1

Pergamon Press Inc., 122 East 55th St., New York 22, N.Y.

Gauthier-Villars, 55 Quai des Grands-Augustins, Paris 6

Pergamon Press GmbH, Kaiserstrasse 75, Frankfurt-am-Main

First English edition 1965

Library of Congress Catalog Card Number 63-22366

This book is a translation of the original *Zadania z olimpiad matematycznych,* Vol. I, published by Państwowe Zakłady Wydawnictw Szkolnych, 1960

Printed in Poland (DRP)

CONTENTS

PREFACE

THIS book is a translation of the second Polish edition, published in 1960, in which various improvements were made.

The contest for secondary school pupils known as the Mathematical Olympiad has been held in Poland every year since 1949/50. It is organized by the Polish Mathematical Society under the supervision of the Ministry of Education, which provides the necessary financial means. Direct control of the contest is in the hands of a Central Committee in Warsaw and of Regional Committees in seven university towns. In each committee university professors collaborate with secondary school teachers.

Participation in the contests is voluntary. They are open to all secondary school pupils in the country but in practice candidates for the Olympiad are recruited from the two senior forms (ages: 16 to 18).

The Olympiad comprises three stages.

Stage one (preparatory) lasts throughout October, November and December. At the beginning of each month the Central Committee sends out to all secondary schools in the country a set of problems which pupils are expected to solve individually, working at home, within the months in question. No check is made to see whether they work entirely on their own. The pupils' solutions are then mailed by their school to the appropriate regional committee, where they are assessed. The authors of the best solutions are then admitted to the second stage (regional) contests. These take place in March—on the same day at all the seven regional centres, and the same problems are set at each. The contestants have to come up in person to their nearest regional centre. The contest lasts two days. Each day three problems have to be solved in the allotted time of four hours. The candidates work in one room under the supervision of members of the local regional committee; they are not allowed to communicate with one another or to receive help from the professors. Their chance to discuss the problems with the professors comes at the social gatherings organized for this purpose after the contest.

The pupils who have produced the best papers are admitted

to the third stage (final) contest, which takes place in Warsaw in the month of April and is organized along the same lines as the regional contest. The papers are assessed by the Central Committee and the authors of the best of them are awarded prizes.

The problems set at the contests require only a knowledge of school mathematics (i.e. elementary algebra, geometry and trigonometry) but are on the whole more difficult than the usual school exercises. Their degree of difficulty, however, is not uniform, for it is considered desirable that not only the most gifted pupils but also those of average ability should —with a certain effort— manage to solve some of the problems and gain a number of points at the contests.

This book contains the problems set at the first five Olympiads. It has been prepared in order to provide secondary school pupils with suitable topics to be worked out on their own, individually or collectively. It aims at extending their knowledge of mathematics and training them in mathematical thinking. Accordingly, the solutions of most of the problems have been given in an extended form, the readers' attention being drawn to various details of the reasoning. In addition, several problems have been provided with commentaries containing generalizations or further development of the topics in question, including various supplementary data of elementary mathematics outside the scope of the school syllabus. These commentaries are given separately, in the form of remarks following the solutions. This arrangement of the contents has been adopted in order to make the book easy to read even for less advanced pupils, who can skip over the material contained in the remarks. More advanced readers, however, will find in them instructive examples of mathematical reasoning, which may stimulate their own initiative in posing problems and seeking solutions. In most cases several solutions of the same problem have been given, which might perhaps seem superfluous. The aim, however, has been to show that the solution of a problem may result from different mental associations and be obtained through different processes of reasoning. It will be observed that several of the solutions given in the book are due to the pupils themselves. Thus for instance problem 15 was solved by means of geometrical illustration (method III) by one of the participants in the Third Olympiad and the original method III of solving problem 153 was found by a participant in the Fourth Olympiad. It has frequently occurred at the Olympiads that pupils have presented different solutions from those expected by the Committee.

S. Straszewicz

PART ONE

Arithmetic and Algebra

PROBLEMS

§ 1. Integers

1. Find two natural numbers a and b given their greatest common divisor $D = 12$ and their least common multiple $M = 432$. Give the method of finding solutions in the general case.

2. Prove that a sum of natural numbers is divisible by 9 if and only if the sum of all the digits of those numbers is divisible by 9.

3. Prove that the number which in the decimal system is expressed by means of 91 unities is a composite number.

4. Prove that if an integer a is not divisible either by 2 or by 3, then the number a^2-1 is divisible by 24.

5. Find when the sum of the cubes of three successive natural numbers is divisible by 18.

6. Prove that if the sum of three natural numbers is divisible by 3, then the sum of the cubes of those numbers is also divisible by 3.

7. Show that if n is an integer, then n^3-3n^2+2n is divisible by 6.

8. Prove that number $2^{55}+1$ is divisible by 11.

9. Prove that the sum of two successive natural numbers and the sum of their squares are relatively prime.

10. Prove that every odd prime number p can be represented as the difference of the squares of two natural numbers, and that this can be done in one way only.

11. What digits should be put instead of zeros in the third and the fifth places in number 3000003 in order to give a number divisible by 13?

12. Prove that if a natural number n is greater than 4 and is not a prime, then the product of the successive natural numbers from 1 to $n-1$ is divisible by n.

13. Prove that if n is an even natural number, then number 13^n+6 is divisible by 7.

14. Prove that among ten successive natural numbers there are always at least one and at most four numbers that are not divisible by any of the numbers 2, 3, 5, 7.

15. Prove that none of the digits 2, 4, 7, 9 can be the last digit of a number equal to the sum

$$1+2+3+\ldots+n$$

where n is an arbitrary natural number.

§ 2. Polynomials, Algebraic Fractions, Irrational Expressions

16. Factorize the expression

$$(x+a)^7-(x^7+a^7).$$

17. Factorize the expression

$$W = x^4(y-z)+y^4(z-x)+z^4(x-y).$$

18. Factorize the expression

$$W = a^4(b^2-c^2)+b^4(c^2-a^2)+c^4(a^2-b^2).$$

19. Factorize the expression

$$a(b-c)^3+b(c-a)^3+c(a-b)^3.$$

20. Determine p and q such that the trinomial x^4+px^2+q is divisible by a given trinomial x^2+ax+b.

21. Factorize the polynomial

$$x^8+x^4+1$$

into factors of at most the second degree.

22. Prove that the polynomial

$$x^{44}+x^{33}+x^{22}+x^{11}+1$$

is divisible by the polynomial $x^4+x^3+x^2+x+1$.

23. Prove that if

$$\frac{a-b}{1+ab}+\frac{b-c}{1+bc}+\frac{c-a}{1+ca} = 0,$$

then at least two of the numbers a, b, c are equal.

24. Prove that if

$$\frac{1}{ab}+\frac{1}{bc}+\frac{1}{ca} = \frac{1}{ab+bc+ca},$$

then two of the numbers a, b, c are opposite numbers.

25. Find the least value of the fraction

$$\frac{x^4+x^2+5}{(x^2+1)^2}.$$

26. Find numbers a, b, c, d for which the equation

$$\frac{2x-7}{4x^2+16x+15}=\frac{a}{x+c}+\frac{b}{x+d} \tag{1}$$

would be an identity.

27. Find $x^{13}+\frac{1}{x^{13}}$ knowing that $x+\frac{1}{x}=a$, where a is a given number.

28. Prove that $\sqrt{2}$, $\sqrt{3}$, $\sqrt{5}$ cannot be terms of the same arithmetical progression.

29. Prove that if $1\leqslant a\leqslant 2$, then we have

$$\sqrt{[a+2\sqrt{(a-1)}]}+\sqrt{[a-2\sqrt{(a-1)}]}=2.$$

30. The expression

$$y=\sqrt{(x-1)}+\sqrt{[x+24-10\sqrt{(x-1)}]}$$

has a constant value in a certain interval. Find that interval.

31. Prove that if n is a natural number, then we have

$$(\sqrt{2}-1)^n=\sqrt{m}-\sqrt{(m-1)},$$

where m is a natural number.

§ 3. Equations

32. Prove that if

$$a(y+z)=x,\quad b(z+x)=y,\quad c(x+y)=z \tag{1}$$

and at least one of the numbers x, y, z is not equal to zero, then we have

$$ab+bc+ca+2abc-1=0. \tag{2}$$

33. Solve the equation

$$|x|+|x-1|+|x-2|=a,$$

where a denotes a given positive number.

34. A motor-boat set off up the river at 9 o'clock and at the same time a ball was dropped from the motor-boat into the river. At 9.15 the motor-boat turned round and travelled down the river. At what time did it overtake the ball?

35. Find for what value of the parameter m the sum of the squares of the roots of the equation

$$x^2+(m-2)x-(m+3)=0$$

has the least value.

36. Prove that the equation

$$(x-a)(x-c)+2(x-b)(x-d)=0, \qquad (1)$$

in which $a<b<c<d$, has two real roots.

37. Prove that if the equations

$$x^2+mx+n=0 \quad \text{and} \quad x^2+px+q=0 \qquad (1)$$

have a common root, then the following relation holds between the coefficients of these equations:

$$(n-q)^2-(m-p)(np-mq)=0. \qquad (2)$$

38. Find the condition which must be satisfied by the coefficients of the trinomials

$$x^2+mx+n \quad \text{and} \quad x^2+px+q$$

for a root of each of these trinomials to lie between the roots of the other trinomial. The letters m, n, p, q denote real numbers.

39. Determine the coefficients of the equation x^3-ax^2+ $+bx-c=0$ in such a way as to have numbers a, b, c as the roots of this equation.

40. Find the necessary and sufficient conditions which must be satisfied by the real numbers a, b, c for the equation

$$x^3+ax^2+bx+c=0$$

to have three real roots forming an arithmetical progression.

41. Prove that the equation

$$\frac{m^2}{a-x}+\frac{n^2}{b-x}=1,$$

where $m\neq 0$, $n\neq 0$, $a\neq b$, has two real roots (m, n, a, b denote real numbers).

42. How many real roots has the equation

$$\frac{x^2}{x^2-a^2}+\frac{x^2}{x^2-b^2}=4$$

(a, b—real numbers)?

43. Solve the equation

$$\frac{1}{a}+\frac{1}{b}+\frac{1}{x}=\frac{1}{a+b+x}$$

(a, b — real numbers $\neq 0$).

44. Find whether the equation

$$\frac{1}{x-a}+\frac{1}{x-b}+\frac{1}{x-c}=0,$$

where a, b, c denote given real numbers, has real roots.

45. Give the conditions under which the equation

$$\sqrt{(x-a)}+\sqrt{(x-b)}=\sqrt{(x-c)}$$

has roots, assuming that there are no equal numbers among the numbers a, b, c.

46. The lengths of the sides of a right-angled triangle are natural numbers. The length of one of the perpendicular sides is 10. Find the remaining sides of this triangle.

47. Find the integral solutions of the equation

$$y^3-x^3=91.$$

48. Solve the system of equations

$$x^2-yz=3, \quad y^2-zx=4, \quad z^2-xy=5.$$

49. Solve the system of equations

$$x^2+x+y=8, \quad y^2+2xy+z=168, \quad z^2+2yz+2xz=12480.$$

50. Solve the system of equations

$$xy=ax+by, \quad yz=ay+bz, \quad zx=az+bx,$$

where a and b denote given real numbers.

51. Solve the system of equations

$$xy(x-y)=ab(a-b), \quad x^3-y^3=a^3-b^3.$$

52. Solve the system of equations

$$x^2+y^2+z^2=19^2, \quad \frac{1}{x}+\frac{1}{y}+\frac{1}{z}=0, \quad x-y+z=11.$$

53. Solve the system of equations

$$x_1x_2=1, \quad x_2x_3=2, \quad x_3x_4=3, \quad \ldots, \quad x_nx_1=n.$$

§ 4. Inequalities

54. Somebody has a balance with arms of unequal length. In order to weigh 2 kg of sugar he proceeds as follows: he puts a 1 kg weight on the left-hand scale and pours sugar on the right-hand one until the two scales are balanced; emptying the two scales, he then puts the 1 kg weight on the right-hand scale and pours sugar on the left-hand one until the two scales are balanced.

Are the two quantities of sugar taken together less than 2 kg, more than 2 kg, or exactly 2 kg?

55. Show that for any numbers a, b, c the following inequality holds:

$$a^2+b^2+c^2 \geqslant ab+bc+ca.$$

56. Prove that for any numbers a, b, c, x, y the following inequality holds:

$$(a^2+b^2+c^2)(x^2+y^2+z^2) \geqslant (ax+by+cz)^2.$$

57. Prove that if a and b are the perpendicular sides of a right-angled triangle, c its hypotenuse and n a number greater than 2, then

$$a^n+b^n < c^n.$$

58. Prove that if $m > 0$, then

$$m+\frac{4}{m^2} \geqslant 3.$$

59. Prove that if $a > b > 0$, then

$$a+\frac{1}{(a-b)b} \geqslant 3.$$

60. Prove that if the sum of positive numbers a, b, c is equal to 1, then

$$\frac{1}{a}+\frac{1}{b}+\frac{1}{c} \geqslant 9.$$

61. Prove that if $a>0$, $b>0$, $c>0$, then the following inequality holds:

$$ab(a+b)+bc(b+c)+ca(c+a) \geqslant 6abc.$$

62. Prove that if $u > 0$, $v > 0$, $w > 0$, then the following inequality holds:

$$u^3+v^3+w^3 \geqslant 3uvw.$$

63. Prove that if $x > 0$, $y > 0$ and $x+y = 1$, then

$$\left(1+\frac{1}{x}\right)\left(1+\frac{1}{y}\right) \geqslant 9.$$

64. Prove that if $a < b < c < d$, then

$$(a+b+c+d)^2 > 8(ac+bd).$$

65. Prove that if n is an integer greater than 2, then

$$2^{\frac{1}{2}n(n-1)} > n!.$$

SOLUTIONS

§ 1. Integers

1. If the greatest common divisor of numbers a and b is 12, then

$$a = 12x, \quad b = 12y,$$

x and y being relatively prime natural numbers. Consequently, the least common multiple of numbers a and b is $12xy$, and therefore

$$12xy = 432, \quad \text{i.e.} \quad xy = 36.$$

Conversely, if $xy = 36$ and numbers x, y are relatively prime, then $12x$ and $12y$ give a solution of the problem.

We shall thus find numbers x and y by decomposing 36 into the product of two relatively prime factors. There exist two such decompositions:

$$36 = 1 \times 36 \quad \text{and} \quad 36 = 4 \times 9.$$

We obtain two solutions:

$$x = 1, \quad y = 36 \quad \text{or} \quad x = 4, \quad y = 9.$$

The required numbers are:

$$12 \times 1 = 12 \quad \text{and} \quad 12 \times 36 = 432$$

or

$$12 \times 4 = 48 \quad \text{and} \quad 12 \times 9 = 108.$$

In the general case, given the greatest common divisor D and the least common multiple M of numbers a and b (M and D being natural numbers), we reason in the same way and obtain the equations

$$a = Dx, \quad b = Dy \quad (x \text{ and } y \text{ being relatively prime}),$$

whence

$$Dxy = M \quad \text{and} \quad xy = \frac{M}{D}.$$

A solution exists provided M is divisible by D. If this condition is satisfied, the problem has as many solutions as there are ways in which the natural number M/D can be decomposed into two relatively prime factors x and y. Numbers D and M always constitute one of the solutions.

2. (a) Let us first prove that the difference between a natural number a and the sum of its digits is divisible by 9. Let C_0, $C_1, C_2, C_3, \ldots$ denote, in succession, the digits indicating unities, tens, hundreds, thousands and further orders of the number in question. Then

$$a = C_0+C_1\times 10+C_2\times 100+C_3\times 1000+\ldots,$$
$$s = C_0+C_1+C_2+C_3+\ldots$$

Subtracting these equalities, we obtain

$$a-s = C_1\times 9+C_2\times 99+C_3\times 999+\ldots$$

Since each component on the right-hand side is divisible by 9, number $a-s$ is also divisible by 9.

(b) Let $a_1, a_2, \ldots, a_n$ denote natural numbers and $s_1, s_2, \ldots, s_n$ the respective sums of their digits.

In the identity

$$a_1+a_2+\ldots+a_n$$
$$= [(a_1-s_1)+(a_2-s_2)+\ldots+(a_n-s_n)]+(s_1+s_2+\ldots+s_n)$$

the component of the right-hand side which is contained in the square brackets is divisible by 9 because, according to (a), it is the sum of numbers which are divisible by 9. Consequently, number $a_1+a_2+\ldots+a_n$ is divisible by 9 if and only if the component $s_1+s_2+\ldots+s_n$, i.e. the sum of all the digits of numbers $a_1, a_2, \ldots, a_n$, is divisible by 9, which is what we were to prove.

3. Since $91 = 7\times 13$, the digits of the given number can be divided into 13 groups containing 7 unities each:

$$1111111 \quad 1111111 \quad \ldots \quad 1111111.$$

This shows that the number in question is divisible by 1111111; the quotient is

$$1\ 0000001 \ \ldots \ 0000001.$$

More precisely, let $1111111 = N$. The given number can be represented as the sum of 13 components,

$$L = N\times 10^{84}+N\times 10^{77}+\ldots+N\times 10^{7}+N$$

corresponding to the 13 groups of digits. By factorizing, we obtain

$$L = N\times(10^{84}+10^{77}+\ldots+10^{7}+1), \tag{1}$$

which shows that the given number is the product of two natural numbers different from 1, i.e. it is not a prime number.

Another factorization of number L will be obtained by dividing its digits into 7 groups containing 13 unities each; it will then be observed that number L is divisible by the number

$$M = 1111111111111.$$

We can also reason in the following way:

$$L = 1+10+10^2+\ldots+10^{90}.$$

Number L is thus the sum of 91 terms of a geometrical progression having 1 as its first term and 10 as the common ratio. According to the well-known formula for the sum of the terms of a geometrical progression, we have

$$L = \frac{10^{91}-1}{10-1}.$$

Hence

$$L = \frac{(10^7)^{13}-1}{10-1} = \frac{(10^7)^{13}-1}{10^7-1} \times \frac{10^7-1}{10-1}. \tag{2}$$

Since for arbitrary $a \neq b$ and a natural n

$$\frac{a^n-b^n}{a-b} = a^{n-1}+a^{n-2}b+\ldots+ab^{n-2}+b^{n-1},$$

we find by substituting in the above formula first $a = 10^7$, $b = 1$, $n = 13$ and then $a = 10$, $b = 1$, $n = 7$ that each factor on the right-hand side of formula (2) is equal to a natural number different from 1. The factors are the same as in (1).

Analogously

$$L = \frac{(10^{13})^7-1}{10-1} = \frac{(10^{13})^7-1}{10^{13}-1} \times \frac{10^{13}-1}{10-1}.$$

Remark. The theorem proved above is a particular case of the following theorem. *If all the digits of number L written in a certain positional (not necessarily decimal) notation are unities, and the number of those unities is composite, then number L is composite.*

This theorem can easily be proved on the lines of the preceding proof. We leave the proof as an exercise for the reader. The inverse theorem is not true.

If the number of digits of L written by means of unities alone is a prime number, then L can be either a prime number or a composite number. For example, in the decimal system

11 is a prime number while 111 is divisible by 3 and 11111 is divisible by 41; in the binary system 11, 111, 11111, 1111111 are prime numbers while 11111111111 is a composite number.

It is not known yet whether in every sequence of numbers whose digits (in a certain positional system) are all unities there are infinitely many prime numbers; neither do we know whether there are infinitely many composite numbers in every sequence of that kind.

4. *Method I.* $a^2-1 = (a-1)(a+1)$. Either of the numbers $a-1$ and $a+1$ is divisible by 2, and, since their difference is 2, one of these numbers is divisible by 4. Thus their product is divisible by 8. On the other hand, since a is not divisible by 3, either the preceding number, $a-1$, or the following number, $a+1$, is divisible by 3; thus the product $(a-1)(a+1)$ is certainly divisible by 3. And if the number $(a-1)(a+1)$ is divisible by either of the relatively prime numbers 3 and 8, it must be divisible by their product, namely by 24.

Method II. If a is not divisible either by 2 or by 3, then, divided by 6, it leaves the remainder 1 or the remainder 5, i.e. it is of the form $6k+1$ or $6k-1$, k being an integer. If $a = 6k+1$, then

$$a^2-1 = (6k+1)^2-1 = 36k^2+12k+1-1 = 12k(3k+1);$$

since of the two numbers k and $3k+1$ one is divisible by 2, the product $12k(3k+1)$ is divisible by 12×2, i.e. by 24.

Now if $a = 6k-1$, then

$$a^2-1 = (6k-1)^2-1 = 36k^2-12k+1-1 = 12k(3k-1),$$

whence, as before, we conclude that a^2-1 is divisible by 24.

5. *Answer.* The sum of the cubes of three successive natural numbers is divisible by 18 if and only if the first of those numbers is odd.

6. *Hint:* $(a^3+b^3+c^3)-(a+b+c) = (a^3-a)+(b^3-b)+(c^3-c)$ $= (a-1)a(a+1)+(b-1)b(b+1)+(c-1)c(c+1)$.

Generalize the theorem to a sum of n numbers.

7. *Hint:* $n^3-3n^2+2n = n(n-1)(n-2)$.

8. We know that if n is a natural number and a and b are arbitrary numbers, then

$$a^n-b^n = (a-b)(a^{n-1}+a^{n-2}b+a^{n-3}b^2+ \ldots +ab^{n-2}+b^{n-1}).$$

Putting $a = 2^5$, $b = -1$, $n = 11$, we obtain

$$2^{55}+1 = (2^5+1)(2^{50}-2^{45}+2^{40}- \ldots -2^5+1) = 33\times C,$$

where C is an integer. $2^{55}+1$ is thus divisible by 33, and hence by 11.

REMARK 1. In the same manner we can prove in general that number $2^{mn}+1$, where n is an odd number, is divisible by 2^m+1.

REMARK 2. The above problem, and many other problems concerning divisibility of numbers, can be solved quickly by using the properties of congruences.

Two integers a and b are said to be *congruent modulo k* (k being a natural number) if the difference $a-b$ is divisible by k; this is expressed by the formula

$$a \equiv b \pmod{k},$$

which is called a *congruence*.

We can also say that two numbers are congruent modulo k if they leave the same remainder when divided by k.

From the definition of congruence we can immediately draw the following conclusions:

(1) $a \equiv a \pmod{k}$ for any integer a and any natural number k.
(2) If $a \equiv b \pmod{k}$, then $b \equiv a \pmod{k}$.
(3) If $a \equiv b \pmod{k}$ and $b \equiv c \pmod{k}$, then $a \equiv c \pmod{k}$.
(4) If $a \equiv b \pmod{k}$ and $c \equiv d \pmod{k}$, then $a+c \equiv b+d \pmod{k}$

The proofs of theorems 1–4 present no difficulty: the reader is invited to carry them out by himself.

(5) If $a \equiv b \pmod{k}$ and $c \equiv d \pmod{k}$, then $ac \equiv bd \pmod{k}$.

To prove this it suffices to observe that $ac-bd \equiv ac-bc+bc-$ $-bd \equiv (a-b)c+(c-d)b$.

Theorem (5) implies theorem

(6) If $a \equiv b \pmod{k}$, then $a^2 \equiv b^2 \pmod{k}$, and generally: $a^n \equiv b^n \pmod{k}$ for any natural n. We reach this last conclusion by induction.

Using congruences we solve problem 8 in the following simple way:

Since $2^5 \equiv -1 \pmod{11}$, we have $(2^5)^{11} \equiv (-1)^{11} \pmod{11}$, whence $(2^5)^{11} \equiv -1 \pmod{11}$, which means that $2^{55}+1$ is divisible by 11.

As an example of the use of congruences, we shall solve one more problem:

Find the last digit of number 2^{1000}.

Since $2^5 \equiv 2 \pmod{10}$, we have

$$2^{1000} \equiv 2^{200} \equiv 2^{40} \equiv 2^8 \equiv 2^5 \times 2^3 \equiv 2 \times 2^3 \equiv 6 \pmod{10}.$$

Thus the last digit of number 2^{1000} is 6.

9. Denote by s the sum of the successive natural numbers n and $n+1$ and by t the sum of their squares:

$$s = n+(n+1) = 2n+1, \quad t = n^2+(n+1)^2 = 2n^2+2n+1.$$

Then

$$s^2 = 4n^2+4n+1 \quad \text{and} \quad 2t-s^2 = 1.$$

The equality obtained shows that every common divisor of s and t is a divisor of unity; consequently, numbers s and t have no common divisor greater than 1, i.e. they are relatively prime.

REMARK. The above theorem can be generalized.

If two natural numbers a and b of different evenness are relatively prime, then the sum $s = a+b$ and the sum of the squares $t = a^2+b^2$ are relatively prime.

Indeed, if numbers s and t were not relatively prime, there would exist a prime number p constituting a common divisor of numbers s and t, and we should have $p > 2$ because if a and b are of different evenness, then numbers s and t are odd. Now

$$2t-s^2 = (a-b)^2,$$

whence p would be a divisor of $a-b$ and consequently also a divisor of number $(a+b)+(a-b) = 2a$ and of number $(a+b)-(a-b) = 2b$. This, however, is impossible because $2a$ and $2b$ have no common divisor greater than 2.

10. Suppose that

$$p = a^2-b^2,$$

where a and b denote natural numbers. Since

$$a^2-b^2 = (a+b)(a-b)$$

and p is prime, one of the factors $a+b$, $a-b$ is equal to p and the other is equal to 1. Since $b > 0$, we have $a+b > a-b$, and we infer that

$$a+b = p, \quad a-b = 1.$$

Solving these equations, we obtain

$$a = \frac{p+1}{2}, \quad b = \frac{p-1}{2}.$$

Since p is odd, a and b are integers. The equality

$$p = \left(\frac{p+1}{2}\right)^2 - \left(\frac{p-1}{2}\right)^2$$

gives the required representation.

11. Denote the required digits by x and y and write the number $30x0y03$ in the form

$$N = 3\times 10^6 + x\times 10^4 + y\times 10^2 + 3.$$

We are to find integers x and y for which N will be divisible by 13 and the following inequalities will be satisfied:

$$0 \leqslant x \leqslant 9, \quad 0 \leqslant y \leqslant 9.$$

Numbers 10^6, 10^4, 10^2 divided by 13 leave the remainders 1, 3 and 9 respectively, whence

$$3\times 10^6 = 13k_1 + 3, \quad x\times 10^4 = 13k_2 + 3x, \quad y\times 10^2 = 13k_3 + 9y,$$

where k_1, k_2, k_3 are natural numbers. In that case

$$N = 13k + 3 + 3x + 9y + 3,$$

i.e.

$$N = 13k + 3(x + 3y + 2),$$

where k is a natural number.

N is divisible by 13 if and only if $x+3y+2$ is divisible by 13, i.e. if x and y satisfy the equation

$$x + 3y + 2 = 13m,$$

where m denotes a natural number.

The inequalities $x \leqslant 9$, $y \leqslant 9$ imply that

$$x + 3y + 2 \leqslant 9 + 3\times 9 + 2, \quad \text{i.e.} \quad x + 3y + 2 \leqslant 38;$$

consequently, m must satisfy the inequality $13m \leqslant 38$, i.e. m must be either 1 or 2.

(i) Taking $m = 1$ we obtain for x and y the equation

$$x + 3y + 2 = 13, \quad \text{i.e.} \quad x = 11 - 3y.$$

With the restrictions $0 \leqslant x \leqslant 9$, $0 \leqslant y \leqslant 9$, this equation has three integral solutions:

$$y = 1, \quad y = 2, \quad y = 3,$$
$$x = 8, \quad x = 5, \quad x = 2.$$

(ii) Taking $m = 2$ we have the equation

$$x + 3y + 2 = 26, \quad \text{i.e.} \quad x = 3(8 - y)$$

and we obtain the solutions:

$$y = 5, \quad y = 6, \quad y = 7, \quad y = 8,$$
$$x = 9, \quad x = 6, \quad x = 3, \quad x = 0.$$

The problem thus has 7 solutions, the corresponding numbers being 3080103, 3050203, 3020303, 3090503, 3060603, 3030703, 3000803.

REMARK. Problem 11 concerned changing the digits of a certain definite number. Let us consider two examples of more general problems of this kind.

I. *We are given a natural number* N *not divisible by a prime number* p *and having the digit* 0 *in the* kth *decimal place. Is it possible to replace the* 0 *by such a digit* x *that the new number will be divisible by* p?

To begin with, it will be observed that if p is equal to 2 or 5, the problem has no solution. Let us therefore suppose that $p \neq 2$ and $p \neq 5$.

If we put the digit x in the kth decimal place of number N, the number obtained, $N(x)$, is expressed by the formula

$$N(x) = N + x \times 10^{k-1}. \tag{1}$$

Consider the values of the function $N(x)$ as x assumes the values $0, 1, 2, \ldots, p-1$

$$N(0),\ N(1),\ N(2),\ \ldots,\ N(p-1). \tag{2}$$

Each of the p numbers of sequence (2) leaves a different remainder when divided by p. Indeed, by formula (1) the difference of two numbers of sequence (2) is expressed by the formula

$$N(i) - N(j) = (i-j) \times 10^{k-1}.$$

If $i \neq j$, this number is not divisible by p because the factor $i-j$ is different from zero and absolutely less than p and the factor 10^{k-1} has only the prime divisors 2 and 5. Accordingly, numbers $N(i)$ and $N(j)$ give different remainders when divided by p.

Since the remainders of the numbers of sequence (2) are all different and there are p such remainders, one and only one of those remainders is equal to 0, i.e. one and only one of the numbers of sequence (2) is divisible by p; suppose that number to be $N(x_0)$.

If $x_0 \leqslant 9$, then x_0 gives the solution of the problem.

If $x_0 > 9$, the problem has no solution.

We conclude that:

(a) If $p = 3$, the problem always has three solutions; the required digits x are either 1, 4, 7 or 2, 5, 8.

(b) If $p = 7$, the problem has either two solutions—the digits 1, 8 or 2, 9, or one solution—one of the digits 3, 4, 5, 6.

(c) If $p \geqslant 11$, the problem can either have one solution or have no solution at all. If $p = 11$, the first case occurs, e.g. for number 10; the second case occurs for number 109 for example.

II. *We are given a natural number N not divisible by a prime number $p \geqslant 11$ and having the digit 0 in the kth and lth decimal places. Is it possible to replace those zeros by such digits x and y that the new number will be divisible by p?*

If we put the digit x in the kth decimal place and the digit y in the lth decimal place of number N, we shall obtain number $N(x, y)$ expressed by the formula

$$N(x, y) = N + x \times 10^{k-1} + y \times 10^{l-1}. \tag{3}$$

Consider those values of the function $N(x, y)$ which are obtained by substituting for x and y the values $0, 1, 2, \ldots, p-1$. We obtain p^2 numbers:

$$\begin{array}{lllll} N(0,0), & N(0,1), & N(0,2), & \ldots, & N(0,p-1), \\ N(1,0), & N(1,1), & N(1,2), & \ldots, & N(1,p-1), \\ \ldots & \ldots & \ldots & \ldots & \ldots \\ N(p-1,0), & N(p-1,1), & N(p-1,2), & \ldots, & N(p-1,p-1). \end{array}$$

We ascertain, as in problem I, that in each row and in each column of this table there is one and only one number divisible by p; we denote such a number in the ith row by the symbol $N(i, y_i)$.

The table thus contains p numbers divisible by p; they are the numbers

$$N(0, y_0),\ N(1, y_1),\ N(2, y_2),\ \ldots,\ N(p-1, y_{p-1}), \tag{4}$$

the sequence of numbers $y_0, y_1, y_2, \ldots, y_{p-1}$ containing the same numbers as the sequence $0, 1, 2, \ldots, p-1$, but not in the same order. Observe for instance that y_0 is certainly not equal to 0 because number $N(0, 0) = N$, by our assumption, is not divisible by p.

Number $N(i, y_i)$ of sequence (4) gives the solution of the problem if and only if $i \leqslant 9$ and $y_i \leqslant 9$. It is easy to count how many such numbers can be contained in sequence (4). They should be sought among the numbers

$$N(0, y_0),\ N(1, y_1),\ \ldots,\ N(9, y_9). \tag{5}$$

If none of the numbers $y_0, y_1, \ldots, y_9$ is greater than 9, the problem has 10 solutions, given by the numbers of sequence (5). That is the greatest possible number of solutions.

If the sequence $y_0, y_1, \ldots, y_9$ includes numbers greater than 9 (there can be at most $p-10$ such numbers), the corresponding terms should be deleted from sequence (5). The problem then has at least $10-(p-10)$, i.e. $20-p$ solutions. Hence the following conclusion:

(a) If $p = 11$, the problem has at least 9 solutions. An example of the maximum number of 10 solution is given by the number 101015.

(b) If $p = 13$, the problem has at least 7 solutions; the case in which there are only 7 solutions has been discussed above.

(c) If $p = 17$, the problem has at least three solutions.

(d) If $p = 19$, the problem has at least one solution.

(e) If $p \geqslant 23$, the problem may have no solution.

A suitable example for the case $p = 23$ can easily be found if we consider that number 10^{22} divided by 23 gives the remainder 1. We suggest this as an exercise.

12. Since n is not prime, there exist natural numbers p and q such that $1 < p < n$, $1 < q < n$ and $n = p \times q$. (For example we can take as p the greatest prime factor of n.)

We shall distinguish two cases.

Case 1. $p \neq q$.

In this case p and q are two different numbers of the sequence $1, 2, \ldots, n-1$, and consequently the product $1 \times 2 \times \ldots \times (n-1)$ is divisible by $n = p \times q$.

Case 2. $p = q$.

In this case $n = p^2$, and since $n > 4$, we have $p^2 > 4$, whence $p > 2$, $p^2 > 2p$, and thus $2p < n$. Numbers p and $2p$ are two different numbers of the sequence $1, 2, \ldots, n-1$, and consequently the product $1 \times 2 \times \ldots \times (n-1)$ is divisible by $p \times 2p = 2n$, and thus of course divisible by n.

REMARK. Given the same assumptions regarding n we can prove more than is required in the problem. We can prove that even the product $1 \times 2 \times \ldots \times (n-3)$ is divisible by n.

For this purpose let us observe that in the equality $n = p \times q$, where p and q are natural numbers greater than 1, none of the numbers p and q can be greater than $n-3$. Indeed, if we had, say, $p > n-3$, then, in view of $q \geqslant 2$, we should obtain $p \times q > 2(n-3)$, whence $n > 2n-6$ and $n < 6$; this, however, is impossible because we have assumed that $n > 4$ and $n \neq 5$.

To prove the proposition given above let us, as before, distinguish two cases:

Case 1. $p \neq q$.

Since $p \leqslant n-3$ and $q \leqslant n-3$, p and q are two different natural numbers of the sequence $1, 2, \ldots, n-3$, and consequently the product $1 \times 2 \times \ldots \times (n-3)$ is divisible by $n = p \times q$.

Case 2. $p = q$.

In this case $n = p^2$. Since $n > 4$, we have $p > 2$, and consequently $p \geqslant 3$. Hence we deduce successively that $p^2 \geqslant 3p$,

$n \geqslant 2p+p$, $n \geqslant 2p+3$ and finally $2p \leqslant n-3$. Numbers p and $2p$ are thus two different numbers of the sequence $1, 2, \ldots, n-3$, and consequently the product $1\times 2\times \ldots \times (n-3)$ is divisible by $p\times 2p = 2n$, and thus of course divisible by n.

It will also be observed that the product $1\times 2\times \ldots \times (n-4)$ is not divisible by n either if $n = 6$ or if $n = 9$.

13. We shall prove that if n is even, then 13^n divided by 7 gives the remainder 1.

Method I. We shall use *induction.*

Let $n = 2k$. If $k = 1$, the theorem holds because

$$13^2 = 169 = 7\times 24+1.$$

Suppose that for a certain k we have the equality

$$13^{2k} = 7m+1,$$

where m is a natural number.

In that case

$$13^{2(k+1)} = 13^{2k}\times 13^2 = (7m+1)(7\times 24+1) = 7\times(169m+24)+1.$$

We conclude that the theorem holds for any natural k.

Method II. We apply the *binomial theorem*

$$(a+b)^n = a^n+\binom{n}{1}a^{n-1}b+\ldots+\binom{n}{i}a^{n-i}b^i+\ldots+b^n.$$

Taking $a = 14$, $b = -1$, we obtain

$$13^n = [14+(-1)]^n = 14^n+\binom{14}{1}\times 14^{n-1}\times(-1)+$$
$$+\ldots+\binom{14}{i}\times 14^{n-i}\times(-1)^i+\ldots+(-1)^n.$$

All the terms on the right side, except the last, are divisible by 7, and $(-1)^n = 1$ because n is even.

Exercise. Prove the theorem by means of congruences.

14. In a sequence consisting of ten successive natural numbers there are five even and five odd numbers. Consequently problem 14 can be reduced to the following problem:

Prove that among five successive odd numbers there are at least one and at most four numbers not divisible by any of the numbers 3, 5 and 7.

It will be observed that in the sequence of odd numbers every third is divisible by three, every fifth by five and every seventh by seven. Consequently:

(a) Among five successive odd numbers there is at least one divisible by 3; thus those numbers can include no more than four numbers that are divisible by none of the numbers 3, 5 and 7. The example of numbers 11, 13, 15, 17, 19, shows that five successive odd numbers can include four numbers which are divisible by none of the numbers 3, 5 and 7.

(b) Among five successive odd numbers there are at most two divisible by 3, at most one divisible by 5 and at most one divisible by 7. It follows that at least one of those five numbers is divisible by none of the numbers 3, 5 and 7.

15. *Method I.* It will be observed that if the final digits of numbers a and b are c and d, then

$$ab = (10k+c)(10l+d) = 10(10kl+kd+cl)+cd,$$

where k and l are integers; thus the last digit of the product ab is the same as the last digit of the product cd.

We know that

$$1+2+3+ \dots +n = \frac{n(n+1)}{2}.$$

If n ends with the digit 0, 1, 2, 3, 4, 5, 6, 7, 8, 9,
then $n+1$ ends with the digit 1, 2, 3, 4, 5, 6, 7, 8, 9, 0,
and thus $n(n+1)$ ends with the digit 0, 2, 6, 2, 0, 0, 2, 6, 2, 0.

If $\frac{n(n+1)}{2}$ ended with the digit 2, 4, 7, 9,

then $n(n+1)$ would end with the digit 4, 8, 4, 8,

which, as we have ascertained before, is impossible.

Method II. If the last digit of the number in question is x, then $\frac{1}{2}n(n+1) = 10k+x$ (k—a natural number) and consequently

$$n^2+n-(20k+2x) = 0.$$

Since n is an integer, the discriminant of the above equation, i.e. number

$$\Delta = 80k+8x+1,$$

is the square of an integer. The last digit of number $8x+1$ constitutes the last digit of the discriminant. If x is equal to 2, 4, 7, 9, then $8x+1$ ends with the digits 7, 3, 7, 3 respectively.

Now the square of an integer ending with digit c has the form $(10a+c)^2 = 100a^2+20ac+c^2$, and thus it has the same final digit as c^2. And since the squares of numbers from 0 to 9 end neither with the digit 7 nor with the digit 3, x can be none of the digits 2, 4, 7 and 9.

Method III. We introduce the notation

$$S_n = 1+2+3+\ldots+n.$$

To begin with, we ascertain that number S_{5k} is divisible by 5. Indeed,

$$S_5 = 1+2+3+4+5 = 15,$$

and

$$S_{5(k+1)} = S_{5k}+(5k+1)+(5k+2)+\ldots+(5k+5) = S_{5k}+25k+S_5.$$

Thus if S_{5k} is divisible by 5, then also $S_{5(k+1)}$ is divisible by 5, whence it follows by induction that for any natural k number S_{5k} is divisible by 5. Thus the last digit of number S_{5k} is either 0 or 5.

Let $n = 5k+r$, where r is one of the numbers 1, 2, 3, 4; then

$$S_{5k+r} = S_{5k}+(5k+1)+\ldots+(5k+r) = S_{5k}+5kr+S_r.$$

Since number $S_{5k}+5kr$ is divisible by 5, its last digit is either 0 or 5. Consequently, the last digit of number S_{5k+r} is equal either to the last digit of number S_r or to the last digit of number S_r+5.

Now: if r is equal to	1, 2, 3, 4,
then the last digit of number S_r is	1, 3, 6, 0,
and the last digit of number S_r+5 is	6, 8, 1, 5.

This implies that for any natural n the last digit of number S_n is one of the digits 0, 1, 3, 5, 6, 8, whence it can be none of the digits 2, 4, 7, 9.

The same argument can be applied to the sum of the squares

$$T_n = 1^2+2^2+3^2+\ldots+n^2.$$

Since

$$T_5 = 1^2+2^2+3^2+4^2+5^2 = 55$$

and

$$T_{5k+r} = T_{5k}+(5k+1)^2+\ldots+(5k+r)^2 = T_{5k}+25k^2r+10kS_r+T_r,$$

we infer as before that number T_{5k} is divisible by 5 for any natural k and that the last digit of number T_{5k+r} is equal either to the last digit of number T_r or to the last digit of number T_r+5.

If r is equal to	1, 2, 3, 4,
then the last digit of number T_r is	1, 5, 4, 0,
and the last digit of number T_r+5 is	6, 0, 9, 5.

Consequently, for any natural n the last digit of number T_n is one of the digits 0, 1, 4, 5, 6, 9, i.e. digits 2, 3, 7, and 8 are excluded.

The same result will be obtained for the sum of the cubes of natural numbers

$$U_n = 1^3+2^3+3^3+ \dots +n^3.$$

The above argument cannot be applied to the sum V_n of the fourth powers of numbers $1, 2, \dots, n$, because number V_5 is not divisible by 5. It is easy to verify that any of the ten digits can be the final digit of number V_n. We suggest that the reader should use the same method to investigate the cases of the next few powers.

The preceding proof could be presented in a visual manner as follows.

Divide a circular disc into 5 sectors A, B, C, D, E (Fig. 1).

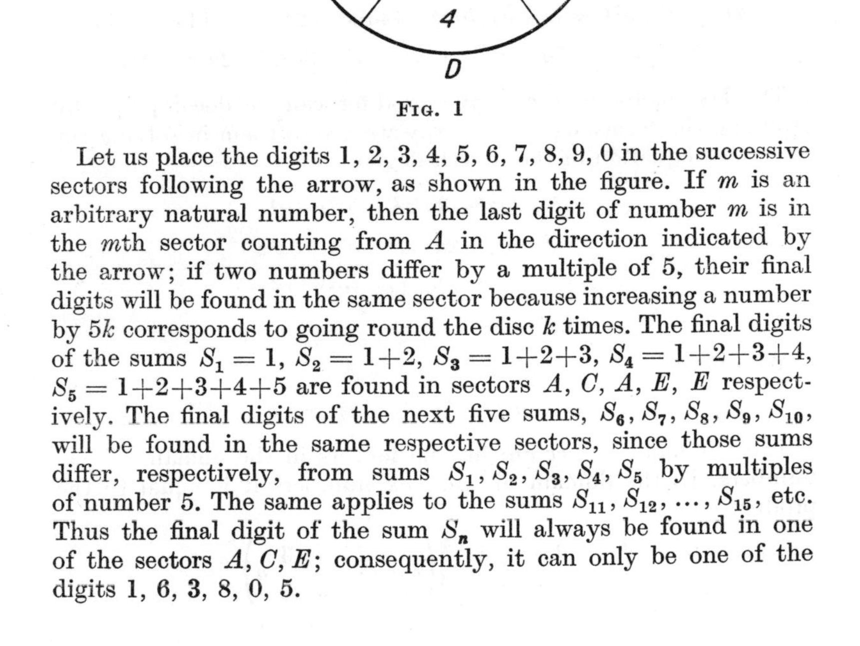

Fig. 1

Let us place the digits 1, 2, 3, 4, 5, 6, 7, 8, 9, 0 in the successive sectors following the arrow, as shown in the figure. If m is an arbitrary natural number, then the last digit of number m is in the mth sector counting from A in the direction indicated by the arrow; if two numbers differ by a multiple of 5, their final digits will be found in the same sector because increasing a number by $5k$ corresponds to going round the disc k times. The final digits of the sums $S_1 = 1$, $S_2 = 1+2$, $S_3 = 1+2+3$, $S_4 = 1+2+3+4$, $S_5 = 1+2+3+4+5$ are found in sectors A, C, A, E, E respectively. The final digits of the next five sums, $S_6, S_7, S_8, S_9, S_{10}$, will be found in the same respective sectors, since those sums differ, respectively, from sums S_1, S_2, S_3, S_4, S_5 by multiples of number 5. The same applies to the sums $S_{11}, S_{12}, \dots, S_{15}$, etc. Thus the final digit of the sum S_n will always be found in one of the sectors A, C, E; consequently, it can only be one of the digits 1, 6, 3, 8, 0, 5.

In the same way we shall ascertain that the last digit of the sum T_n of the squares and of the sum U_n of the cubes of natural numbers from 1 to n can only be found in one of the sectors A, D, E, i.e. that it can only be one of the digits 1, 6, 4, 9, 0, 5.

REMARK. The question which digits can be the final digits of the sums T_n, U_n, etc. can also be solved by method I. We must then make use of the formulas

$$1^2+2^2+ \dots +n^2 = \frac{n(n+1)(2n+1)}{6},$$

$$1^3+2^3+ \dots +n^3 = \frac{n^2(n+1)^2}{4},$$

etc. Method III does not require these formulas.

§ 2. Polynomials, Algebraic Fractions, Irrational Expressions

16. Since

$$(x+a)^7 = (x+a)(x^6+6ax^5+15a^2x^4+20a^3x^3+15a^4x^2+6a^5x+a^6),$$
$$x^7+a^7 = (x+a)(x^6-ax^5+a^2x^4-a^3x^3+a^4x^2-a^5x+a^6),$$

we have

$$(x+a)^7-(x^7+a^7) = (x+a)(7ax^5+14a^2x^4+21a^3x^3+14a^4x^2+7a^5x)$$
$$= 7ax(x+a)(x^4+2ax^3+3a^2x^2+2a^3x+a^4).$$

The last factor of the above product can be decomposed by grouping the terms, e.g. in the way we group them in solving the reciprocal equation of the fourth degree:

$$x^4+2ax^3+3a^2x^2+2a^3x+a^4$$
$$= (x^4+a^4)+2ax(x^2+a^2)+3a^2x^2$$
$$= (x^2+a^2)^2+2ax(x^2+a^2)+a^2x^2$$
$$= (x^2+a^2+ax)^2.$$

We obtain

$$(x+a)^7-(x^7+a^7) = 7ax(x+a)(x^2+ax+a^2)^2.$$

The trinomial x^2+ax+a^2 has no factors in the domain of real numbers. In the domain of complex numbers it is equal to the product

$$\left(x+\frac{1+i\sqrt{3}}{2}a\right)\left(x+\frac{1-i\sqrt{3}}{2}a\right)$$

and the given polynomial is a product of factors of the first degree:

$$(x+a)^7-(x^7+a^7) = 7ax(x+a)\left(x+\frac{1+i\sqrt{3}}{2}a\right)^2\left(x+\frac{1-i\sqrt{3}}{2}a\right)^2.$$

17. To begin with, it will be observed that the expression W is invariant under the *cyclic substitution* of the variables x, y, z, i.e. it remains unchanged when we replace in it, simultaneously, x by y, y by z and z by x.

If we substitute $x = y$ in the expression W, which is a polynomial of the fourth degree with respect to x, we shall obtain $W = 0$; the polynomial W is thus divisible by $x-y$, and since it is invariant under cyclic substitution, it is also divisible by $y-z$ and by $z-x$, and consequently by $(x-y)(y-z)(z-x)$. In view of the fact that W is a homogeneous polynomial of the fifth degree with respect to x, y, z and $(x-y)(y-z)(z-x)$ is a homogeneous polynomial of the third degree with respect to these variables, the quotient of these polynomials must be a homogeneous polynomial of the second degree with respect to x, y, z, and the polynomial W has the form

$$W = (x-y)(y-z)(z-x)(ax^2+by^2+cz^2+dxy+eyz+fzx).$$

Since both W and $(x-y)(y-z)(z-x)$ are invariant under the cyclic substitution of x, y, z, their quotient must have the same property; therefore $a = b = c$, $d = c = f$ and

$$W = (x-y)(y-z)(z-x)[a(x^2+y^2+z^2)+d(xy+yz+zx)].$$

In order to determine the coefficients a and d it is sufficient to find the values of the two sides of the above equality substituting for x, y and z arbitrary, but different, numerical values. For example substituting $x = 0$, $y = 1$, $z = -1$, we obtain $-2 = 2(2a-d)$ or $2a-d = -1$; and substituting $x = 0$, $y = 1$, $z = 2$, we obtain $-14 = 2(5a+2d)$ or $(5a+2d) = -7$; these relations give us $a = -1$, $d = -1$. Consequently

$$W = (x-y)(y-z)(x-z)(x^2+y^2+z^2+xy+yz+zx).$$

REMARK. The polynomial of the second degree occurring in the decomposition of polynomial W which we have obtained is itself irreducible, i.e. it is not a product of two polynomials of the first degree. In order to prove this, it will be observed that such polynomials would have to be homogeneous with respect to x, y and z, and consequently for any values of x, y and z the following equality would hold:

$$x^2+y^2+z^2+xy+yz+zx = (ax+by+cz)(a_1x+b_1y+c_1z).$$

The coefficients a, b, c, a_1, b_1, c_1 would then satisfy the system of equations

$$aa_1 = 1, \quad (1)$$

$$bb_1 = 1, \quad (2)$$

$$cc_1 = 1, \quad (3)$$

$$ab_1 + a_1 b = 1, \quad (4)$$

$$bc_1 + b_1 c = 1, \quad (5)$$

$$ca_1 + c_1 a = 1. \quad (6)$$

Now, by squaring equations (4) and (5) and taking into consideration (1), (2) and (3), we obtain

$$(ab_1)^2 + (a_1 b)^2 = -1, \quad (7)$$

$$(bc_1)^2 + (b_1 c)^2 = -1. \quad (8)$$

This shows that the system of equations (1)–(6) has no real solutions. But neither has it any complex solutions. Indeed, by multiplying equations (4) and (5) and taking into consideration (2) and (6), we obtain

$$ab_1^2 c + a_1 b^2 c_1 = 0; \quad (9)$$

next, by multiplying (9) and (6) and taking into consideration (1) and (3), we obtain

$$(ab_1)^2 + (a_1 b)^2 + (bc_1)^2 + (b_1 c)^2 = 0,$$

whence, by (7) and (8), follows the contradiction $-2 = 0$.

18. *Answer*:

$$W = (a+b)(b+c)(c+a)(b-a)(c-b)(a-c).$$

19. *Answer*:

$$(a-b)(b-c)(c-a)(a+b+c).$$

20. The divisibility of the fourth degree polynomial $x^4 + px^2 + q$ by the quadratic trinomial $x^2 + ax + b$ denotes the existence of a quadratic trinomial $x^2 + mx + n$ such that for every value of x we have the equality

$$x^4 + px^2 + q = (x^2 + ax + b)(x^2 + mx + n),$$

i.e.

$$x^4 + px^2 + q = x^4 + (a+m)x^3 + (am+b+n)x^2 + (an+bm)x + bn.$$

This equality holds if and only if on both sides the coefficients of equal powers of x are equal. We thus obtain for the unknowns m, n, p, q the following system of equations:

$$a + m = 0, \quad (1)$$

$$am+b+n = p, \tag{2}$$

$$an+bm = 0, \tag{3}$$

$$bn = q. \tag{4}$$

From equation (1) we find $m = -a$, which, substituted in equation (3), gives

$$a(n-b) = 0. \tag{5}$$

(a) If $a \neq 0$, then $n = b$. From equations (2) and (4) we obtain

$$p = 2b-a^2, \qquad q = b^2.$$

In this case the problem has the following solution:

$$x^4+(2b-a^2)x^2+b^2 = (x^2+ax+b)(x^2-ax+b).$$

(b) If $a = 0$, then equation (5) is an identity and n can be any number. From equations (2) and (4) we obtain

$$p = b+n, \qquad q = bn.$$

The solution has the form

$$x^4+(b+n)x^2+bn = (x^2+b)(x^2+n)$$

where n is an arbitrary number.

21. We transform the given polynomial in the following way:

$$\begin{aligned} x^8+x^4+1 &= x^8+2x^4+1-x^4 \\ &= (x^4+1)^2-x^4 \\ &= (x^4+x^2+1)(x^4-x^2+1) \\ &= (x^4+2x^2+1-x^2)(x^4+2x^2+1-3x^2) \\ &= [(x^2+1)^2-x^2][(x^2+1)^2-(x\sqrt{3})^2] \\ &= (x^2+x+1)(x^2-x+1)(x^2+x\sqrt{3}+1)(x^2-x\sqrt{3}+1). \end{aligned}$$

The polynomial has been decomposed into four factors of the second degree. Since none of those factors has real roots, the factors cannot be further decomposed in the domain of real numbers and the above result gives the required factorization.

In the domain of complex numbers each of the above quadratic functions has two conjugate complex roots: they are, respectively,

$$a_1 = \frac{-1+i\sqrt{3}}{2} \quad \text{and} \quad a_2 = \frac{-1-i\sqrt{3}}{2},$$

$$a_3 = \frac{1+i\sqrt{3}}{2} \quad \text{and} \quad a_4 = \frac{1-i\sqrt{3}}{2},$$

$$a_5 = \frac{-\sqrt{3}+i}{2} \quad \text{and} \quad a_6 = \frac{-\sqrt{3}-i}{2},$$

$$a_7 = \frac{\sqrt{3}+i}{2} \quad \text{and} \quad a_8 = \frac{\sqrt{3}-i}{2}.$$

The given polynomial is the product of eight linear factors:

$$(x-a_1)(x-a_2)(x-a_3)(x-a_4)(x-a_5)(x-a_6)(x-a_7)(x-a_8).$$

The above decomposition into complex factors can be obtained by a shorter method. We know that the equation

$$x^n-1 = 0$$

has n roots defined by the equation

$$x = \cos\frac{2k\pi}{n}+i\sin\frac{2k\pi}{n},$$

where k runs over the successive values $0, 1, 2, \ldots, n-1$.

These roots can be represented geometrically in the plane of the complex variable x as the vertices of a regular n-gon inscribed in a circle with centre 0 and radius 1.

Now

$$x^8+x^4+1 = \frac{x^{12}-1}{x^4-1}.$$

The roots of the equation $x^8+x^4+1 = 0$ will thus be obtained by writing down the twelve roots of the equation $x^{12}-1 = 0$ and rejecting those four which are also roots of the equation $x^4-1 = 0$. Geometrically, this amounts to choosing eight of the vertices of a regular dodecagon, as shown in Fig. 2, in which the rejected vertices are marked by dots.

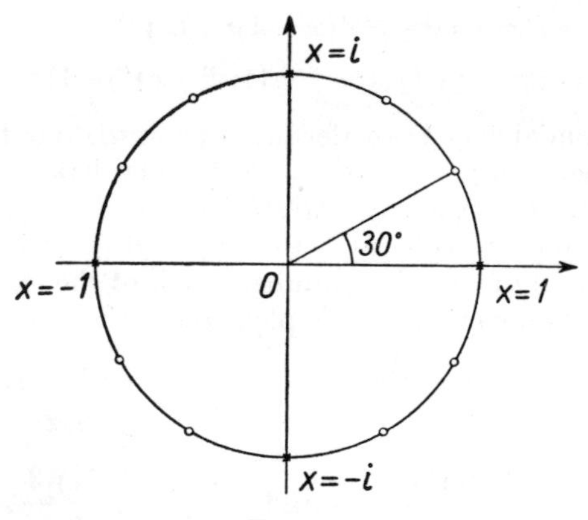

Fig. 2

The roots of the equation $x^8+x^4+1 = 0$ are thus the numbers

$$\cos 30^\circ+i\sin 30^\circ, \qquad \cos 60^\circ+i\sin 60^\circ,$$
$$\cos 120^\circ+i\sin 120^\circ, \qquad \cos 150^\circ+i\sin 150^\circ,$$
$$\cos 210^\circ+i\sin 210^\circ, \qquad \cos 240^\circ+i\sin 240^\circ,$$
$$\cos 300^\circ+i\sin 300^\circ, \qquad \cos 330^\circ+i\sin 330^\circ,$$

i.e. the numbers

$$\frac{\sqrt{3}+i}{2}, \quad \frac{1+i\sqrt{3}}{2}, \quad \frac{-1+i\sqrt{3}}{2}, \quad \frac{-\sqrt{3}+i}{2},$$
$$\frac{-\sqrt{3}-i}{2}, \quad \frac{-1-i\sqrt{3}}{2}, \quad \frac{1-i\sqrt{3}}{2}, \quad \frac{\sqrt{3}-i}{2},$$

in accordance with the result obtained before.

22. *Method I.* We shall assume the well-known theorem of algebra which states that the binomial a^n-1 (where n is a natural number) is divisible by the binomial $a-1$, i.e. that

$$a^n-1 = (a-1)(a^{n-1}+a^{n-2}+\ldots+1). \tag{1}$$

The second factor on the right-hand side of formula (1) is a geometrical progression, of which only the first two terms and the last term are written down, the remaining terms being replaced by dots.

The solution of the problem can be obtained by applying formula (1) to the binomial $x^{55}-1$ in two ways. First, if we substitute $a = x^{11}$, $n = 5$ in formula (1), we obtain the equality

$$x^{55}-1 = (x^{11}-1)(x^{44}+x^{33}+x^{22}+x^{11}+1),$$

and applying formula (1) to the factor $x^{11}-1$ we have

$$x^{55}-1 = (x-1)(x^{10}+x^9+\ldots+1)(x^{44}+x^{33}+\ldots+1). \tag{2}$$

If, on the other hand, we substitute $a = x^5$, $n = 11$ in formula (1), we obtain the equality

$$x^{55}-1 = (x^5-1)(x^{50}+x^{45}+\ldots+1),$$

which, on the application of formula (1) to the factor x^5-1, gives the equality

$$x^{55}-1 = (x-1)(x^4+x^3+x^2+x+1)(x^{50}+x^{45}+\ldots+1). \tag{3}$$

Equalities (2) and (3) imply

$$(x^{10}+x^9+\ldots+1)(x^{44}+x^{33}+\ldots+1)$$
$$= (x^4+x^3+\ldots+1)(x^{50}+x^{45}+\ldots+1). \tag{4}$$

From this equality we shall derive the required theorem; we first transform the first factor of the left-hand side:

$$\begin{aligned} &x^{10}+x^9+\dots+1 \\ &= (x^{10}+x^9+\dots+x^6)+(x^5+x^4+\dots+x)+1 \\ &= x^6(x^4+x^3+\dots+1)+x(x^4+x^3+\dots+1)+1 \\ &= (x^4+x^3+\dots+1)(x^6+x)+1. \end{aligned}$$

Formula (4) can therefore be written in the form

$$\begin{aligned} (x^4+x^3+\dots+1)(x^6+x)(x^{44}+x^{33}+\dots+1)+(x^{44}+x^{33}+\dots+1) \\ = (x^4+x^3+\dots+1)(x^{50}+x^{45}+\dots+1). \end{aligned}$$

Let us subtract the first term of the equality from each side, and then let us factorize the right-hand side; we obtain

$$\begin{aligned} x^{44}+x^{33}+\dots+1 = (x^4+x^3+\dots+1)[x^{50}+x^{45}+\dots+1- \\ -(x^6+x)(x^{44}+x^{33}+\dots+1)]. \quad (5) \end{aligned}$$

Equality (5) shows that the polynomial $x^{44}+x^{33}+x^{22}+x^{11}+1$ is divisible by the polynomial $x^4+x^3+x^2+x+1$.

The polynomial appearing in the square brackets on the right-hand side of formula (5) could be ordered, on opening the round brackets, according to the powers of x; this, however, is not necessary for the proof of the theorem. We shall only remark that the terms containing x^{50} and x^{45} are reduced and consequently formula (5) can be written in a simpler form:

$$\begin{aligned} x^{44}+x^{33}+\dots+1 = (x^4+x^3+\dots+1)[x^{40}+{}^{35}+\dots+1- \\ -(x^6+x)(x^{33}+x^{22}+x^{11}+1)]. \quad (6) \end{aligned}$$

Method II. Let us multiply each of the given polynomials

$$f(x) = x^{44}+x^{33}+x^{22}+x^{11}+1$$

and

$$g(x) = x^4+x^3+x^2+x+1$$

by $x-1$; we shall obtain the polynomials

$$\begin{aligned} F(x) &= (x-1)(x^{44}+x^{33}+x^{22}+x^{11}+1) \\ &= x^{45}-x^{44}+x^{34}-x^{33}+x^{23}-x^{22}+x^{12}-x^{11}+x-1 \end{aligned}$$

and

$$G(x) = (x-1)(x^4+x^3+x^2+x+1) = x^5-1.$$

In order to prove that the polynomial $f(x)$ is divisible by the polynomial $g(x)$ it is sufficient to show that the polynomial $F(x)$ is divisible by the polynomial $G(x)$, i.e. by x^5-1.

Now

$$x^{45}-x^{44}+x^{34}-x^{33}+x^{23}-x^{22}+x^{12}-x^{11}+x-1$$
$$=(x^{45}-1)-x^{34}(x^{10}-1)-x^{23}(x^{10}-1)-x^{12})(x^{10}-1)-$$
$$-x(x^{10}-1)=[(x^5)^9-1]-(x^{10}-1)(x^{34}+x^{23}+x^{12}+x)$$
$$=(x^5-1)(x^{40}+x^{35}+\ldots+1)-(x^5-1)(x^5+1)(x^{34}+x^{23}+x^{12}+x)$$
$$=(x^5-1)[x^{40}+x^{35}+\ldots+1-(x^5+1)(x^{34}+x^{23}+x^{12}+x)].$$

Consequently

$$F(x)=G(x)[x^{40}+x^{35}+\ldots+1-(x^6+x)(x^{33}+x^{22}+x^{11}+1)].$$

Thus the polynomial $F(x)$ is indeed divisible by the polynomial $G(x)$, which is what we wanted to prove.

Dividing both sides of the last equality by $x-1$, we obtain the formula

$$x^{44}+x^{33}+x^{22}+x^{11}+1=(x^4+x^3+x^2+x+1)[x^{40}+x^{35}+\ldots+$$
$$+1-(x^6+x)(x^{33}+x^{22}+x^{11}+1)],$$

i.e. the formula (6) obtained by method I.

Method III. A very short and simple solution of the problem will be obtained by the use of complex numbers and their geometrical representation on a plane. If x denotes a complex variable, then the roots of the equation

$$x^n-1=0$$

can be represented as the vertices of a regular n-gon $A_1A_2\ldots A_n$ inscribed in a unit circle C, drawn in the plane of complex numbers x from the origin 0, the vertex A_n of the polygon lying at point $x=1$.

Let us write, as in method I:

$$x^{55}-1=(x^{11}-1)(x^{44}+x^{33}+x^{22}+x^{11}+1),$$
$$x^5-1=(x-1)(x^4+x^3+x^2+x+1).$$

The roots of the binomial $x^{55}-1$ correspond to the vertices A_1, $A_2, \ldots, A_{55}$ of a regular 55-gon inscribed in the above-mentioned circle C, the root $x=1$ corresponding to the vertex A_{55} (Fig. 3). Similarly, the roots of the binomial $x^{11}-1$ correspond to the vertices of a regular 11-gon; those vertices are to be found among the vertices of that 55-gon; namely they are the points $A_5, A_{10}, A_{15}, \ldots, A_{55}$. The roots of the polynomial $x^{44}+x^{33}+$ $+x^{22}+x^{11}+1$ correspond to those vertices of the 55-gon which will remain after we have rejected the vertices of the 11-gon; there are 44 of them and the vertices $A_{11}, A_{22}, A_{33}, A_{44}$ are among them.

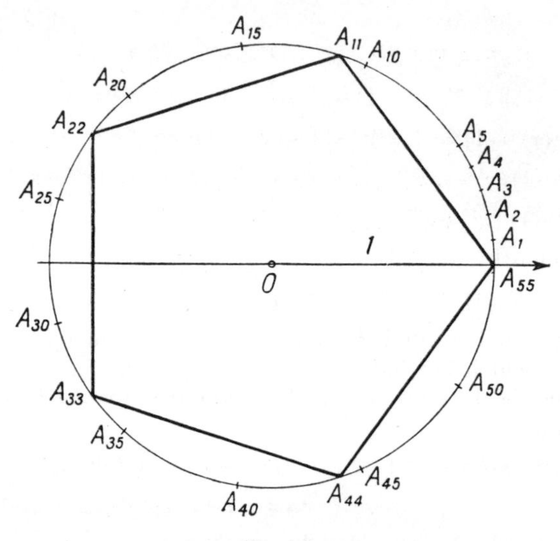

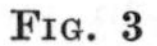
Fig. 3

On the other hand, to the roots of the binomial x^5-1 correspond the vertices of a regular 5-gon, which are also to be found among the vertices of the 55-gon: they are the points A_{11}, A_{22}, A_{33}, A_{44}, A_{55}; to the roots of the polynomial $x^4+x^3+x^2+x+1$ correspond only the vertices A_{11}, A_{22}, A_{33}, A_{44}, because we must reject the vertex A_{55}, corresponding to the number $x = 1$.

Apparently the roots of the polynomial $x^4+x^3+x^2+x+1$ are at the same time roots of the polynomial $x^{44}+x^{33}+x^{22}+x^{11}+1$, which implies that the latter polynomial is divisible by the former.

23. *Method I.* Let us multiply both sides of the equality

$$\frac{a-b}{1+ab}+\frac{b-c}{1+bc}+\frac{c-a}{1+ca}=0 \tag{1}$$

by the product $(1+ab)(1+bc)(1+ca)$; on the left side we shall obtain the expression

$$W = (a-b)(1+bc)(1+ca)+$$
$$+(b-c)(1+ab)(1+ca)+(c-a)(1+ab)(1+bc). \tag{2}$$

We could perform the multiplications indicated in this expression, obtaining 24 terms; we could then reduce similar terms and finally, by a suitable grouping of the terms and by factorizing, give the expression W the following form:

$$W = (a-b)(b-c)(c-a). \tag{3}$$

If $W = 0$, then one of the factors of this product is equal to zero; consequently at least two of the numbers a, b, c are equal, which is what was to be proved.

However, formula (3) could be obtained without a lengthy algebraic manipulation. In mathematics we give preference to methods based on reasoning and not on calculations, which are often tedious and cumbersome. In this case we can use a very simple argument.

It will be observed that with respect to any of the letters a, b, c, e.g. the letter a, W is a polynomial of the second degree. If we substitute b for a in equality (2), we shall obtain $W = 0$; by a well-known theorem, the polynomial W is thus divisible by $a-b$. In the same way we can ascertain that the polynomial W is divisible by $b-c$ and by $c-a$; actually this results from the fact that W remains unchanged if we apply a cyclic substitution, i.e. if we replace a by b, b by c and c by a; thus if W is divisible by $a-b$, it must be divisible by the binomials arising from the binomial $a-b$ through a cyclic substitution, i.e. by the binomials $b-c$ and $c-a$. Thus

$$W = (a-b)(b-c)(c-a)k. \tag{4}$$

Since both W and $(a-b)(b-c)(c-a)$ are polynomials of the second degree with respect to a, the factor k must be a polynomial of degree zero with respect to a, and likewise with respect to b and with respect to c, i.e. k is simply a numerical coefficient. The value of k will be found by substituting for a, b, c in equality (4) some definite numbers, e.g. $a = 1$, $b = -1$, $c = 0$; equality (4) then gives $2 = 2k$ and $k = 1$. We have thus proved formula (3), from which the theorem follows as before.

Method II. The structure of the components of the left side of equation (1) brings to mind the formula for the tangent of the difference of two angles, which leads us to the solution of the problem with the use of trigonometry. We can write

$$a = \tan\alpha, \qquad b = \tan\beta, \qquad c = \tan\gamma,$$

where α, β, γ are definite angles contained in the open interval from $-90°$ to $90°$. Equation (1) assumes the form

$$\frac{\tan\alpha-\tan\beta}{1+\tan\alpha\tan\beta}+\frac{\tan\beta-\tan\gamma}{1+\tan\beta\tan\gamma}+\frac{\tan\gamma-\tan\alpha}{1+\tan\gamma\tan\alpha}=0$$

or

$$\tan(\alpha-\beta)+\tan(\beta-\gamma)+\tan(\gamma-\alpha) = 0, \tag{5}$$

each of the angles $\alpha-\beta$, $\beta-\gamma$, $\gamma-\alpha$ being contained in the open interval from $-180°$ to $180°$.

Applying to the sum of the first two components in (5) the formula

$$\tan m+\tan n = \tan(m+n)(1-\tan m \tan n),$$

we obtain

$$\tan(\alpha-\gamma)[1-\tan(\alpha-\beta)\tan(\beta-\gamma)]+\tan(\gamma-\alpha) = 0,$$

and therefore

$$\tan(\alpha-\gamma)\tan(\alpha-\beta)\tan(\beta-\gamma) = 0. \quad (6)$$

One of the tangents in formula (6) must be equal to zero, and since, as has been pointed out before, the corresponding angle is greater than $-180°$ and less than $180°$, that angle must also be equal to zero. At least two of the angles α, β, γ are thus equal, and this means that at least two of the numbers a, b, c, are equal.

24. *Hint.* See the solution of the preceding problem, method I.

25. *Method I.* Denote the value of the fraction by the letter y and perform the transformation

$$y = \frac{x^4+x^2+5}{(x^2+1)^2} = \frac{(x^2+1)^2-(x^2+1)+5}{(x^2+1)^2} = 1-\frac{1}{x^2+1}+\frac{5}{(x^2+1)^2}.$$

Let $1/(1+x^2) = u$; then y is a quadratic function of the variable u:

$$y = 5u^2-u+1.$$

We know from algebra that the quadratic function au^2+bu+c where $a > 0$ has its minimum when $u = -b/2a$. Thus in our case the least value of y corresponds to the value $u = \frac{1}{10}$, which gives $y_{\min} = \frac{19}{20}$. The corresponding value of x is obtained from the equation $1/(x^2+1) = \frac{1}{10}$, whence $x^2 = 9$, and consequently $x = 3$ or $x = -3$.

Method II. Let $x^2 = z$; then

$$y = \frac{z^2+z+5}{(z+1)^2};$$

since $z+1 = x^2+1 > 0$, the above equation is equivalent to

$$(z+1)^2 y = z^2+z+5,$$

which can be written as

$$(y-1)z^2+(2y-1)z+(y-5) = 0.$$

If certain values of y and z satisfy this equation, then either $y = 1$ or the equation is quadratic with respect to z and its discriminant is non-negative, i.e.

$$(2y-1)^2-4(y-1)(y-5) \geqslant 0$$

whence

$$20y-19 \geqslant 0 \quad \text{or} \quad y \geqslant \frac{19}{20}.$$

Hence the least value of y is $y_{\min} = \frac{19}{20}$. From the given relation between y and z we obtain the corresponding value $z = 9$, and thus $x^2 = 9$, i.e. $x = 3$ or $x = -3$.

REMARK. We shall solve a more general problem: find the least and the greatest values of the function

$$y = \frac{x^2+mx+n}{x^2+px+q} \tag{1}$$

under the assumption that the trinomials x^2+mx+n and x^2+ $+px+q$ have no root in common.

If the numbers x, y satisfy equation (1), then they also satisfy the equation

$$(x^2+px+q)y = x^2+mx+n. \tag{2}$$

Conversely, if the numbers x, y satisfy equation (2), then $x^2+px+q \neq 0$; otherwise equality $x^2+px+q = 0$ would, by (2), imply the equality $x^2+mx+n = 0$, which would contradict the assumption that the given trinomials have no root in common. Consequently x and y satisfy equation (1). Equations (1) and (2) are thus equivalent.

We shall write equation (2) in the form

$$(y-1)x^2+(py-m)x+(qy-n) = 0. \tag{3}$$

Our problem consists in finding, among those values of y for which equation (3) has roots, the least value $y_{\min}$ and the greatest value $y_{\max}$.

If $y = 1$, equation (3) is linear with respect to x and has a solution for $p \neq m$. If $y \neq 1$, equation (3) is quadratic with respect to x.

Let us find its discriminant

$$\delta(y) = (py-m)^2-4(y-1)(qy-n)$$

and write it in the form

$$\delta(y) = (p^2-4q)y^2-2(pm-2q-2n)y+(m^2-4n).$$

Equation (3) has roots for those values of $y \neq 1$ for which $\delta(y) \geqslant 0$. We shall distinguish three cases.

(i) $p^2-4q > 0$. The quadratic function $\delta(y)$ is then positive provided $|y|$ is sufficiently large. Thus in this case there exists neither a greatest nor a least value y of fraction (1).

(ii) $p^2-4q < 0$. The quadratic function $\delta(y)$ is then negative for sufficiently large values of $|y|$.

Since the function y of x expressed by formula (1) assumes a definite value for every x and is not constant, we must have $\delta(y) \geqslant 0$ for infinitely many values of y; we could also prove this by showing that the discriminant of the quadratic function $\delta(y)$ is positive in case (ii).

Thus in this case the quadratic function $\delta(y)$ has two roots and is positive if the value of y is contained between those roots. Since $\delta(1) = (p-m)^2 \geqslant 0$, the roots are the required values $y_{\min}$ and $y_{\max}$.

(iii) $p^2-4q = 0$. The function $\delta(y)$ has then the form

$$\delta(y) = -2(pm-2q-2n)y+(m^2-4n).$$

It will be observed that in this case $pm-2q-2n \neq 0$; for the equality $p^2-4q = 0$ implies that the trinomial x^2+px+q is equal to $(x+\frac{1}{2}p)^2$ and has a double root $-\frac{1}{2}p$. By hypothesis this root is not a root of the trinomial x^2+mx+n, and therefore

$$\frac{p^2}{4}-\frac{pm}{2}+n = q+n-\frac{pm}{2} = \frac{1}{2}(2q+2n-pm) \neq 0.$$

Consequently the solution of the problem in case (iii) is as follows.

(a) If $pm-2q-2n > 0$, then $\delta(y) \geqslant 0$ for

$$y \leqslant \frac{m^2-4n}{2(mp-2q-2n)}.$$

Since under assumption (a)

$$\frac{m^2-4n}{2(pm-2q-2n)} \geqslant 1,$$

in view of $m^2-4n-2(pm-2q-2n) = (p-m)^2 \geqslant 0$, we have

$$y_{\max} = \frac{m^2-4n}{2(pm-2q-2n)},$$

and $y_{\min}$ does not exist.

(b) If $pm-2q-2n < 0$, then $\delta(y) \geqslant 0$ for

$$y \geqslant \frac{m^2-4n}{2(pm-2q-2n)},$$

and

$$\frac{m^2-4n}{2(pm-2q-2n)} \leqslant 1;$$

thus

$$y_{\min} = \frac{m^2-4n}{2(pm-2q-2n)},$$

and $y_{\max}$ does not exist.

We suggest that the reader should illustrate geometrically the above cases (i), (ii), and (iii) choosing suitable numerical values of m, n, p, q.

26. First, it may be observed that $4x^2+16x+15 = 4(x^2+ +4x+\frac{15}{4}) = 4(x+\frac{5}{2})(x+\frac{3}{2})$; thus the left-hand side of equation (1) has a numerical value only for values of x other than $-\frac{5}{2}$ or $-\frac{3}{2}$. The right-hand side of equation (1) has a numerical meaning only for values of x other than $-c$ or $-d$. If equation (1) is to be an identity, both its sides must have a numerical meaning for the same values of x; therefore the denominators of the fractions on the right side must be the binomials $x+\frac{5}{2}$, $x+\frac{3}{2}$ and the problem is reduced to finding numbers a and b for which the equation

$$\frac{2x-7}{4x^2+16x+15} = \frac{a}{x+\frac{5}{2}} + \frac{b}{x+\frac{3}{2}} \tag{2}$$

is an identity.

Suppose that such numbers exist, i.e. that equation (2), in which a and b denote definite numbers, is an identity. Then both sides assume the same numerical values for every value of x other than $-\frac{5}{2}$ and than $-\frac{3}{2}$.

Let us multiply both sides of equation (2) by $4x^2+16x+15$

$$2x-7 = 4a\left(x+\frac{3}{2}\right)+4b\left(x+\frac{5}{2}\right). \tag{3}$$

Equation (3) is also true for every value of x other than $-\frac{5}{2}$ or $-\frac{3}{2}$. And we know that if two first degree functions of variable x assume equal values, even if that occurs only for two different values of x, then they have equal coefficients of x and equal constant terms (i.e. they are identically equal).

Consequently

$$\begin{aligned} 4a+4b &= 2, \\ 6a+10b &= -7, \end{aligned} \tag{4}$$

and therefore

$$a = 3, \quad b = -\frac{5}{2}. \tag{5}$$

Substituting these values in equation (2) we obtain

$$\frac{2x-7}{4x^2+16x+15} = \frac{3}{x+\frac{5}{2}} + \frac{-\frac{5}{2}}{x+\frac{3}{2}}. \tag{6}$$

It does not yet follow that equation (6) is an identity; we have only proved that *if there exists a solution* of the problem, then equation (6) constitutes that solution. We must still verify whether the two sides of equation (6) are indeed identically equal. For this purpose we might transform identically the right-hand side of (6), i.e. reduce its components to a common denominator, perform the addition and simplify the result; we should then obtain the same fraction as the one appearing on the left-hand side of (6). Such a procedure, however, is unnecessary, since it can be replaced by the following short argument:

If a and b have values (5), equations (4) are valid; then both sides of equation (3) are linear functions of x with identical coefficients and identical constant terms, and consequently equation (3) is valid for every value of x. Dividing both sides of (3) by $4(x+\frac{5}{2})(x+\frac{3}{2})$, we obtain (for $a = 3$ and $b = -\frac{5}{2}$) equation (6); thus equation (6) is valid for every value of x other than $-\frac{5}{2}$ or $-\frac{3}{2}$, i.e. equation (6) is an identity.

REMARK 1. The values of a and b can also be found in the following manner. Equation (3) must be valid for every value of x, i.e. unlike equality (2), also for $x = -\frac{5}{2}$ and $x = -\frac{3}{2}$. Substituting these values in equation (3), we immediately obtain $-12 = -4a$ and $-10 = 4b$, whence $a = 3$, $b = -\frac{5}{2}$.

REMARK 2. The theorem stating that two first degree functions which assume equal values for two (different) values of x have coefficients respectively equal is a particular case of the following theorem:

If two nth *degree polynomials in* x *are equal,*

$$a_0x^n+a_1x^{n-1}+ \dots +a_{n-1}x+a_n = b_0x^n+b_1x^{n-1}+ \dots +b_{n-1}x+b_n$$

for $n+1$ *different values of* x, *then the corresponding coefficients of those polynomials are equal, i.e. the equality is an identity.*

The above equation can be replaced by

$$(a_0-b_0)x^n+(a_1-b_1)x^{n-1}+ \dots +(a_{n-1}-b_{n-1})x+(a_n-b_n) = 0;$$

and we obtain the following theorem:

If an nth *degree polynomial in* x *assumes the value* 0 *for* $n+1$ *different values of* x (*in other words: if it has* $n+1$ *different roots*), *then all the coefficients of that polynomial are zeros, i.e. the polynomial is identically equal to zero.*

This theorem can be proved as follows. Assume that the theorem is valid for polynomials of degree $n-1$. Let the polynomial

$$c_0x^n+c_1x^{n-1}+\ldots+c_{n-1}x+c_n \tag{7}$$

have $n+1$ different roots $x_1, x_2, \ldots, x_{n+1}$. We can then determine the numbers $d_0, d_1, \ldots, d_{n-1}$ in such a way that the following identity holds:

$$c_0x^n+c_1x^{n-1}+\ldots+c_{n-1}x+c_n \\ = (x-x_1)(d_0x^{n-1}+d_1x^{n-2}+\ldots+d_{n-2}x+d_{n-1}). \tag{8}$$

Indeed it is sufficient to find such values $d_0, d_1, \ldots, d_{n-1}$ that the coefficients of the identical powers of x will be the same on both sides of equation (8), i.e. that $d_1, d_2, \ldots, d_{n-1}$ will satisfy the system of equations

$$\begin{aligned} c_0 &= d_0, \\ c_1 &= d_1-d_0x_1, \\ c_2 &= d_2 - d_1x_1, \\ &\cdots\cdots\cdots\cdots \\ c_{n-1} &= d_{n-1}-d_{n-2}x_1, \\ c_n &= -d_{n-1}x_1. \end{aligned} \tag{9}$$

From the first $n-1$ equations of system (9) we obtain successively

$$\begin{aligned} d_0 &= c_0, \\ d_1 &= c_1+d_0x_1 = c_1+c_0x_1, \\ d_2 &= c_2+d_1x_1 = c_2+c_1x_1+c_0x_1^2, \\ &\cdots\cdots\cdots\cdots\cdots\cdots\cdots\cdots \\ d_{n-1} &= c_{n-1}+d_{n-2}x_1 = c_{n-1}+c_{n-2}x_1+\ldots+c_1x_1^{n-2}+c_0x_1^{n-1}. \end{aligned}$$

The above values satisfy also the last equation of system (9), since by substituting the value of d_{n-1} in this equation we get

$$c_n = -(c_{n-1}+c_{n-2}x_1+\ldots+c_1x_1^{n-2}+c_0x_1^{n-1})x_1$$

or

$$c_0x_1^n+c_1x_1^{n-1}+\ldots+c_{n-1}x_1+c_n = 0,$$

which is valid since x_1 is a root of polynomial (7).

Identity (8) implies that numbers $x_2, x_3, \ldots, x_{n+1}$, which by hypothesis are roots of the polynomial appearing on the left-hand side of this identity, are also roots of the polynomial on the right-hand side, and since the factor $x-x_1$ is not equal to zero for any of the n numbers $x_2, x_3, \ldots, x_{n+1}$, those numbers are the roots of the polynomial $d_0x^{n-1}+d_1x^{n-2}+ \ldots +d_{n-2}x+d_{n-1}$ of degree $n-1$. Thus by the assumption we made at the beginning, $d_0 = d_1 = \ldots = d_{n-2} = d_{n-1} = 0$, and therefore equations (9) give

$$c_0 = c_1 = c_2 = \ldots = c_{n-1} = c_n = 0.$$

We have proved that if the theorem is valid for polynomials of degree $n-1$, then it is also valid for polynomials of degree n. Clearly, the theorem is valid for polynomials of degree zero, since, if the polynomial c_0 is equal to zero for a value $x = x_1$, it means that $c_0 = 0$.

By the induction principle, we infer from the above two premises that the theorem is valid for polynomials of any degree.

REMARK 3. In the above arguments we used the notion of identical equality of two algebraic expressions. This notion is usually defined as follows:

Suppose we are given two expressions $A(x, y, \ldots)$ and $B(x, y, \ldots)$ containing the same variables $x, y, \ldots$ We say that $A(x, y, \ldots)$ and $B(x, y, \ldots)$ are identically equal or that the equation $A(x, y, \ldots) = B(x, y, \ldots)$ is an *identity* if for every system of numbers $x_0, y_0, \ldots$ for which one of these expressions has a definite value the other expression also has a definite numerical value, both values being equal, i.e. $A(x_0, y_0, \ldots) = B(x_0, y_0, \ldots)$. Two expressions identically equal to a third are of course identically equal to each other. Passing from one expression to an expression identically equal to it is called an identical transformation of the given expression.

It is often convenient to use a more general notion of identity. Suppose that we are given two expressions, $A(x, y, \ldots)$ and $B(x, y, \ldots)$, and a set Z of numbers. The equation $A(x, y, \ldots) = B(x, y, \ldots)$ is said to be an *identity in the set* Z if it is valid for any values of $x, y, \ldots$ belonging to the set Z. An equation which is an identity in one set of numbers is not necessarily an identity in another set. For example the equation

$$\log xy = \log x + \log y$$

is an identity in the set of positive numbers but is not an identity in the set of all real numbers, since, if $x < 0$ and $y < 0$, then the left-hand side of this equality has a numerical meaning and the right-hand side has not. Similarly, the equation

$$\sqrt{ab} = \sqrt{a} \times \sqrt{b}$$

is an identity in the set of non-negative numbers, and the equation

$$\frac{a^2-1}{a-1} = a+b$$

is an identity in the set of all real numbers different from 1.

27. We shall approach the problem in a more general manner and show that for any natural k we can find $x^k+\frac{1}{x^k}$ given $x+\frac{1}{x}$. It will be observed that

$$\left(x^m+\frac{1}{x^m}\right)\left(x^n+\frac{1}{x^n}\right) = \left(x^{m+n}+\frac{1}{x^{m+n}}\right)+\left(x^{m-n}+\frac{1}{x^{m-n}}\right),$$

whence

$$x^{m+n}+\frac{1}{x^{m+n}} = \left(x^m+\frac{1}{x^m}\right)\left(x^n+\frac{1}{x^n}\right)-\left(x^{m-n}+\frac{1}{x^{m-n}}\right). \qquad (1)$$

Equation (1) allows us to reduce the evaluation of $x^k+\frac{1}{x^k}$ to the evaluation of expressions of the same form but with a lower exponent. Formulas of this kind are called *recursive formulas*.

If $n = 1$, (1) gives

$$x^{m+1}+\frac{1}{x^{m+1}} = \left(x^m+\frac{1}{x^m}\right)\left(x+\frac{1}{x}\right)-\left(x^{m-1}+\frac{1}{x^{m-1}}\right), \qquad (2)$$

and if $n = m$, we obtain

$$x^{2m}+\frac{1}{x^{2m}} = \left(x^m+\frac{1}{x^m}\right)^2-2. \qquad (3)$$

We shall use the above formulas to find $x^{13}+\frac{1}{x^{13}}$ given $x+\frac{1}{x}=a$.

By formula (3)

$$x^2+\frac{1}{x^2} = \left(x+\frac{1}{x}\right)^2-2 = a^2-2,$$

$$x^4+\frac{1}{x^4} = \left(x^2+\frac{1}{x^2}\right)^2-2 = a^4-4a^2+2$$

By formula (2)

$$x^3+\frac{1}{x^3} = \left(x^2+\frac{1}{x^2}\right)\left(x+\frac{1}{x}\right)-\left(x+\frac{1}{x}\right) = a^3-3a\,;$$

hence by formula (3)

$$x^6+\frac{1}{x^6} = \left(x^3+\frac{1}{x^3}\right)^2-2 = a^6-6a^4+9a^2-2.$$

Then, using formula (1), we obtain

$$x^7+\frac{1}{x^7} = \left(x^4+\frac{1}{x^4}\right)\left(x^3+\frac{1}{x^3}\right)-\left(x+\frac{1}{x}\right)$$

$$= (a^4-4a^2+2)(a^3-3a)-a = a^7-7a^5+14a^3-7a,$$

and finally

$$x^{13}+\frac{1}{x^{13}} = \left(x^7+\frac{1}{x^7}\right)\left(x^6+\frac{1}{x^6}\right)-\left(x+\frac{1}{x}\right)$$

$$= (a^7-7a^5+14a^3-7a)(a^6-6a^4+9a^2-2)-a$$

$$= a^{13}-13a^{11}+65a^9-156a^7+182a^5-91a^3+13a.$$

28. If numbers $\sqrt{2}$, $\sqrt{3}$, $\sqrt{5}$ were terms of an arithmetical progression with difference d, they would differ from one another by a multiple of number d, i.e. there would exist integers m and n, different from 0 and such that

$$\sqrt{3}-\sqrt{2} = md,$$

$$\sqrt{5}-\sqrt{2} = nd.$$

Eliminating d, we should obtain

$$m(\sqrt{5}-\sqrt{2}) = n(\sqrt{3}-\sqrt{2}),$$

whence

$$m\sqrt{5}-n\sqrt{3} = (m-n)\sqrt{2}.$$

Squaring both sides, we should obtain

$$5m^2+3n^2-2mn\sqrt{15} = 2(m-n)^2$$

and finally

$$\sqrt{15} = \frac{5n^2+3n^2-2(m-n)^2}{2mn}.$$

This equality, however, is contradictory since the left-hand side is an irrational number whereas the right-hand side is a rational number. Numbers $\sqrt{2}$, $\sqrt{3}$, $\sqrt{5}$ thus cannot be terms of the same arithmetical progression.

29. Since $1 \leqslant a \leqslant 2$, $\sqrt{(a-1)}$ denotes a definite number, and $0 \leqslant \sqrt{(a-1)} \leqslant 1$. Further, we have

$$a+2\sqrt{(a-1)} = a-1+2\sqrt{(a-1)}+1 = [\sqrt{(a-1)}+1]^2,$$

$$a-2\sqrt{(a-1)} = a-1-2\sqrt{(a-1)}+1 = [\sqrt{(a-1)}-1]^2.$$

The expression

$$u = \sqrt{[a+2\sqrt{(a-1)}]}+\sqrt{[a-2\sqrt{(a-1)}]}$$

thus has a definite numerical value, namely

$$u = |\sqrt{(a-1)}+1|+|\sqrt{(a-1)}-1| = \sqrt{(a-1)}+1+1-\sqrt{(a-1)} = 2.$$

30. *Hint.*

$$\sqrt{[x+24-10\sqrt{(x-1)}]} = \sqrt{[\sqrt{(x-1)}-5]^2} = |\sqrt{(x-1)}-5|;$$

the expression y has a constant value equal to 5 if $1 \leqslant x \leqslant 26$.

31. *Method I.* First, we shall prove that for any natural n there exist natural numbers a and b such that

$$(1-\sqrt{2})^n = \sqrt{a^2}-\sqrt{(2b^2)},$$

and

$$a^2-2b^2 = (-1)^n.$$

Proof. For $n = 1$ the theorem is valid, namely $a = b = 1$. Suppose that the theorem is valid for a certain n; then

$$\begin{aligned}(1-\sqrt{2})^{n+1} &= (1-\sqrt{2})^n(1-\sqrt{2}) = [\sqrt{a^2}-\sqrt{(2b^2)}](1-\sqrt{2}) \\ &= (a-b\sqrt{2})(1-\sqrt{2}) = (a+2b)-(a+b)\sqrt{2} \\ &= \sqrt{[(a+2b)^2]}-\sqrt{[2(a+b)^2]} \\ &= \sqrt{a_1^2}-\sqrt{(2b_1^2)},\end{aligned}$$

where a_1 and b_1 are natural numbers and

$$\begin{aligned}a_1^2-2b_1^2 &= (a+2b)^2-2(a+b)^2 = -a^2+2b^2 \\ &= -(a^2-2b^2) = (-1)^{n+1}.\end{aligned}$$

Thus the theorem is also valid for the exponent $n+1$. Hence we infer by induction that the theorem is valid for any natural n.

The theorem involved in the problem is an immediate conclusion from the theorem proved above; for, if n is an even number, then

$$(\sqrt{2}-1)^n = (1-\sqrt{2})^n = \sqrt{a^2}-\sqrt{(2b^2)},$$

where a and b, and therefore also a^2 and $2b^2$, are natural numbers and $a^2-2b^2 = 1$. If n is an odd number, then

$$(\sqrt{2}-1)^n = -(1-\sqrt{2})^n = \sqrt{(2b^2)}-\sqrt{a^2},$$

where $2b^2$ and a^2 are natural numbers and

$$2b^2-a^2 = -(a^2-2b^2) = -(-1) = 1.$$

Method II. Since

$$(\sqrt{2}-1)^n = \frac{(\sqrt{2}+1)^n+(\sqrt{2}-1)^2}{2} - \frac{(\sqrt{2}+1)^n-(\sqrt{2}-1)^n}{2},$$

we have

$$(\sqrt{2}-1)^n = \sqrt{m}-\sqrt{k},$$

where

$$m = \left[\frac{(\sqrt{2}+1)^n+(\sqrt{2}-1)^n}{2}\right]^2 = \frac{(\sqrt{2}+1)^{2n}+(\sqrt{2}-1)^{2n}+2}{4},$$

$$k = \left[\frac{(\sqrt{2}+1)^n-(\sqrt{2}-1)^n}{2}\right]^2 = \frac{(\sqrt{2}+1)^{2n}+(\sqrt{2}-1)^{2n}-2}{4},$$

and consequently

$$m-k = 1, \quad \text{i.e.} \quad k = m-1.$$

It remains to prove that m is a natural number. According to Newton's binomial formula we have

$$(\sqrt{2}+1)^n = (\sqrt{2})^n+\binom{n}{1}(\sqrt{2})^{n-1}+\binom{n}{2}(\sqrt{2})^{n-2}+\binom{n}{3}(\sqrt{2})^{n-3}+\ldots$$

$$(\sqrt{2}-1)^n = (\sqrt{2})^n-\binom{n}{1}(\sqrt{2})^{n-1}+\binom{n}{2}(\sqrt{2})^{n-2}-\binom{n}{3}(\sqrt{2})^{n-3}+\ldots,$$

and thus

$$\sqrt{m} = \frac{(\sqrt{2}+1)^n+(\sqrt{2}-1)^n}{2} = (\sqrt{2})^n+\binom{n}{2}(\sqrt{2})^{n-2}+\ldots$$

The above equality implies that, if n is even, $\sqrt{m}$ is a sum of natural numbers, whence $\sqrt{m}$ and m are natural numbers. If n is odd, then $\sqrt{m}$ is a sum of numbers of the form $a\sqrt{2}$, where a is a natural number; consequently $\sqrt{m}$ is the product of a natural number and $\sqrt{2}$, whence m is a natural number. The theorem is thus proved.

We shall give two more solutions of the problem, not so concise as the above two but having the advantage of suggesting themselves quite naturally.

Method III. We shall use the induction method. The theorem stating that for a natural n we have the equality

$$(\sqrt{2}-1)^n = m-\sqrt{(m-1)} \qquad (m\text{—natural number})$$

is valid if $n = 1$; in this case $m = 2$. Suppose that it is valid for a certain natural n; then

$$\begin{aligned}(\sqrt{2}-1)^{n+1} &= (\sqrt{2}-1)^n(\sqrt{2}-1) = [(\sqrt{m}-\sqrt{(m-1)}](\sqrt{2}-1)\\ &= \sqrt{(2m)}+\sqrt{(m-1)}-\sqrt{[2(m-1)]}-\sqrt{m}\\ &= \sqrt{[\sqrt{(2m)}+\sqrt{(m-1)}]^2}-\sqrt{\{\sqrt{[2(m-1)]}+\sqrt{m}\}^2},\end{aligned}$$

and

$$\begin{aligned}&[\sqrt{(2m)}+\sqrt{(m-1)}]^2-\{\sqrt{[2(m-1)]}+\sqrt{m}\}^2\\ &= \{3m-1+2\sqrt{[2m(m-1)]}\}-\{3m-2+2\sqrt{[2m(m-1)]}\} = 1.\end{aligned}$$

We shall prove that $[\sqrt{2m}+\sqrt{(m-1)}]^2$ is a natural number. Since

$$[\sqrt{(2m)}+\sqrt{(m-1)}]^2 = 3m-1+2\sqrt{[2m(m-1)]},$$

it is sufficient to prove that $2m(m-1)$ is a square of a natural number. It will be observed that $(\sqrt{2}-1)^n$ is a number of the form $a\sqrt{2}+b$, where a and b are integers, since each term of the expansion of $(\sqrt{2}-1)^n$ according to Newton's formula is either an integer or the product of an integer and $\sqrt{2}$. Consequently, by the induction hypothesis

$$\sqrt{m}-\sqrt{(m-1)} = a\sqrt{2}+b.$$

By squaring, we obtain

$$2m-1-2\sqrt{[m(m-1)]} = 2a^2+b^2+2ab\sqrt{2}.$$

It follows that

$$-2\sqrt{[m(m-1)]} = 2ab\sqrt{2}, \qquad 2m(m-1) = 4a^2b^2,$$

and thus the number $2m(m-1)$ is the square of the natural number $2|ab|$.

We have shown that the theorem is valid for the exponent $n+1$ if it is valid for the exponent n. And since it is valid for $n = 1$, it is valid for any natural n.

REMARK. In the end part of the above proof we assumed the following theorem:

If

$$A+B\sqrt{C} = K+L\sqrt{M},$$

where A, B, C, K, L, M are rational numbers, $L \neq 0$, and M is not a square of a rational number, then

$$A = K \quad \text{and} \quad B\sqrt{C} = L\sqrt{M}.$$

The proof of this theorem is simple. The equality assumed implies that

$$B\sqrt{C} = K-A+L\sqrt{M}.$$

Hence

$$B^2C = (K-A)^2+2(K-A)L\sqrt{M}+L^2M,$$

$$2(K-A)L\sqrt{M} = B^2C-L^2M-(K-A)^2.$$

The right-hand side of the last equation represents a rational number, and thus the left-hand side must also be equal to a rational number; under the assumption made regarding L and M, this occurs only for $K = A$.

$$\begin{aligned}(ab-1)x+(a+ab)z &= 0,\\ (bc+c)x+(bc-1)z &= 0,\\ bx-y+bz &= 0,\end{aligned} \tag{8}$$

consisting of equations (4) and the second equation of system (3). Indeed, if equations (3) hold, then equations (8) also hold; conversely, equations (8) imply the first and the third equations of system (3) if from the first and the second equations of (8) we subtract the third equation, having first multiplied it once by a and once by c.

Now, if equation (2), which can be written in the form

$$(bc+c)(a+ab)-(ab-1)(bc-1) = 0,$$

is satisfied, then we can satisfy the first two equations of (8) by setting, say,

$$x = a+ab, \quad z = 1-ab;$$

substituting these values in the third equation of system (8), we shall obtain $y = b+ab$.

If at least one of the numbers $a+ab$, $1-ab$, $b+ab$ is different from 0, then they constitute a non-zero solution of the system of equations (8).

If they are each equal to zero,

$$a+ab = 0, \quad 1-ab = 0, \quad b+ab = 0,$$

then $a = -1$, $b = -1$; in this case system (8) assumes the form

$$\begin{aligned}0\times x+0\times z &= 0,\\ 0\times x-(c+1)z &= 0,\\ -x-y-z &= 0,\end{aligned}$$

and it is obvious that this system has non-zero solutions, e.g. $x = 1$, $y = -1$, $z = 0$.

The theorems proved above—the direct one and the inverse one—can be jointly expressed as follows:

The necessary and sufficient condition that the system of equations (1) *have non-zero solutions is equality* (2).

REMARK 2. The above theorem is a particular case of an important theorem of algebra which states that a system of n homogeneous linear equations has non-zero solutions if and only if the so-called *determinant* of the system is equal to zero. In our case this determinant is

$$\Delta = \begin{vmatrix} -1 & a & a \\ b & -1 & b \\ c & c & -1 \end{vmatrix} = ab+bc+ca+2abc-1.$$

33. Let us recall the definition of the *absolute value* $|l|$ of a real number l:

$$\text{if } l \geqslant 0 \quad \text{then} \quad |l| = l,$$
$$\text{if } l < 0 \quad \text{then} \quad |l| = -l.$$

In order to solve the equation

$$|x|+|x-1|+|x-2| = a, \tag{1}$$

it is best to replace it with an equation which does not contain the symbol of absolute value. We shall seek the solutions of the equation successively in the intervals $(-\infty, 0)$, $(0, 1)$, $(1, 2)$, $(2, \infty)$.

1. If $x \leqslant 0$, then equation (1) assumes the form

$$-x-(x-1)-(x-2) = a, \quad \text{i.e.} \quad -3x+3 = a.$$

Hence

$$x = \frac{3-a}{3}.$$

The condition $x \leqslant 0$ is satisfied if $a \geqslant 3$.

2. If $0 \leqslant x \leqslant 1$, equation (1) assumes the form

$$x-(x-1)-(x-2) = a, \quad \text{i.e.} \quad -x+3 = a.$$

Hence

$$x = 3-a.$$

The condition $0 \leqslant x \leqslant 1$ is satisfied if $2 \leqslant a \leqslant 3$.

3. If $1 \leqslant x \leqslant 2$, equation (1) assumes the form

$$x+x-1-(x-2) = a, \quad \text{i.e.} \quad x+1 = a.$$

Hence

$$x = a-1.$$

The condition $1 \leqslant x \leqslant 2$ is satisfied if $2 \leqslant a \leqslant 3$.

4. If $x \geqslant 2$, equation (1) assumes the form

$$x+x-1+x-2 = a, \quad \text{i.e.} \quad 3x-3 = a.$$

Hence

$$x = \frac{a+3}{3}.$$

The condition $x \geqslant 2$ is satisfied if $a \geqslant 3$.

Let us list the results obtained:

If $a < 2$, equation (1) has no solutions.

If $a = 2$, equation (1) has 1 solution: $x = 1$.

If $2 < a < 3$, equation (1) has 2 solutions: $x = 3-a$,
$x = a-1$.

If $a \geqslant 3$, equation (1) has 2 solutions: $x = (3-a)/3$,
$x = (a+3)/3$.

The dependence of x upon a is represented graphically in Fig. 4.

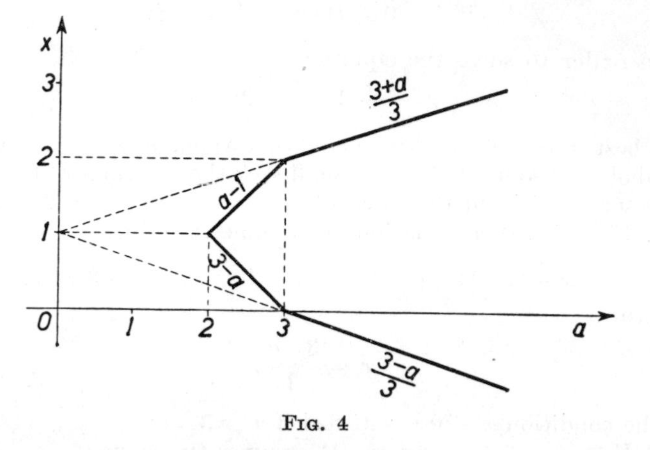

FIG. 4

34. On stagnant water, the motor-boat would reach the ball after another 15 minutes, i.e. at 9·30. It will be the same on the river because the current "carries" the motor-boat at the same rate as it carries the ball. The motor-boat overtook the ball at 9·30.

If one did not hit upon the above simple reasoning, it would be possible to solve the problem by means of equations.

Seemingly there are too few data in the problem because we are told neither the velocity proper of the motor-boat, i.e. its speed on stagnant water, nor the velocity of the river current.

In spite of that let us denote by x the number of hours elapsing between 9 o'clock and the moment at which the motor-boat overtook the ball. Let us introduce the velocity u of the current expressed in kilometres per hour and the velocity v of the motor-boat in stagnant water also expressed in km/hr. To form an equation we must express by means of these quantities the distance covered by the ball and that covered by the motor-boat.

In the course of x hr the ball travelled xu km with the current of the river.

The motor-boat sailed against the current for the first quarter of an hour; its velocity was then $v-u$ km/hr and the distance it covered was therefore $\frac{1}{4}(v-u)$ km. For the remaining $x-\frac{1}{4}$ hr the motor-boat sailed with the current, and thus with a velocity of $u+v$ km/hr; it covered $(x-\frac{1}{4})(x+v)$ km, travelling first the

$\frac{1}{4}(v-u)$ km back to the starting point and then sailing the xu km covered by the ball down to the point where it was overtaken by the boat.

Thus

$$\left(x-\frac{1}{4}\right)(u+v)=\frac{1}{4}(v-u)+xu.$$

When we simplify this equation by opening the brackets and grouping the terms containing x on one side and the remaining terms on the other side, we obtain the equation

$$xv=\frac{1}{2}v.$$

Since the specific velocity v of the motor-boat is certainly not equal to zero, we can divide both sides of the equation by v and obtain

$$x=\frac{1}{2}.$$

Thus the motor-boat will overtake the ball in $\frac{1}{2}$ hr from the moment it started, i.e. at 9·30.

The unknown quantities u and v have proved to be needed only to form the equation. In the calculations they disappeared, i.e.—as we say in mathematics—they were eliminated.

35. If the roots of the equation

$$x^2+(m-2)x-(m+3)=0 \tag{1}$$

are numbers x_1 and x_2, then

$$x_1^2+(m-2)x_1-(m+3)=0,$$
$$x_2^2+(m-2)x_2-(m+3)=0.$$

Adding these equalities we obtain

$$x_1^2+x_2^2+(m-2)(x_1+x_2)-2(m+3)=0.$$

Hence

$$x_1^2+x_2^2=-(m-2)(x_1+x_2)+2(m+3),$$

and since $x_1+x_2=-(m-2)$, we have

$$x_1^2+x_2^2=(m-2)^2+2(m+3)=m^2-2m+10=(m-1)^2+9.$$

The expression $(m-1)^2+9$ has its minimum value when $m=1$. Equation (1) then has the form $x^2-x-4=0$; its roots are

$$x_1=\frac{1+\sqrt{17}}{2}, \quad x_2=\frac{1-\sqrt{17}}{2}$$

and

$$x_1^2+x_2^2 = \left(\frac{1+\sqrt{17}}{2}\right)^2+\left(\frac{1-\sqrt{17}}{2}\right)^2 = 9.$$

REMARK. If we take only real numbers into consideration, then a quadratic equation has roots if and only if its discriminant Δ is non-negative.

In our problem the discriminant of equation (1) is

$$\Delta = (m-2)^2+4(m+3) = m^3+16$$

and is positive for any value of m; thus equation (1) always has two roots.

If we pose the same problem for the equation

$$x^2-(m+2)x+(m+5) = 0 \tag{2}$$

for example, then a procedure analogous to the preceding one gives

$$x_1^2+x_2^2 = (m+1)^2-7.$$

The expression $(m+1)^2-7$ has its least value when $m = -1$; that least value is -7. This result seems false since the sum of the squares $x_1^2+x_2^2$ cannot have a negative value.

The point is that the discriminant of equation (2) is

$$\Delta = (m+2)^2-4(m+5) = m^2-16,$$

and is non-negative only if $m \leqslant -4$ or if $m \geqslant 4$. The equation has roots only for such values of m; for other values of m (including $m = -1$) there are no roots.

It is thus necessary to seek the least value of the sum of the squares of the roots of the equation, equal to $(m+1)^2-7$, under the assumption that $|m| \geqslant 4$.

Now if m increases from $-\infty$ to -4, the value of the expression $(m+1)^2-7$ decreases from ∞ to 2; if m increases from 4 to ∞, the value of the expression increases from 18 to ∞.

Hence the conclusion that the sum of the squares of the roots of equation (2) is the least when $m = -4$; then equation (2) assumes the form $x^2+2x+1 = 0$ and has roots $x_1 = x_2 = -1$; we then have $x_1^2+x_2^2 = 2$.

The position is different if we consider the quadratic equation x^2+px+q in the domain of complex numbers; then the roots always exist and—whether they are real or imaginary—the following formula is valid:

$$x_1^2+x_2^2 = (x_1+x_2)^2-2x_1x_2 = p^2-2q.$$

If the coefficients p and q in the equation are real, then according to this formula $x_1^2+x_2^2$ is also a real number, even if x_1 and x_2 are complex numbers.

E.g., for equation (2), in which m denotes any real number, the sum of the squares of the roots is equal to the real number $(m+1)^2-7$ and has its least value -7 when $m=-1$. For this value of m equation (2) is of the form $x^2-x+4=0$; its roots are

$$x_1 = \frac{1+i\sqrt{15}}{2} \quad \text{and} \quad x_2 = \frac{1-i\sqrt{15}}{2}$$

and it can easily be verified that

$$x_1^2+x_2^2 = \left(\frac{1+i\sqrt{15}}{2}\right)^2+\left(\frac{1-i\sqrt{15}}{2}\right)^2 = -7.$$

36. Consider the left-hand side of the equation

$$(x-a)(x-c)+2(x-b)(x-d) = 0, \tag{1}$$

i.e. the quadratic function

$$(x-a)(x-c)+2(x-b)(x-d). \tag{2}$$

We know from algebra that a quadratic function has two real roots (in other words two zeros) if and only if it assumes both positive and negative values. Now if x is greater than any of the numbers, a, b, c, d, then each of the differences $x-a$, $x-b$, $x-c$, $x-d$ is positive and function (2) has a positive value.

It is easy to indicate a value of x for which this function is negative; e.g. if $x=b$, it has the value

$$(b-a)(b-c),$$

which is negative, because by hypothesis $b-a>0$ and $b-c<0$. The theorem is thus proved.

Remark 1. It can be seen from the above proof that, instead of making the assumption $a<b<c<d$, it is sufficient to assume that $a<b<c$.

Similarly, it would be sufficient to assume that $b<c<d$, since then, by substituting the value $x=c$ in (2), we should obtain a negative number $(c-b)(c-d)$. Moreover, it will be observed that the value of expression (2) remains the same if we interchange the letters a and c, and also if we interchange the letters b and d. We can therefore state a theorem which is stronger than the preceding one:

If either of the numbers b and d lies between the numbers a and c or if either of the numbers a and c lies between the numbers b and d, then equation (1) *has two real roots.*

This theorem remains valid if we replace in equation (1) the coefficient 2 by any positive number k; the above proof requires no alterations. It will be different if, instead of the coefficient 2, we take a negative number. We suggest that the reader should investigate this case and find a sufficient condition for the roots of the equation to be real.

REMARK 2. Another, considerably longer, proof of the theorem can be derived from the consideration of the discriminant of equation (1):

$$\Delta = (a+c+2b+2d)^2-12(ac+2bd).$$

Following the pattern of the solution of problem 64, we can easily show that if either of the numbers a and c lies between the numbers b and d or if either of the numbers b and d lies between the numbers a and c, then $\Delta > 0$.

37. *Method I.* Suppose that equations (1) have a root in common, i.e. that there exists a number x for which equalities (1) hold. From (1) we must infer equality (2), in which number x does not appear, i.e. we must eliminate the variable x (cf. problem 32).

Subtracting equations (1) we obtain

$$(m-p)x+(n-q) = 0. \tag{3}$$

(i) If $m-p = 0$, then it follows from (3) that $n-q = 0$; consequently equation (2) is satisfied, because each of the terms on the left-hand side is equal to zero.

(ii) If $m-p \neq 0$, then equation (3) implies

$$x = -\frac{n-q}{m-p}. \tag{4}$$

We substitute value (4) in one of the equations (1), say in the first of them, and obtain

$$\left(\frac{n-q}{m-p}\right)^2 -m\frac{n-q}{m-p}+n = 0,$$

whence

$$(n-q)^2-m(n-q)(m-p)+n(m-p)^2 = 0;$$

finally we have

$$(n-q)^2-(m-p)(np-mq) = 0,$$

i.e. the required condition (2).

REMARK. In eliminating the unknown x we have distinguished two cases. It is possible, however, to perform the elimination in such a way that no distinctions are necessary; that is preferable because in mathematics we always strive for arguments that are as general as possible.

Multiplying both sides of equation (3) by x, we obtain

$$(m-p)x^2+(n-q)x = 0, \tag{5}$$

and from the first of the given equations (1) we have

$$x^2 = -(mx+n). \tag{6}$$

We substitute the expression for x^2 from formula (6) in equation (5):

$$-(m-p)(mx+n)+(n-q)x = 0.$$

Hence

$$[-m(m-p)+(n-q)]x-n(m-p) = 0. \tag{7}$$

Now we eliminate x from the linear equations (3) and (7). Accordingly we multiply equation (3) by $[-m(m-p)+(n-q)]$ and equation (7) by $-(m-p)$ and obtain by addition:

$$(n-q)[-m(m-p)+(n-q)]+n(m-p)^2 = 0.$$

Hence

$$(n-q)^2-(m-p)[m(n-q)-n(m-p)] = 0$$

and finally

$$(n-q)^2-(m-p)(np-mq) = 0.$$

Method II. If one of the roots x_1, x_2 of the first equation of (1) is equal to one of the roots of the second equation, then one of the differences x_1-x_3, x_2-x_3, x_1-x_4, x_2-x_4 is equal to zero. This holds if and only if

$$(x_1-x_3)(x_2-x_3)(x_1-x_4)(x_2-x_4) = 0.$$

If we perform the multiplication on the left-hand side of this equality and make use of the relations between the coefficients and the roots of equations (1), namely $x_1+x_2 = -m$, $x_1x_2 = n$, $x_3+x_4 = -p$, $x_3x_4 = q$, we shall obtain

$$(n-q)^2-(m-p)(np-mq) = 0.$$

Method III. If one of the roots x_1, x_2 of the first equation of (1) is at the same time a root of the second equation, then one of the numbers $x_1^2+px_1+q$, $x_2^2+px_2+q$ is equal to zero, and consequently

$$(x_1^2+px_1+q)(x_2^2+px_2+q) = 0.$$

Performing the multiplication on the left-hand side and taking into account the relations $x_1+x_2 = -m$, $x_1x_2 = n$, we obtain

$$(n-q)^2-(m-p)(np-mq) = 0.$$

REMARK. The theorem proved in this problem states that equality (2) is a necessary condition for equations (1) to have a root in common. It can easily be shown that it is also a sufficient condition, i.e. that the inverse theorem is also true: if equality (2) holds, equations (1) have a root in common.

Indeed, if we substitute in equation (2) $m = -(x_1+x_2)$, $n = x_1x_2$, $p = -(x_3+x_4)$, $q = x_3x_4$, we shall obtain (see method II above):

$$(x_1-x_3)(x_2-x_3)(x_1-x_4)(x_2-x_4) = 0,$$

which implies that one of the differences x_1-x_3, x_2-x_3, x_1-x_4, x_2-x_4 is equal to 0, i.e. one of the roots of the first equation of (1) is equal to one of the roots of the second equation.

The above argument assumes the existence of the roots of equations (1), and is valid in the domain of complex numbers. If we consider only the real roots of equations (1) with real coefficients, the theorem must be modified. Although it follows from (2) that one of the roots x_3, x_4 is equal to one of the roots x_1, x_2, yet those roots might be imaginary numbers. We know, however, that the roots of each of the equations (1) would then be conjugate† complex numbers. Thus if we had, for instance, $x_3 = x_1$, then the conjugate numbers x_4, x_2 would also be equal, $x_4 = x_2$, and the given equations (1) would be identical.

Thus, in the set of real numbers, the inverse theorem under consideration reads:

If equality (2) *holds and equations* (1) *are not identical, then those equations have a common root.*

38. The problem is this: from the assumption that the trinomials

$$x^2+mx+n \quad \text{and} \quad x^2+px+q \qquad (1)$$

have real roots such that the first pair of roots separate the other pair we are to derive the relation which then holds between the coefficients m, n, p and q.

---

† The equation $x^2+mx+n = 0$ with real coefficients has imaginary roots if $m^2-4n < 0$; those roots are conjugate complex numbers

$$x_{1,2} = \frac{-m \pm i\sqrt{(4n-m^2)}}{2}.$$

Denote the roots of the first trinomial by x_1, x_2 and the roots of the second trinomial by x_3, x_4. Let x_3 lie inside the interval (x_1, x_2) and x_4 outside this interval.

Method I. We consider the differences between the roots of one trinomial and those of the other.

By hypothesis the differences x_3-x_1 and x_3-x_2 have opposite signs and the differences x_4-x_1 and x_4-x_2 have identical signs, whence

$$(x_3-x_1)(x_3-x_2) < 0 \quad \text{and} \quad (x_4-x_1)(x_4-x_2) > 0. \tag{2}$$

Since $x_1+x_2 = -m$, $x_1x_2 = n$, we have

$$(x_3-x_1)(x_3-x_2) = x_3^2-(x_1+x_2)x_3+x_1x_2 = x_3^2+mx_3+n,$$
$$(x_4-x_1)(x_4-x_2) = x_4^2-(x_1+x_2)x_4+x_1x_2 = x_4^2+mx_4+n.$$

Inequalities (2) assume the form

$$x_3^2+mx_3+n < 0 \quad \text{and} \quad x_4^2+mx_4+n > 0. \tag{3}$$

It follows that

$$(x_3^2+mx_3+n)(x_4^2+mx_4+n) < 0. \tag{4}$$

Let us perform the multiplication on the left-hand side of inequality (4) taking into account that $x_3+x_4 = -p$, $x_3x_4 = q$, and $x_3^2+x_4^2 = (x_3+x_4)^2-2x_3x_4 = p^2-2q$. We shall obtain

$$\begin{aligned}
&(x_3^2+mx_3+n)(x_4^2+mx_4+n)\\
&= x_3^2x_4^2+mx_3x_4(x_3+x_4)+m^2x_3x_4+n(x_3^2+x_4^2)+mn(x_3+x_4)+n^2\\
&= q^2-mpq+m^2q+n(p^2-2q)-mnp+n^2\\
&= (n-q)^2+mq(m-p)+np(p-m)\\
&= (n-q)^2+(m-p)\ (mq-np).
\end{aligned}$$

Inequality (4) assumes the form

$$(n-q)^2+(m-p)(mq-np) < 0. \tag{5}$$

We have obtained the following theorem:

If the pairs of the roots of trinomials (1) *separate each other, then the coefficients of these trinomials satisfy condition* (5).

Method II. We shall apply a well-known theorem on the sign of the quadratic function

$$f(x) = x^2+mx+n,$$

which reads:

If the function $f(x) = x^2+mx+n$ has real roots x_1 and x_2 and $x_1 < x_2$, then for $x_1 < x < x_2$ we have $f(x) < 0$ and for $x < x_1$ or $x > x_2$ we have $f(x) > 0$.

Thus if number x_3 lies inside and number x_4 outside the interval (x_1, x_2), then the first of the values

$$f(x_3) = x_3^2+mx_3+n, \quad f(x_4) = x_4^2+mx_4+n \tag{6}$$

is negative and the other positive, whence

$$(x_3^2+mx_3+n)(x_4^2+mx_4+n) < 0,$$

which, as shown in method I, leads to condition (5).

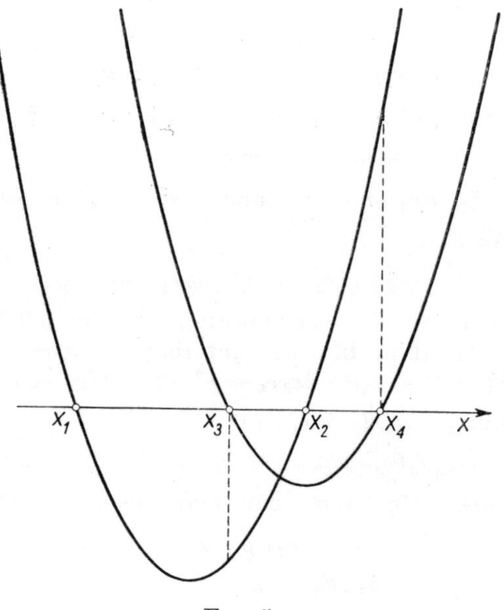

FIG. 5

Figure 5 illustrates the respective positions of the parabolas which are the graphs of the given quadratic trinomials

$$y = x^2+mx+n \quad \text{and} \quad y = x^2+px+q$$

with mutually separated roots.

REMARK 1. We have proved that if trinomials (1) have real roots and if the pairs of those roots separate each other, then inequality (5) is satisfied; this inequality thus gives a necessary condition for the pairs of the roots of trinomials (1) to separate each other.

We shall prove that this condition is sufficient, i.e. that the inverse theorem holds:

If inequality (5) *holds, then trinomials* (1) *have real roots and the pairs of the roots of those trinomials separate each other.*

Proof. Substitute in inequality (5) the values

$$p = -(x_3+x_4), \; q = x_3x_4.$$

We obtain inequality (4), which, by formulas (6), can be written in the form

$$f(x_3)\,f(x_4) < 0. \tag{7}$$

From inequality (7) we shall first draw the conclusion that the roots x_3 and x_4 are real numbers.

Suppose that x_3 and x_4 are imaginary numbers. We know from algebra that, if the roots of a quadratic function are imaginary, then they are conjugate complex numbers, i.e. that

$$x_3 = \alpha+i\beta, \qquad x_4 = \alpha-i\beta,$$

where α and β are real numbers and i denotes the imaginary unit ($i^2 = -1$).

Let us find the values of the function $f(x) = x^2+mx+n$ by substituting for x the numbers x_3 and x_4

$$\begin{aligned} f(x_3) &= (\alpha+i\beta)^2+m(\alpha+i\beta)+n \\ &= (\alpha^2-\beta^2+m\alpha+n)+i(2\alpha\beta+m\beta), \\ f(x_4) &= (\alpha-i\beta)^2+m(\alpha-i\beta)+n \\ &= (\alpha^2-\beta^2+m\alpha+n)-i(2\alpha\beta+m\beta). \end{aligned}$$

We can see that if x_3 and x_4 are conjugate complex numbers, then $f(x_3)$ and $f(x_4)$ are also conjugate complex numbers. A product of conjugate complex numbers is a non-negative real number, as can be seen from the equality $(a+bi)(a-bi) = a^2+b^2$; consequently

$$f(x_3)\,f(x_4) \geqslant 0. \tag{8}$$

The assumption that x_3 and x_4 are imaginary numbers has led to inequality (8), which contradicts inequality (7); thus the numbers x_3 and x_4 are real.

Therefore it follows from (7) that $f(x_3)$ and $f(x_4)$ are real numbers with opposite signs. Consequently the function $f(x) = x^2+mx+n$ has real roots x_1 and x_2, one of them lying between x_3 and x_4 and the other outside the interval (x_3, x_4), which is what was to be proved.

Finally we have the following theorem:

A necessary and sufficient condition for the trinomials

$$x^2+mx+n \quad \textit{and} \quad x^2+px+q$$

(m, n, p, q—real numbers) to have real and mutually separating pairs of roots is the inequality

$$(n-q)^2+(m-p)(mq-np) > 0.$$

REMARK 2. If the pairs of numbers (x_1, x_2) and (x_3, x_4) separate each other, then, according to which of the numbers x_3, x_4 lies between the numbers x_1 and x_2, either the differences x_3-x_1, x_3-x_2 have opposite signs and the differences x_4-x_1, x_4-x_2 have identical signs or *vice versa*. In both cases

$$(x_3-x_1)(x_3-x_2)(x_4-x_1)(x_4-x_2) < 0. \tag{9}$$

Conversely: if inequality (9) holds, then one of the products $(x_3-x_1)(x_3-x_2)$ and $(x_4-x_1)(x_4-x_2)$ is positive and the other is negative; hence the pairs (x_1, x_2) and (x_3, x_4) separate each other.

Condition (9) can be replaced by the following equivalent condition:

$$\frac{x_3-x_1}{x_3-x_2} : \frac{x_4-x_1}{x_4-x_2} < 0. \tag{10}$$

Inequality (9) or (10) thus expresses a *necessary and sufficient condition for the pairs of numbers (x_1, x_2) and (x_3, x_4) to separate each other.*

The expression appearing on the left side of inequality (10) plays a considerable part in mathematical considerations: it is called the *cross-ratio of an ordered quadruple of numbers* x_1, x_2, x_3, x_4 and is denoted by the symbol (x_1, x_2, x_3, x_4).

Thus

$$(x_1, x_2, x_3, x_4) = \frac{x_3-x_1}{x_3-x_2} : \frac{x_4-x_1}{x_4-x_2}.$$

Let us take, on the number axis, points A_1, A_2, A_3, A_4 with abscissae x_1, x_2, x_3, x_4.

The cross-ratio of four numbers (x_1, x_2, x_3, x_4) is also termed the *cross-ratio of four points* A_1, A_2, A_3, A_4 and denoted by (A_1, A_2, A_3, A_4).

Since the differences of abscissae $x_3-x_1, x_3-x_2, x_4-x_1, x_4-x_2$ are the relative measures of the vectors $A_1A_3, A_2A_3, A_1A_4, A_2A_4$ lying on the axis (cf. problem 66), we have:

$$(A_1, A_2, A_3, A_4) = \frac{A_1A_3}{A_2A_3} : \frac{A_1A_4}{A_2A_4}.$$

Thus the cross-ratio (A_1, A_2, A_3, A_4) is the ratio of the ratios in which the points A_3 and A_4 divide the directed segment A_1A_2.

If $(A_1, A_2, A_3, A_4) < 0$, the pairs of points (A_1, A_2) and (A_3, A_4) separate each other.

If $(A_1, A_2, A_3, A_4) > 0$, those pairs do not separate each other, i.e. either one pair lies within the other, or the two pairs lie outside each other.

In the particular case of $(A_1, A_2, A_3, A_4) = -1$, i.e. if the division ratios $\frac{A_1A_3}{A_2A_3}$ and $\frac{A_1A_4}{A_2A_4}$ are opposite numbers, we say that the pairs (A_1, A_2) and (A_3, A_4) separate each other *harmonically* or that the *quadruple of points* A_1, A_2, A_3, A_4 *is harmonic.*

This case is shown in Fig. 6, in which

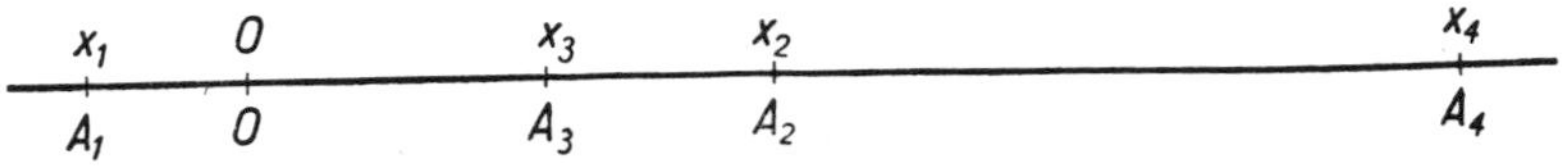

x_1 0 x_3 x_2 x_4

A_1 O A_3 A_2 A_4

FIG. 6

$$\frac{A_1A_3}{A_2A_3} = -2, \qquad \frac{A_1A_4}{A_2A_4} = 2.$$

39. *Method I.* We know from algebra that the numbers x_1, x_2, x_3 are the roots of the equation $x^3+mx^2+nx+p = 0$ if and only if the following conditions are satisfied:

$$x_1+x_2+x_3 = -m,$$
$$x_1x_2+x_2x_3+x_3x_1 = n,$$
$$x_1x_2x_3 = -p.$$

The numbers a, b, c are thus the roots of the equation

$$x^3-ax^2+bx-c = 0 \tag{1}$$

if and only if they satisfy the equations

$$\begin{aligned} a+b+c &= a, \\ ab+bc+ca &= b, \\ abc &= c. \end{aligned} \tag{2}$$

The first of these equations gives $b+c = 0$, which permits us to reduce the left-hand side of the second equation to the term bc and to replace the system of equations (2) by the equivalent system

$$\begin{aligned} b+c &= 0, \\ b(c-1) &= 0, \\ c(ab-1) &= 0. \end{aligned} \tag{3}$$

of the remaining factors of the second equation we can replace system (7) by an alternative of two systems of equations

$$\begin{aligned} c &= ab, \\ b+1 &= 0, \\ (b-1)(a^2b+1) &= 0, \end{aligned} \tag{7a}$$

or

$$\begin{aligned} c &= ab, \\ b-a &= 0, \\ (b-1)(a^2b+1) &= 0. \end{aligned} \tag{7b}$$

System (7a) has two solutions:

IV. $a = 1, \quad b = -1, \quad c = -1,$

V. $a = -1, \quad b = -1, \quad c = 1.$

Solution IV does not satisfy the conditions of the problem because the roots of the equation $x^3-x^2-x+1 = 0$, i.e. $(x-1)^2(x+1) = 0$, are the numbers $1, 1, -1$ and not the numbers $1, -1, -1$.

Solution V satisfies the conditions of the problem (see method I).

We replace system (7b) by an equivalent system by substituting a for b in the first and in the third equations; we shall obtain the system of equations

$$\begin{aligned} a^2-c &= 0, \\ a-b &= 0, \\ (a-1)(a^3+1) &= 0. \end{aligned} \tag{8}$$

System (8) has the solutions:

VI. $a = 1, \quad b = 1, \quad c = 1,$

VII. $a = -1, \quad b = -1, \quad c = 1,$

VIII. $a = \dfrac{1+i\sqrt{3}}{2} = \varepsilon, \quad b = \varepsilon, \quad c = \varepsilon^2,$

IX. $a = \dfrac{1-i\sqrt{3}}{2} = \dfrac{1}{\varepsilon}, \quad b = \dfrac{1}{\varepsilon}, \quad c = \dfrac{1}{\varepsilon^2}.$

Solution VI does not satisfy the conditions of the problem since the roots of the equation x^3-x^2+x-1 are the numbers $1, i, -i$, and not the numbers $1, 1, 1$.

Solution VII is identical with solution V.

Solution VIII does not satisfy the conditions of the problem since the roots of the equation $x^2-\varepsilon x^2+\varepsilon x-\varepsilon^2=0$ are the numbers $\varepsilon, \varepsilon^2, -\varepsilon^2$ and not the numbers $\varepsilon, \varepsilon, \varepsilon^2$.

Similarly, we ascertain that solution IX does not satisfy the conditions of the problem.

Finally our problem has the solutions:

(A) a—arbitrary, $b=0$, $c=0$,

(B) $a=-1$, $b=-1$, $c=1$.

40. The roots of the equation

$$x^3+ax^2+bx+c=0 \tag{1}$$

form an arithmetical progression if and only if the sum of two of them equals twice the third, i.e. if the sum of all three roots equals three times one of them; since the sum of all the roots of equation (1) is $-a$, one of them is $-\frac{1}{3}a$. Therefore a necessary and sufficient condition for the roots of equation (1) to form an arithmetic progression is that the number $-\frac{1}{3}a$ should satisfy the equation.

Substituting $-\frac{1}{3}a$ for x in (1), we obtain this condition in the form

$$\left(-\frac{1}{3}a\right)^3+a\left(-\frac{1}{3}a\right)^2+b\left(-\frac{1}{3}a\right)+c=0, \tag{2}$$

or in the simplified form

$$2a^3-9ab+27c=0. \tag{3}$$

Assume that the coefficients a, b, c satisfy equation (3), i.e. that equation (1) has a root $-\frac{1}{3}a$. It remains to find a necessary and sufficient condition for the other two roots of equation (1) to be real. Those roots are the roots of the quadratic equation which will be obtained by dividing both sides of (1) by $x+\frac{1}{3}a$. It is easiest to perform the division by subtracting equations (1) and (2)

$$x^3+\left(\frac{1}{3}a\right)^3+a\left[x^2-\left(\frac{1}{3}a\right)^2\right]+b\left(x+\frac{1}{3}a\right)=0.$$

By factorizing and simplifying we obtain

$$\left(x+\frac{1}{3}a\right)\left[x^2+\frac{2}{3}ax+\left(b-\frac{2}{9}a^2\right)\right]=0.$$

The roots of the quadratic equation

$$x^2+\frac{2}{3}ax+\left(b-\frac{2}{9}a^2\right)=0 \tag{4}$$

are real numbers if and only if

$$\Delta = \frac{4}{9}a^2 - 4\left(b - \frac{2}{9}a^2\right) \geqslant 0,$$

i.e.

$$a^2 - 3b \geqslant 0$$

or

$$b \leqslant \frac{1}{3}a^2. \tag{5}$$

The required necessary and sufficient conditions are thus relations (3) and (5). It will be observed that if the equality

$$b = \frac{1}{3}a^2$$

holds, then equality (3) gives

$$c = \frac{1}{27}a^2$$

and equation (1) assumes the form

$$x^3 + ax^2 + \frac{1}{3}a^2x + \frac{1}{27}a^3 = 0,$$

i.e.

$$\left(x + \frac{1}{3}a\right)^3 = 0.$$

This equation has three roots $-\frac{1}{3}a$, $-\frac{1}{3}a$, $-\frac{1}{3}a$, forming an arithmetical progression with difference 0.

Remark. The quadratic equation (4) can be obtained in a somewhat simpler way by using the following relation between the roots x_1, x_2, x_3 of equation (1):

$$x_1x_2 + x_2x_3 + x_3x_1 = b.$$

If $x_3 = -\frac{1}{3}a$, then $x_1 + x_2 = -\frac{2}{3}a$, and the above relation gives $x_1x_2 = b - \frac{2}{9}a^2$, in view of which x_1 and x_2 are the roots of equation (4).

41. Let us write the given equation in the form

$$\frac{m^2}{a-x} + \frac{n^2}{b-x} - 1 = 0, \tag{1}$$

and multiply this equation by $(a-x)(b-x)$

$$m^2(b-x) + n^2(a-x) - (a-x)(b-x) = 0. \tag{2}$$

Equations (1) and (2) are equivalent, i.e. they are satisfied by the same values of x. Namely, (2) results from (1) when multiplied by the number $(a-x)(b-x)$. And if a number x satisfies equation (2), then certainly $x \neq a$ and $x \neq b$, because for $x = a$ the left side of equation (2) assumes the value $m^2(b-a)$, which is different from 0, and for $x = b$ it assumes the value $n^2(a-b)$, which is also different from 0; thus $(a-x)(b-x)$ is different from 0; consequently number x satisfies equation (1), which is obtained from (2) through division by $(a-x)(b-x)$.

Accordingly, we can replace equation (1) by equation (2).

Method I. Equation (2) can be written in the form

$$x^2+[(m^2+n^2)-(a+b)]x+[ab-(m^2b+n^2a)] = 0.$$

Let us find the discriminant of this equation:

$$\Delta = [(m^2+n^2)-(a+b)]^2-4[ab-(m^2b+n^2a)].$$

By means of a suitable transformation we shall prove that $\Delta > 0$.

$$\begin{aligned}\Delta &= (m^2+n^2)^2-2(m^2+n^2)(a+b)+(a+b)^2-4ab+4m^2b+4n^2a\\ &= (m^2+n^2)^2-2(m^2-n^2)(a-b)+(a-b)^2\\ &= (m^2-n^2)^2-2(m^2-n^2)(a-b)+(a-b)^2+4m^2n^2\\ &= [(m^2-n^2)-(a-b)]^2+4m^2n^2.\end{aligned}$$

Since $[(m^2-n^2)-(a-b)]^2 \geqslant 0$ and $4m^2n^2 > 0$, we have $\Delta > 0$; thus equation (2) has two different real roots.

In the above calculation we have twice applied the formula

$$(a+b)^2 = (a-b)^2+4ab.$$

Method II. The preceding method requires rather long calculation. We can avoid it by applying the following simple argument. The left-hand side of equation (2) is a quadratic function of the variable x

$$\varphi(x) = m^2(b-x)+n^2(a-x)-(a-x)(b-x).$$

Function $\varphi(x)$ assumes for $x = a$ and $x = b$ the values

$$\varphi(a) = m^2(b-a) \quad \text{and} \quad \varphi(b) = n^2(a-b)$$

respectively.

These values are of opposite signs; it follows that the function $\varphi(x)$, and thus also equation (2), has one real root lying between a and b and another real root lying outside this interval.

Remark. To those readers who are acquainted with analytic geometry we shall now explain the relation which holds between the above problem and a property of curves of the second degree.

We shall rewrite equation (1) using different letters:

$$\frac{x^2}{a-\lambda}+\frac{y^2}{b-\lambda}-1=0. \tag{I}$$

Let x and y denote the orthogonal coordinates of a point in a plane, a and b two unequal numbers, and λ a variable parameter.

Let us investigate what curves are represented by equation (I) for different values of λ. Assume that $a > b$ (if $a < b$, the argument is analogous).

(1) If $\lambda < b$, then $a-\lambda > 0$ and $b-\lambda > 0$. Equation (I) is then the axial equation of an *ellipse* with semi-axes $\sqrt{(a-\lambda)}$ and $\sqrt{(b-\lambda)}$.

(2) If $b < \lambda < a$, then $a-\lambda > 0$, $b-\lambda < 0$. Equation (I) is then the axial equation of a *hyperbola* with semi-axes $\sqrt{(a-\lambda)}$ and $\sqrt{(\lambda-b)}$.

(3) If $\lambda > a$, equation (I) is not satisfied by any real values of x and y; we then say that the equation represents an *imaginary curve*.

In the sequel we shall consider only the values $\lambda < a$.

The curves represented by equation (I) have one interesting property: *all these curves have the same foci.* Indeed, in an ellipse half the distance between the foci, which we shall denote by c, is equal to the square root of the differences of the squared semi-axes, and in a hyperbola this distance is equal to the square root of the sum of the squared semi-axes. In case (1) we thus have

$$c = \sqrt{[(a-\lambda)-(b-\lambda)]} = \sqrt{(a-b)},$$

and in case (2) we also have

$$c = \sqrt{[(a-\lambda)+(\lambda-b)]} = \sqrt{(a-b)},$$

which means that c has a constant value, independent of λ.

We say that equation (I) with a variable parameter λ represents a *family of confocal conics.*

Let us assign to x and y definite (non-zero) values and let us regard equation (I) as an equation with the unknown λ. Since $a \neq b$, the assumptions of problem 41 are satisfied, and we know from the solution of that problem that equation (I) then has two different real roots, λ_1 and λ_2.

We thus obtain the following theorem as a geometrical interpretation of the algebraic theorem of problem 41.

Through every point of a plane lying neither on the x-axis nor on the y-axis there pass two curves of the family (I) *of confocal conics.*

It is easy to ascertain that one of those curves is an ellipse and the other a hyperbola.

Indeed, if we replace equation (I) by the equivalent equation

$$x^2(b-\lambda)+y^2(a-\lambda)-(a-\lambda)(b-\lambda) = 0,$$

we shall find that the left side of this equation assumes for $\lambda = a$ a negative value $x^2(b-a)$ and for $\lambda = b$ a positive value $y^2(a-b)$; consequently one of the roots λ_1, λ_2 of this equation is less than b (this root gives an ellipse), and the other root is contained between b and a (and gives a hyperbola).

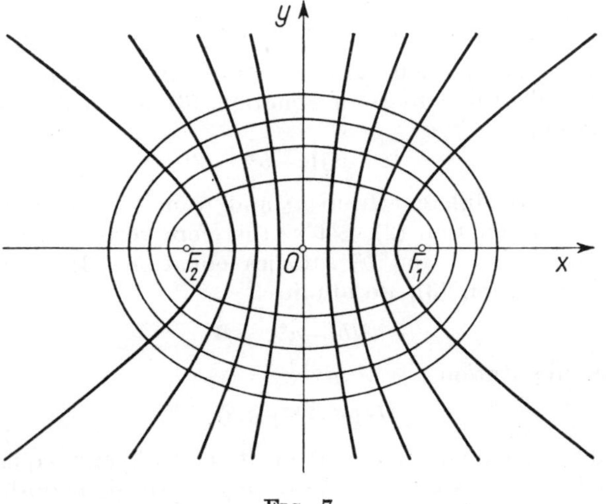

FIG. 7

Figure 7 represents several curves of a family of this kind. Their common foci are the points F_1 and F_2.

An ellipse and a hyperbola of family (I) which pass through the same point of the plane intersect at that point at right angles. The proof of this theorem will present no difficulty to those readers who know how to write the equation of a tangent to curve (I) at a given point of the curve. We say that the family of confocal conics is an *orthogonal net* of curves.

42. *Method I.* Since 0 is not a root of the given equation

$$\frac{x^2}{x^2-a^2}+\frac{x^2}{x^2-b^2} = 4, \tag{1}$$

the number of its roots is even; if α is a root of this equation, so is $-\alpha$. Introduce the notation $x^2 = z$; equation (I) will assume the form

$$\frac{z}{z-a^2}+\frac{z}{z-b^2} = 4. \tag{2}$$

The number of real roots of equation (1) is twice the number of positive roots of equation (2); we shall therefore investigate equation (2).

Let us multiply both sides of (2) by $(z-a^2)(z-b^2)$. We obtain

$$z(z-b^2)+z(z-a^2) = 4(z-a^2)(z-b^2);$$

which, when rearranged, gives

$$2z^2-3(a^2+b^2)z+4a^2b^2 = 0. \tag{3}$$

Each root of equation (2) is a root of equation (3); conversely, however, only those roots of equation (3) which do not satisfy the equation

$$(z-a^2)(z-b^2) = 0,$$

(i.e. which are different from a^2 and from b^2) are at the same time roots of equation (2). Let us therefore verify whether either of the numbers a^2 and b^2 satisfies equation (3). Substituting $z = a^2$ in equation (3), we obtain

$$a^2(b^2-a^2) = 0, \tag{4}$$

and the substitution $z = b^2$ gives

$$b^2(a^2-b^2) = 0. \tag{5}$$

Equality (4) holds if $a = 0$ or if $a^2 = b^2$, and equation (5) holds if $b = 0$ or if $a^2 = b^2$. Accordingly, we distinguish the following cases.

(i) $a \neq 0$, $b \neq 0$, $a^2 \neq b^2$. In this case neither a^2 nor b^2 is a root of equation (3); equations (2) and (3) are thus equivalent. The discriminant of equation (3)

$$\Delta = 9(a^2+b^2)^2-32a^2b^2 = 9a^4-14a^2b^2+9b^4 = 9(a^2-b^2)^2+4a^2b^2$$

is positive, and consequently, equation (3) has two (different) roots. They are both positive since they have a positive sum $\frac{3}{2}(a^2+b^2)$, and a positive product $2a^2b^2$.

Thus in case (i) equation (2) has two positive roots and equation (1) four real roots.

(ii) $a = 0$, $a^2 \neq b^2$, whence $b \neq 0$. Equation (3) is not equivalent to equation (2); it has the roots $z = a^2 = 0$ and $z = \frac{3}{2}b^2$, of which only the second satisfies equation (2) and is positive. Equation (1) has two real roots.

(iii) $b = 0$, $a^2-b^2 \neq 0$, whence $a \neq 0$. This case is analogous to case (ii): equation (1) has two real roots.

(iv) $a^2 = b^2 \neq 0$. Equation (3) has, as in case (i), two (different) positive roots, but is not equivalent to equation (2), because one of the roots is $z = a^2 = b^2$. Consequently, equation (2) has one positive root and equation (1) has two positive roots.

(v) $a^2 = b^2 = 0$. Equation (3) has a double root $z = 0$, equations (2) and (1) have no roots.

Method II. Equation (1) can be written in the form

$$\frac{x^2-a^2+a^2}{x^2-a^2}+\frac{x^2-b^2+b^2}{x^2-b^2}-4 = 0$$

or in the form

$$\frac{a^2}{x^2-a^2}+\frac{b^2}{x^2-b^2}-2 = 0. \tag{1a}$$

The left-hand side of equation (1a) is a function of the variable x: we shall denote it by $f(x)$; we are to find the values of x for which $f(x) = 0$.

We shall confine ourselves to the case where $a \neq 0$, $b \neq 0$, $a^2 < b^2$; we can assume that $a > 0$ and $b > 0$ because only the squares of a and b appear in the equation. Other cases can be investigated in an analogous manner: we leave that to the reader as an exercise.

Since $f(0) = -4$, and $f(-x) = f(x)$, it is sufficient to take care of the positive roots of function $f(x)$; its negative roots are the numbers opposite to the positive roots.

Each positive root of function $f(x)$ must satisfy one of the inequalities $0 < x < a$, $a < x < b$, $x > b$ or, as we usually say, it must belong to one of the intervals $(0, a)$, (a, b), (b, ∞).

In the interval $(0, a)$ the fractions $a^2/(x^2-a^2)$ and $b^2/(x^2-b^2)$ are both negative; thus in this interval there are no roots of $f(x)$.

In the interval (a, b) the fraction $a^2/(x^2-a^2)$ is positive and decreases with the increasing value of x; the fraction $b^2/(x^2-b^2)$ is negative and also decreasing when x increases since then its absolute value increases. Consequently function $f(x)$ is decreasing in the interval (a, b). If x is sufficiently near a, function $f(x)$ is positive because the fraction $a^2/(x^2-a^2)$ assumes arbitrarily large values and $b^2/(x^2-b^2)-2$ assumes values arbitrarily near the number $b^2/(a^2-b^2)-2$. If x is sufficiently near b, function $f(x)$ is negative because then $b^2/(x^2-b^2)$ assumes negative values with arbitrarily large absolute values, and $a^2/(x^2-a^2)-2$ assumes values arbitrarily near the number $a^2/(b^2-a^2)-2$. We have ascertained that in the interval (a, b) function $f(x)$ decreases and assumes both positive and negative values. Since the function is continuous, it has one and only one root in the interval (a, b).

In the interval (b, ∞), the fractions $a^2/(x^2-a^2)$ and $b^2/(x^2-b^2)$ are positive and decreasing while x increases; consequently the same applies to function $f(x)$. If x is sufficiently near b, function $f(x)$ is positive, which we can ascertain in the same manner as above. If x is sufficiently large, the fractions $a^2/(x^2-a^2)$ and $b^2/(x^2-b^2)$ assume values arbitrarily near 0, whence $f(x)$ has a negative value. It follows, as before, that in the interval (b, ∞) function $f(x)$ has one and only one root.†

We conclude from the above that in the case where $a \neq 0$, $b \neq 0$ and $a^2 \neq b^2$ the equation (1) has two positive and two negative roots.

43. *Answer.* If $a+b \neq 0$, then numbers $-a$ and $-b$ are the roots of the equation, and if $a+b = 0$, then all numbers except zero are roots of the equation.

44. Every number x satisfying the equation

$$\frac{1}{x-a}+\frac{1}{x-b}+\frac{1}{x-c}=0 \tag{1}$$

must also satisfy the equation

$$(x-b)(x-c)+(x-c)(x-a)+(x-a)(x-b)=0, \tag{2}$$

obtained from equation (1) by multiplying both sides by $(x-a)(x-b)(x-c)$.

Conversely, every root x of equation (2) different from a, b and c is a root of equation (1), since if we divide (2) by the number $(x-a)(x-b)(x-c)$, which is different from 0, we obtain (1). Considering that none of the numbers a, b, c is a root of equation (1), we can see that the roots of (1) are those roots of (2) which are equal to none of the numbers a, b, c.

We shall therefore investigate (2) taking into consideration three cases.

Case 1. Numbers a, b, c are different from one another; e.g. let $a < b < c$. The left side of (2) is a quadratic function $f(x)$ of variable x:

$$f(x)=(x-b)(x-c)+(x-c)(x-a)+(x-a)(x-b). \tag{3}$$

Let us find the values of function $f(x)$ if x equals a, b, or c. Formula (3) gives

† In order to understand the above reasoning the reader needs rudimentary information on continuous functions.

$$f(a) = (a-b)(a-c),$$
$$f(b) = (b-c)(b-a),$$
$$f(c) = (c-a)(c-b).$$

Numbers $f(a)$ and $f(c)$ are positive since they are the products of numbers of the same sign; number $f(b)$ is negative since $b-c < 0$, and $b-a > 0$. Consequently the quadratic function $f(x)$ has two real roots, one of them lying between a and b and the other between b and c.

Equation (1) has the same roots. Thus equations (1) and (2) are equivalent in this case.

Case 2. Two of the numbers a, b and c are equal; e.g. let $a = b$ and $a \neq c$. Equation (2) assumes the form

$$(x-a)(x-c+x-c+x-a) = 0,$$

i.e.

$$(x-a)(3x-2c-a) = 0,$$

whence we find its roots, a and $(a+2c)/3$. In view of $a \neq c$, the second root equals neither a nor c. In this case equation (1) has only one root, namely the number $(a+2c)/3$.

Case 3. Numbers a, b, c are equal. Equation (2) assumes the form

$$3(x-a)^2 = 0$$

and has a double root $x = a$. Equation (1) has no roots in this case.

REMARK. The proof that in case 1 equation (2) has two real roots can also be carried out by showing that its discriminant Δ is positive. Accordingly, we write equation (2) in the form

$$3x^2-2(a+b+c)x+(ab+bc+ca) = 0$$

and find

$$\frac{1}{4}\Delta = (a+b+c)^2-3(ab+bc+ca) = a^2+b^2+c^2-ab-bc-ca$$

$$= \frac{1}{2}(2a^2+2b^2+2c^2-2ab-2bc-2ca)$$

$$= \frac{1}{2}[(a-b)^2+(b-c)^2+(c-a)^2] > 0.$$

45. *Method I.* From the assumption that no two of the numbers a, b, c are equal it follows that none of them is a root of the given equation. Indeed, substituting in that equation, say, $x = a$ we obtain the equality $\sqrt{(a-b)} = \sqrt{(a-c)}$, which is false because $b \neq c$. Suppose that number x is a root of the given equation. Then $x > a$, $x > b$, $x > c$; besides, $\sqrt{(x-a)} < \sqrt{(x-c)}$, whence $x-a < x-c$; consequently $c < a$, and analogously $c < b$. Thus the equation can have roots if and only if numbers a, b, c satisfy the inequalities

$$c < a, \quad c < b.$$

We shall show that, if this condition is satisfied, the equation has one and only one root. Accordingly, we shall form a rational equation which must be satisfied by the roots of the given equation. We might obtain that equation by squaring the given equation twice. It will be more convenient, however, to follow another procedure. We write the following 4 equations (the first of them equivalent to the given equation):

$$\sqrt{(x-a)}+\sqrt{(x-b)}-\sqrt{(x-c)} = 0, \tag{1}$$

$$\sqrt{(x-a)}-\sqrt{(x-b)}+\sqrt{(x-c)} = 0, \tag{2}$$

$$-\sqrt{(x-a)}+\sqrt{(x-b)}+\sqrt{(x-c)} = 0, \tag{3}$$

$$\sqrt{(x-a)}+\sqrt{(x-b)}+\sqrt{(x-c)} = 0. \tag{4}$$

We multiply these equations, using the easily verifiable formula

$$-(\alpha+\beta-\gamma)(\alpha-\beta+\gamma)(-\alpha+\beta+\gamma)(\alpha+\beta+\gamma)$$
$$= \alpha^4+\beta^4+\gamma^4-2\alpha^2\beta^2-2\beta^2\gamma^2-2\gamma^2\alpha^2,$$

and obtain the equation

$$(x-a)^2+(x-b)^2+(x-c)^2-2(x-a)(x-b)-$$
$$-2(x-b)(x-c)-2(x-c)(x-a) = 0;$$

with the brackets removed and the terms rearranged it assumes the form

$$3x^2-2(a+b+c)x-(a^2+b^2+c^2-2ab-2bc-2ca) = 0. \tag{5}$$

Denoting by Δ the discriminant of equation (5), we have

$$\frac{1}{4}\Delta = (a+b+c)^2+3(a^2+b^2+c^2-2ab-2bc-2ca)$$
$$= 4(a^2+b^2+c^2-ab-bc-ca)$$
$$= 2[(a-b)^2+(b-c)^2+(c-a)^2],$$

whence we can see that $\Delta > 0$; thus equation (5) has two real roots.

We must now investigate whether those roots satisfy equation (1). It will be observed that if x is a root of any of the equations (1), (2), (3), (4) then it satisfies also equation (5), and in that case equation (5) has a root greater than any of the numbers a, b, c. Conversely, if equation (5) has a root x satisfying the conditions $x > a$, $x > b$, $x > c$, then x is a root of one of the equations (1)–(4), since the left side of equation (5) can then be represented in the form of the product of the left sides of the equations (1)–(4). Therefore we must find out whether any of the roots of equation (5) is greater than a, b and c. Let $f(x)$ denote the left-hand side of equation (5); an easy calculation gives

$$f(a) = -(b-c)^2, \quad f(b) = -(a-c)^2, \quad f(c) = -(a-b)^2.$$

By the well-known theorem on the sign of a quadratic trinomial, we infer from the above that the numbers a, b, c lie between the roots of equation (5); thus the required condition is satisfied only by the greater of the roots of equation (5), namely by the number

$$x_1 = \frac{1}{3}[(a+b+c)+2\sqrt{(a^2+b^2+c^2-ab-bc-ca)}]. \tag{6}$$

Number x_1 expressed by formula (6) satisfies, as has been observed above, one of the equations (1)–(4). Now, if $c < a$ and $c < b$, then number x_1 can satisfy none of the equations (2), (3), (4) since then $\sqrt{(x_1-c)} > \sqrt{(x_1-a)}$ and $\sqrt{(x_1-c)} > \sqrt{(x_1-b)}$, and consequently the value of the left side of each of the equations (2), (3), (4) is positive. Thus number x_1 is a root of equation (1).

Method II. Let us introduce the notation

$$\sqrt{(x-a)} = u, \quad \sqrt{(x-b)} = v, \quad \sqrt{(x-c)} = w;$$

then

$$u^2-w^2 = c-a, \tag{7}$$

$$v^2-w^2 = c-b, \tag{8}$$

and equation (1) assumes the form

$$u+v = w. \tag{9}$$

If x satisfies equation (1), then u, v, w are positive numbers satisfying equations (7), (8), (9). Conversely, if positive numbers u, v, w satisfy equations (7), (8), (9), then the number $x = u^2+a = v^2+b = w^2+c$ satisfies equation (1). The problem is thus reduced to the determination of positive solutions of the system of equations (7), (8), (9).

If numbers $u > 0$, $v > 0$, $w > 0$ satisfy this system of equations, then (9) implies that $u < w$ and $v < w$, in view of which equations (7) and (8) imply that $c < a$ and $c < b$. The system of equations (7), (8), (9) can thus have positive solutions only if

$$c < a, \quad c < b.$$

We shall prove that if this condition is satisfied then the system of equations (7), (8), (9) has one and only one positive solution.

Substituting in (7) and (8) the value w from equation (9) gives

$$\begin{aligned} u^2+2uv &= b-c, \\ v^2+2uv &= a-c. \end{aligned} \tag{10}$$

We multiply the first of these equations by $a-c$ and the second by $c-b$ and then add them:

$$(a-c)u^2+2(a-b)uv+(c-b)v^2 = 0.$$

Since $v \neq 0$, we can replace this equation by

$$(a-c)\left(\frac{u}{v}\right)^2+2(a-b)\left(\frac{u}{v}\right)+c-b = 0. \tag{11}$$

Equation (11) is quadratic with respect to u/v. Since $a-c > 0$ and $c-b < 0$, equation (11) has two real roots of which one and only one is positive; let us denote it by α. From the equation $u = \alpha v$ and from the equations (10) and (9) we obtain:

$$u = \sqrt{\frac{(b-c)\alpha}{\alpha+2}}, \quad v = \frac{1}{\alpha}u, \quad w = u+v$$

as the only positive solution of system (7), (8), (9).

The given equation (1) has one root

$$x = \frac{(b-c)\alpha}{\alpha+2}+a. \tag{12}$$

Equation (11) gives

$$\alpha = \frac{b-a+\sqrt{(\frac{1}{4}\Delta)}}{a-c}.$$

If we substitute this value of α in formula (12) and make the necessary transformations, we obtain formula (6).

We have obtained the following result:

A necessary and sufficient condition for the equation

$$\sqrt{(x-a)}+\sqrt{(x-b)} = \sqrt{(x-c)},$$

where a, b, c are pairwise different, to have roots are the inequalities

$$c < a, \quad c < b.$$

If this condition is satisfied, the equation has one root

$$x = \frac{1}{3}[a+b+c+2\sqrt{(a^2+b^2+c^2-ab-bc-ca)}].$$

46. Let x denote the length of the hypotenuse and y the length of the third side. Then

$$x^2-y^2 = 100 \quad \text{or} \quad (x+y)(x-y) = 100.$$

Writing

$$x+y = p, \quad x-y = q$$

we have

$$x = \frac{p+q}{2}, \quad y = \frac{p-q}{2}, \quad p\times q = 100.$$

Since x and y are natural numbers and 100 is an even number, p and q are even numbers and $p > q$. Hence $p = 50$, $q = 2$, and consequently $x = 26$, $y = 24$.

47. Let us write the given equation in the form

$$(y-x)(y^2+xy+x^2) = 13\times 7. \tag{1}$$

It will be observed that the trinomial y^2+xy+x^2 has a non-negative value for any x, y since

$$y^2+xy+x^2 = \frac{1}{2}[(x+y)^2+x^2+y^2] \geqslant 0.$$

It follows that if the integers x and y satisfy equation (1), then both factors on the left-hand side have positive integral values. Since the right-hand side is divisible by 13 and by 7, the left-hand side must also be divisible by 13 and by 7. And since 13 and 7 are prime numbers, only the following cases are possible:

(i) Number $y-x$ is divisible both by 13 and by 7; then x and y must satisfy the equations

$$y-x = 91, \quad y^2+xy+x^2 = 1.$$

This system of equations, however, has no real solutions and thus of course no integral solutions.

(ii) Number x^2+xy+x^2 is divisible by 13 and by 7. Then x and y satisfy the equations

$$y-x = 1, \quad y^2+xy+x^2 = 91.$$

This system has the solutions:

$$x = 5, \quad y = 6,$$

$$x = -6, \quad y = -5.$$

(iii) Number $y-x$ is divisible by 13 and number y^2+xy+x^2 is divisible by 7; this assumption leads to the system of equations

$$y-x = 13, \quad y^2+xy+x^2 = 7.$$

This system has no real solutions.

(iv) Number $y-x$ is divisible by 7 and number y^2+xy+x^2 is divisible by 13. Then

$$y-x = 7, \quad y^2+xy+x^2 = 13.$$

Solving this system of equations we obtain the solutions

$$x = -3, \quad y = 4,$$

$$x = -4, \quad y = 3.$$

Thus equation (1) has only the following integral solutions:

$$x = 5, \quad x = -6, \quad x = -3, \quad x = -4,$$
$$y = 6, \quad y = -5, \quad y = 4, \quad y = 3.$$

REMARK. We shall solve a more general problem:

Find the integral solutions of the equation

$$x^3-y^3 = c \tag{1}$$

where c is a given integer.

It will be observed that if $c = 0$, then the solution of the equation is any pair of equal integers, $x = y$. If $c < 0$, then the equation can be written as

$$y^3-x^3 = -c,$$

where the right-hand side is positive.

Thus it is sufficient to consider the case of $c > 0$. Then we must have $x > y$.

Suppose that there exist integers x and y which satisfy equation (1). Let us write it in the form

$$(x-y)(x^2+xy+y^2) = c.$$

Let

$$x-y = u, \tag{2}$$

$$x^2+xy+y^2 = v. \tag{3}$$

Numbers u and v are positive integers and satisfy the condition

$$uv = c. \tag{4}$$

Equations (2) and (3) can be solved in terms of u and v. Equation (2) gives

$$x = y+u. \tag{5}$$

Substituting this value in equation (3) we obtain the quadratic equation

$$y^2+uy+\frac{u^2-v}{3} = 0. \tag{6}$$

The discriminant of this equation must be a square of an integer, i.e.

$$\frac{4v-u^2}{3} = t^2, \tag{7}$$

where t denotes an integer.

Considering formulas (5), (6) and (7) we can see that numbers x and y must satisfy the equations

$$x = \frac{u+t}{2}, \quad y = \frac{-u+t}{2} \tag{I}$$

or the equations

$$x = \frac{u-t}{2}, \quad y = \frac{-u-t}{2}. \tag{II}$$

Conversely, let u, v and t be integers satisfying equations (4) and (7); u and t are either both even or both odd: this follows from formula (7). Consequently x and y defined by formulas (I) and (II) are integers; it is obvious from the above that they satisfy equation (1).

We have thus obtained the following result:

In order to find the integral solutions of equation (1) for $c > 0$ it is necessary to decompose number c into two natural factors u and v such that the number

$$\frac{4v-u^2}{3}$$

is the square of a non-negative integer t; each factorization of this kind is linked with two solutions of equation (1) represented by formulas (I) and (II).

We should therefore try out all decompositions of c into two factors. The number of those trials can be reduced on the strength of the following observation: since $4v-u^2 \geqslant 0$, we have $u^3 \leqslant 4uv$, i.e. $u^3 \leqslant 4c$, whence

$$u \leqslant \sqrt[3]{(4c)}.$$

For example, let $c = 91 = 7\times 13$. Then $u < 8$; therefore only two possibilities should be tried out:

(a) If $u = 1$, $v = 91$, then

$$\frac{4v-u^2}{3} = 121 = 11^2, \quad \text{whence} \quad t = 11,$$

and formulas (I) and (II) give the solutions:

$$x = 6, \ y = 5 \quad \text{and} \quad x = -5, \ y = -6.$$

(b) If $u = 7$, $v = 13$, then

$$\frac{4v-u^2}{3} = 1, \quad \text{whence} \quad t = 1$$

and formulas (I) and (II) give the solutions:

$$x = 4, \ y = -3 \quad \text{and} \quad x = 3, \ y = -4.$$

EXERCISE. Prove the following theorems, which reduce the number of trials necessary to solve equation (1).

(α) If number c is divisible by 3, then both u and v must be divisible by 3 (and thus to ensure the existence of the solutions number c must be divisible by 9).

(β) If number c is of the form $3k+1$ where k is an integer, then u must also be of the form $3k+1$, and if c is of the form $3k+2$, then u must also be of the form $3k+2$.

48. Let us multiply both sides of each of the given equations by y, z, x respectively and then let us add these equations; we shall obtain the equation

$$5x+3y+4z = 0. \tag{1}$$

Multiplying the given equations by z, x, y and adding them, we obtain

$$4x+5y+3z = 0. \tag{2}$$

From equations (1) and (2) we can find the ratios $x : y : z$. By eliminating, say, z we obtain:

$$x = -11y, \tag{3}$$

and by eliminating x we obtain

$$z = 13y. \tag{4}$$

We substitute these values, for example in the first of the given equations:

$$(11^2-13)\,y^2 = 3, \quad \text{whence} \quad y = \tfrac{1}{6} \ \text{or} \ y = -\tfrac{1}{6}.$$

In view of (3) and (4) we obtain two systems of values:

$$x = -\tfrac{11}{6}, \quad y = \tfrac{1}{6}, \quad z = \tfrac{13}{6},$$

$$x = \tfrac{11}{6}, \quad y = -\tfrac{11}{6}, \quad z = -\tfrac{13}{6}.$$

Both the first and the second of these systems satisfy not only the first of the given equations but also the remaining two, which can be verified by substitution.

REMARK. We shall consider a more general system of equations

$$\begin{aligned} x^2 - yz &= a, \\ y^2 - zx &= b, \\ z^2 - xy &= c, \end{aligned} \tag{1}$$

where a, b, c denote any real numbers.

Treating equations (1) in the same way as the preceding ones, we obtain two equations of the first degree.

$$\begin{aligned} cx + ay + bz &= 0, \\ bx + cy + az &= 0. \end{aligned} \tag{2}$$

It is now necessary to distinguish two cases:

Case I. The coefficients c, a, b of the first equation of system (2) are not proportional to the coefficients b, c, a of the second equation, i.e. the three equalities

$$a^2 - bc = 0, \quad b^2 - ac = 0, \quad c^2 - ab = 0 \tag{3}$$

do not hold simultaneously.

This condition is equivalent to the assumption that numbers a, b, c are not all equal. Indeed, if $a = b = c$, then equations (3) are obviously true. Conversely, if equations (3) are true, then we also have

$$a(a^2 - bc) = 0, \quad b(b^2 - ac) = 0, \quad c(c^2 - ab) = 0, \tag{4}$$

i.e.,

$$a^3 = abc, \quad b^3 = abc, \quad c^3 = abc,$$

and consequently

$$a^3 = b^3 = c^3, \quad \text{whence} \quad a = b = c.$$

In view of the fact that in the case in question relations (4) cannot all hold simultaneously, we can assume that, for example, $a(a^2 - bc) \neq 0$, i.e. that

$$a \neq 0 \quad \text{and} \quad a^2 - bc \neq 0.$$

From system (2) we then obtain

$$y = \frac{b^2 - ac}{a^2 - bc} x, \quad z = \frac{c^2 - ab}{a^2 - bc} x. \tag{5}$$

We now substitute expressions (5) for y and z in one of the equations of system (1), say in the first equation:

$$[(a^2-bc)^2-(b^2-ac)(c^2-ab)]\,x^2 = a(a^2-bc)^2.$$

Transforming the coefficient of x^2 we obtain

$$a(a^3+b^3+c^3-3abc)\,x^2 = a(a^2-bc)^2,$$

and since $a \neq 0$,

$$(a^3+b^3+c^3-3abc)\,x^2 = (a^2-bc)^2. \tag{6}$$

If

$$a^3+b^3+c^3-3abc \neq 0$$

then we obtain from equation (6)

$$x = \pm\frac{a^2-bc}{\sqrt{(a^3+b^3+c^3-3abc)}}$$

for $a^3+b^3+c^3-3abc > 0$, and

$$x = \pm\frac{a^2-bc}{\sqrt{(3abc-a^3-b^3-c^3)}}\,i$$

if $a^3+b^3+c^3-3abc < 0$. We then find from equation (5) the corresponding values of y and z.

If

$$a^3+b^3+c^3-3abc = 0,$$

then equation (6) is contradictory since its left-hand side equals zero whereas its right-hand side—by hypothesis—does not equal zero.

In this case the given system of equations has no solutions.

Case II. The coefficients of equations (2) are proportional, which has been shown to occur if $a = b = c$.

It is again necessary to distinguish two cases.

Let $a \neq 0$. Then each of the equations of system (2) gives (on being divided by a) the same equation

$$x+y+z = 0,$$

whence

$$z = -(x+y). \tag{7}$$

Substituting this expression in any of the equations (1) we obtain the equation

$$y^2+xy+x^2-a = 0,$$

and consequently

$$y = \frac{-x \pm \sqrt{(4a-3x^2)}}{2}. \tag{8}$$

Formulas (8) and (7) are the solution of system (1). The value of x can be chosen arbitrarily; the values of y and z are then determined by formulas (8) and (7).

Let $a = 0$; thus also $b = 0$ and $c = 0$. System (1) assumes the form

$$x^2 = yz,$$
$$y^2 = zx,$$
$$z^2 = xy.$$

Hence

$$x^3 = xyz, \quad y^3 = xyz, \quad z^3 = xyz,$$

and consequently

$$x^3 = y^3 = z^3.$$

In the domain of real numbers we obtain hence the equality

$$x = y = z,$$

and the solution of system (1) is

$$x = m, \quad y = m, \quad z = m,$$

where m denotes an arbitrary real number.

In the domain of complex numbers the equality $y^3 = x^3$ implies that $y = \varepsilon x$, where ε denotes any of the numbers

$$1, \quad \frac{-1+i\sqrt{3}}{2}, \quad \frac{-1-i\sqrt{3}}{2},$$

which are the roots of the equation $u^3 = 1$. Substituting the value $y = \varepsilon x$ in the equation $y^2 = zx$ and considering that if $x = 0$, then also $z = 0$, we obtain $z = \varepsilon^2 x$.

The solution of system (1) is then

$$x = m, \quad y = \varepsilon m, \quad z = \varepsilon^2 m,$$

where m denotes an arbitrary complex number.

Remark. As we know

$$a^3+b^3+c^3-3abc = (a+b+c)(a^2+b^2+c^2-ab-bc-ca),$$

and if a, b, c are not all simultaneously equal, we have the inequality

$$a^2+b^2+c^2-ab-bc-ac > 0$$

(see problem 55); consequently, in case I the condition for the existence of solutions can be written as $a+b+c \neq 0$.

If $a+b+c > 0$, the solutions are real, and if $a+b+c < 0$, the solutions are imaginary.

49. The given system of equations has eight solutions:

1. $x = 0$, $y = 8$, $z = 104$ } a double solution,
2. $x = 0$, $y = 8$, $z = 104$ }
3. $x = 4$, $y = -12$, $z = 120$,
4. $x = -4$, $y = -4$, $z = 120$,
5. $x = 3i$, $y = 17-3i$, $z = -130$,
6. $x = -3i$, $y = 17+3i$, $z = -130$,
7. $x = 5$, $y = -22$, $z = -96$,
8. $x = -5$, $y = -12$, $z = -96$.

50. Let us divide the equations (1) by xy, yz and zx respectively:

$$\frac{b}{x}+\frac{a}{y}=1,$$

$$\frac{b}{y}+\frac{a}{z}=1, \tag{2}$$

$$\frac{b}{z}+\frac{a}{x}=1.$$

The solutions of system (2) are those solutions x, y, z of system (1) in which $x \neq 0$, $y \neq 0$, $z \neq 0$.

Adding equations (2) we obtain

$$(a+b)\left(\frac{1}{x}+\frac{1}{y}+\frac{1}{z}\right)=3. \tag{3}$$

If $a+b = 0$, equation (3) expresses a contradiction: $0 = 3$; in this case system (2) has no solutions.

If $a+b \neq 0$, system (2) can be solved in the following way. We multiply equations (2) by b^2, $-ab$ and a^2 respectively, and then add them:

$$(a^3+b^3)\frac{1}{x}=a^2-ab+b^2. \tag{4}$$

Since $a \neq -b$, we have $a^3 \neq -b^3$, i.e. $a^3+b^3 \neq 0$, and since $a^3+b^3 = (a+b)(a^2-ab+b^2)$, we have also $a^2-ab+b^2 \neq 0$; then equation (4) gives $x = a+b$. Analogously $y = a+b$, $z = a+b$.

Thus it is only for $a+b \neq 0$ that system (1) has a solution

x, y, z in which none of the unknowns is equal to zero. This solution is

$$x = a+b, \quad y = a+b, \quad z = a+b.$$

It remains to determine those solutions of system (1) in which at least one of the unknowns is equal to zero.

For example, let $z = 0$. From system (1) we obtain for x and y the system of equations

$$\begin{aligned} xy &= ax+by, \\ ay &= 0, \\ bx &= 0. \end{aligned} \tag{5}$$

If either a or b is different from zero, system (5) has only the solution $x = 0$, $y = 0$. For example if $a \neq 0$, the second equation of (5) gives $y = 0$, whence the first equation gives $x = 0$.

If $a = b = 0$, then the second and the third equation of (5) are satisfied by any values of x and y and the first equation assumes the form

$$xy = 0.$$

Thus one of the unknowns x, y must be equal to zero, and then the other unknown can have any value.

In the above reasoning we have assumed that $z = 0$. We shall obtain analogous results by assuming that $x = 0$ or that $y = 0$.

The results obtained can be listed as follows:

(a) If at least one of the numbers a and b is not equal to zero, system (1) has the solutions:

1. $x = a+b, \quad y = a+b, \quad z = a+b;$
2. $x = 0, \quad y = 0, \quad z = 0.$

These solutions coincide if $a+b = 0$.

(b) If $a = b = 0$, system (1) has the solutions:

1. $x = 0$, $y = 0$, z —arbitrary;
2. $x = 0$, y —arbitrary, $z = 0$;
3. x —arbitrary, $y = 0$, $z = 0$.

51. Let us multiply by -3 the first equation of (1), and then let us add it to the second equation; we obtain

$$(x-y)^3 = (a-b)^3,$$

which is equivalent (in the domain of real numbers) to the equation

$$x-y = a-b.$$

System (1) is equivalent to the system of equations

$$\begin{aligned} xy(x-y) &= ab(a-b), \\ x-y &= a-b \end{aligned} \tag{2}$$

and to the simpler system

$$\begin{aligned} xy(a-b) &= ab(a-b), \\ x-y &= a-b. \end{aligned} \tag{3}$$

We must distinguish two cases:

I. $a-b \neq 0$. System (3) is equivalent to

$$\begin{aligned} xy &= ab, \\ x-y &= a-b, \end{aligned} \tag{4}$$

which has two solutions: $x = a$, $y = b$ and $x = -b$, $y = -a$.

II. $a-b = 0$. The first equation of system (3) is an identity, $0 = 0$, and system (3) reduces to one equation

$$x-y = 0,$$

having the solution $x = m$, $y = m$, where m is any number.

REMARK. In the domain of complex numbers the equation $(x-y)^3 = (a-b)^3$ is equivalent to a disjunction of three equations: $x-a = a-b$, $x-y = (a-b)\varepsilon$ or $x-y = (a-b)\varepsilon^2$, where

$$\varepsilon = \frac{-1+i\sqrt{3}}{2}$$

whence system (1) is equivalent to a disjunction of three systems of equations of which one is system (2) and the other two arise from system (2) by replacing the second equation by the equation $x-y = (a-b)\varepsilon$ or by the equation $x-y = (a-b)\varepsilon^2$. The solving of each of these systems proceeds in the same way as that of system (2). The result is the following:

I. If $a-b \neq 0$, the given system (1) has 6 solutions x, y in the domain of complex numbers, namely

$$\begin{matrix} (a, -b), & (a\varepsilon, b\varepsilon), & (a\varepsilon^2, b\varepsilon^2), \\ (-b, -a), & (-b\varepsilon, -a\varepsilon), & (-b\varepsilon^2, -a\varepsilon^2). \end{matrix}$$

II. If $a = b$, the solution of system (1) is

$$x = m, \quad y = m,$$

where m is any complex number.

52. We shall replace the given system of equations (1) by an equivalent system, which is obtained in the following way.

Multiplying the second of equations (1),

$$\frac{1}{x}+\frac{1}{y}+\frac{1}{z}=0, \tag{1b}$$

by the product xyz, we obtain the equation

$$xy+yz+zx=0. \tag{1b'}$$

Replacing (1b) by (1b′) in system (1) we obtain

$$\begin{aligned} x^2+y^2+z^2 &= 19^2,\\ xy+yz+zx &= 0,\\ x-y+z &= 11. \end{aligned} \tag{2}$$

We shall prove that system (2) is equivalent to system (1). Every triple x, y, z satisfying (1b) also satisfies (1b′) and thus every triple x, y, z satisfying system (1) also satisfies system (2). Is the converse statement true? A solution of (1b′) is given by any triple x, y, z which satisfies (1b) and, moreover, by three numbers of which two are equal to 0 and the third is arbitrary; since triples of the second kind do not satisfy the remaining two equations of system (2), we conclude that every three numbers x, y, z satisfying system (2) also satisfy system (1). We have thus proved the equivalence of systems (1) and (2).

Let us transform system (2). Multiplying the second equation by 2 and adding it to the first equation, we obtain the system

$$\begin{aligned} (x+y+z)^2 &= 19^2,\\ xy+yz+zx &= 0,\\ x-y+z &= 11, \end{aligned} \tag{3}$$

which is equivalent to (2).

System (3) can be replaced by the disjunction:

$$\begin{aligned} x+y+z &= 19,\\ xy+yz+zx &= 0.\\ x-y+z &= 11, \end{aligned} \tag{3a}$$

or

$$\begin{aligned} x+y+z &= -19,\\ xy+yz+zx &= 0,\\ x-y+z &= 11. \end{aligned} \tag{3b}$$

(A) Let us consider system (3a). Subtracting the third equation from the first, we obtain $y = 4$; substituting this value in the second and third equations, we obtain the system of equations

$$\begin{aligned} y &= 4, \\ 4(x+z)+xz &= 0, \\ x+z &= 15, \end{aligned} \tag{4}$$

equivalent to (3a). System (4) can be replaced in turn by the system

$$\begin{aligned} y &= 4, \\ xz &= -60, \\ x+z &= 15. \end{aligned} \tag{5}$$

This system has two solutions:

$$x = \frac{15-\sqrt{465}}{2}, \quad y = 4, \quad z = \frac{15+\sqrt{465}}{2}, \tag{5a}$$

or

$$x = \frac{15+\sqrt{465}}{2}, \quad y = 4, \quad z = \frac{15-\sqrt{465}}{2}. \tag{5b}$$

(B) Let us consider system (3b). Proceeding as before we reduce it to the system

$$\begin{aligned} y &= -15, \\ xz &= -60, \\ x+z &= -4. \end{aligned} \tag{6}$$

This system has two solutions:

$$x = -10, \quad y = -15, \quad z = 6 \tag{6a}$$

or

$$x = 6, \quad y = -15, \quad z = -10. \tag{6b}$$

53. In solving the system of equations

$$\begin{aligned} x_1x_2 &= 1, \\ x_2x_3 &= 2, \\ x_3x_4 &= 3, \\ &\cdots\cdots \\ x_nx_1 &= n, \end{aligned} \tag{1}$$

we must distinguish two cases.

(a) Number n is even, $n = 2m$. Multiplying the equations: first, third, ... up to the $(2m-1)$th, we obtain

$$x_1x_2x_3x_4 \ldots x_{2m-1}x_{2m} = 1\times 3\times \ldots \times(2m-1). \tag{2}$$

Multiplying the equations: second, fourth, ... up to the $2m$th, we obtain:

$$x_2x_3x_4x_5 \ldots x_{2m}x_1 = 2\times 4\times \ldots \times 2m. \tag{3}$$

Equations (2) and (3) are inconsistent since their left sides represent the same product of the unknowns whereas their right sides are different. Hence if n is even, system (1) has no solutions.

(b) Number n is odd, $n = 2m+1$.

To begin with, it will be observed that if the numbers $x_1, x_2, \ldots, x_n$ satisfy system (1) then they are either all positive or all negative and, moreover, $-x_1, -x_2, \ldots, -x_n$ also satisfy system (1). It is thus sufficient to find the positive solutions.

We multiply the equations of system (1):

$$(x_1x_2x_3 \ldots x_{2m+1})^2 = 1\times2\times3\times \ldots \times(2m+1),$$

whence

$$x_1x_2x_3 \ldots x_{2m+1} = \sqrt{[1\times2\times3\times \ldots \times(2m+1)]}. \tag{4}$$

We multiply every other equation: first, third, fifth, ... up to the $(2m+1)$th, i.e. the last:

$$x_1x_2x_3x_4x_5x_6 \ldots x_{2m+1}x_1 = 1\times3\times5\times \ldots \times(2m+1). \tag{5}$$

We divide equation (5) by equation (4):

$$x_1 = \frac{1\times3\times5\times \ldots \times(2m+1)}{\sqrt{[1\times2\times3\times \ldots \times(2m+1)]}}. \tag{6}$$

From the first $2m$ equations of system (1) we obtain

$$x_2 = \frac{1}{x_1}, \quad x_3 = \frac{2}{x_2}, \quad x_4 = \frac{3}{x_3}, \quad x_5 = \frac{4}{x_4}, \quad \ldots, \quad x_{2m+1} = \frac{2m}{x_{2m}}. \tag{7}$$

Hence we can find successively the unknowns $x_2, x_3, \ldots, x_n$ in terms of x_1:

$$x_2 = \frac{1}{x_1}, \quad x_3 = 2x_1, \quad x_4 = \frac{3}{2x_1}, \quad x_5 = \frac{2\times4}{3}x_1, \quad \ldots \tag{8}$$

It is easy to prove that starting from x_4 the above values of the unknowns are expressed by the formulas

$$\begin{aligned} x_{2k} &= \frac{3\times5\times \ldots \times(2k-1)}{2\times4\times \ldots \times(2k-2)}\times\frac{1}{x_1}, \\ x_{2k+1} &= \frac{2\times4\times \ldots \times2k}{1\times3\times \ldots \times(2k-1)}\times x_1, \end{aligned} \tag{9}$$

where k assumes the values 2, 3, ..., m.

We shall use induction. Substituting $k = 2$ in formulas (9) we obtain for x_4 and x_5 the expressions given in (8). Suppose now that formulas (9) are true for a certain $k \geqslant 2$; we shall prove that they will remain true if we replace k by $k+1$. Indeed, by (7) we have $x_{r+1} = r/x_r$, whence

$$x_{2k+2} = \frac{2k+1}{x_{2k+1}} = \frac{1\times 3\times \ldots \times (2k-1)}{2\times 4\times \ldots \times 2k} \times \frac{2k+1}{x_1},$$

$$x_{2k+3} = \frac{2k+2}{x_{2k+2}} = \frac{2\times 4\times \ldots \times 2k}{1\times 3\times \ldots \times (2k-1)} \times \frac{2k+2}{2k+1} \times x_1,$$

and these are the very formulas resulting from formulas (9) if we replace k by $k+1$.

In this way we have found the values of all the unknowns; namely x_1 is expressed by formula (6), $x_2 = 1/x_1$, $x_3 = 2x$, and the remaining unknowns are expressed by formulas (9).

It remains to verify whether the values obtained satisfy the equations (1). As regards the first $2m$ equations, this fact is undoubted because our method of finding the unknowns x_2, $x_1, \ldots, x_{2m+1}$ makes it clear that they satisfy equations (7). We shall verify whether the last equation is satisfied.

The second formula of (9) gives

$$x_{2m+1} = \frac{2\times 4\times \ldots \times 2m}{1\times 3\times \ldots \times (2m-1)}\, x_1,$$

whence

$$x_{2m+1}x_1 = \frac{2\times 4\times \ldots \times 2m}{1\times 3\times \ldots \times (2m-1)}\, x_1^2,$$

and thus, by (6)

$$x_{2m+1}x_1 = \frac{2\times 4\times \ldots \times 2m}{1\times 3\times \ldots \times (2m-1)} \times \frac{1^2\times 3^2\times 5^2\times \ldots \times (2m+1)^2}{2\times 3\times 4\times \ldots \times (2m+1)}$$
$$= 2m+1.$$

The last equation is thus satisfied. We have obtained the following result:

If n is an odd number, system (1) has two solutions. The first solution is defined by formula (6), by the formulas $x_2 = 1/x_1$, $x_3 = 2x_1$, and by formulas (9), in which k assumes the values $2, 3, \ldots, \frac{1}{2}(n-1)$; the second solution consists of these numbers multiplied by -1.

Remark. In the same way as above we can solve (for an odd n) a more general system of equations, replacing the right sides in equations (1) by any real numbers whose product is positive. We can also use the following, slightly different, method of solution.

Suppose we are given the system of equations

$$\begin{aligned} x_1 x_2 &= a_1, \\ x_2 x_3 &= a_2, \\ &\dots\dots\dots \\ x_{2m} x_{2m+1} &= a_{2m}, \\ x_{2m+1} x_1 &= a_{2m+1} \end{aligned} \tag{10}$$

with $a_1 a_2 \dots a_{2m+1} > 0$. Let us write

$$R = \varepsilon \sqrt{(a_1 a_2 \dots a_{2m+1})} \quad \text{where} \quad \varepsilon = 1 \text{ or } \varepsilon = -1.$$

Multiplying all the given equations and extracting the square root from both sides, we obtain

$$x_1 x_2 x_3 \dots x_{2m+1} = R.$$

On the other hand, by multiplying the equations containing the numbers $a_1, a_3, \dots, a_{2m-1}, a_{2m+1}$ on their right-hand sides, we obtain

$$x_1 x_2 x_3 \dots x_{2m+1} x_1 = a_1 a_3 a_5 \dots a_{m+1}.$$

Hence

$$x_1 = \frac{a_1 a_3 a_5 \dots a_{2m+1}}{R}. \tag{11}$$

If we perform in equations (10) a cyclic substitution of indices,

$$\begin{pmatrix} 1, 2, 3, \dots, 2m, & 2m+1 \\ 2, 3, 4, \dots, 2m+1, & 1 \end{pmatrix},$$

i.e. if we replace each of the indices $1, 2, \dots, 2m$ by an index greater by one, and the index $2m+1$ by the index 1, then system (10) will undergo no change, and formula (11) will become

$$x_2 = \frac{a_2 a_4 a_6 \dots a_{2m} a_1}{R}. \tag{12}$$

Applying a cyclic substitution again, we obtain from (12) the formula

$$x_3 = \frac{a_3 a_5 \dots a_{2m+1} a_2}{R}. \tag{13}$$

Continuing this procedure we obtain generally

$$x_{2k} = \frac{a_{2k} a_{2k+2} \dots a_{2m} a_1 a_3 \dots a_{2k-1}}{R},$$

$$x_{2k+1} = \frac{a_{2k+1} a_{2k+3} \dots a_{2m+1} a_2 a_4 \dots a_{2k}}{R},$$

where $k = 1, 2, 3, \ldots, m$. It is easy to verify that the values of $x_1, x_2, \ldots, x_{2m+1}$ obtained in this way satisfy all the equations of system (10); we have here two solutions of this system because in expression R we can assume either $\varepsilon = 1$ or $\varepsilon = -1$.

§ 4. Inequalities

54. Let us denote the number of kilogrammes of sugar weighed at the first trial by x and that weighed at the second trial by y. The lengths of the arms of the balance will be denoted by p and q, $p \neq q$.

By the law of the lever we have

$$q \times x = p \times 1, \quad \text{whence} \quad x = \frac{p}{q},$$

$$p \times y = q \times 1, \quad \text{whence} \quad y = \frac{q}{p}.$$

Consequently

$$x+y = \frac{p}{q}+\frac{q}{p} = \frac{p^2+q^2}{pq} > 2$$

since

$$\frac{p^2+q^2}{pq}-2 = \frac{p^2+q^2-2pq}{pq} = \frac{(p-q)^2}{pq} > 0 \quad \text{for} \quad p \neq q.$$

Therefore the amount of sugar weighed is more than 2 kg.

55. *Method I.* Since

$$a^2-2ab+b^2 = (a-b)^2 \geqslant 0,$$

we have

$$a^2+b^2 \geqslant 2ab,$$

and also

$$b^2+c^2 \geqslant 2bc$$

and

$$c^2+a^2 \geqslant 2ca.$$

Adding these three inequalities and dividing by 2 we obtain

$$a^2+b^2+c^2 \geqslant ab+bc+ca.$$

Method II. We shall use the method of *reductio ad absurdum.* Suppose that for certain numbers a, b, c we have the inequality

$$a^2+b^2+c^2 < ab+bc+ca.$$

In that case

$$a^2+b^2+c^2-ab-bc-ca < 0,$$
$$2a^2+2b^2+2c^2-2ab-2bc-2ca < 0,$$
$$(a-b)^2+(b-c)^2+(c-a)^2 < 0.$$

We have obtained a false inequality since each of the squares on the left-hand side is a non-negative number and consequently the sum of these squares is a non-negative number. Thus for any numbers a, b, c we have the inequality

$$a^2+b^2+c^2 \geqslant ab+bc+ca.$$

REMARK. Methods I and II differ only in the approach since both use premises which are essentially the same:

(1) a square of a number is a non-negative number,

(2) $2(a^2+b^2+c^2-ab-bc-ca) = (a-b)^2+(b-c)^2+(c-a)^2$.

Premise (2) implies that $a^2+b^2+c^2-ab-bc-ca$ is equal to zero only if $a = b = c$.

In the same way as above, we can prove that for any numbers $a_1, a_2, \ldots, a_n$ we have the inequality

$$a_1^2+a_2^2+\ldots+a_n^2 \geqslant a_1a_2+a_2a_3+\ldots a_{n-1}a_n+a_na_1,$$

equality occurring only if

$$a_1 = a_2 = \ldots = a_n.$$

56. We are to prove that the number

$$L = (a^2+b^2+c^2)(x^2+y^2+z^2)-(ax+by+cz)^2$$

is non-negative.

Method I. If we perform in expression L the required multiplication, reduce similar terms and suitably group the remaining ones, we shall obtain

$$L = (b^2x^2-2abxy+a^2y^2)+(a^2z^2-2aczx+c^2x^2)+(c^2y^2-2bcyx+b^2z^2),$$

whence

$$L = (bx-ay)^2+(az-cx)^2+(cy-bz)^2. \tag{1}$$

This shows that number L, as a sum of three squares, is not negative.

Method II. From the form of expression L we can see that $4L$ is equal to the discriminant of the following quadratic function $f(X)$:

$$f(X) = (a^2+b^2+c^2)X^2-2(ax+by+cz)X+(x^2+y^2+z^2).$$

This function can be given the form

$$f(X) = (aX-x)^2+(bX-y)^2+(cX-z)^2, \tag{2}$$

from which we infer that function $f(X)$ can assume only non-negative values; thus its discriminant is not positive since a quadratic function with a positive discriminant assumes both positive and negative values. Hence the conclusion

$$(ax+by+cz)^2-(a^2+b^2+c^2)(x^2+y^2+z^2) \leqslant 0,$$

i.e. $L \geqslant 0$, which is what was to be proved.

REMARK 1. We shall solve an additional question: in what case does the following equality hold:

$$(a^2+b^2+c^2)(x^2+y^2+z^2)-(ax+by+cz)^2 = 0\,?$$

For this purpose we can use method I or method II.

(a) Formula (1) implies that the equality $L = 0$ holds if and only if

$$bx-ay = 0, \quad az-cx = 0, \quad cy-bz = 0,$$

i.e.

$$bx = ay, \quad az = cx, \quad cy = bz.$$

These equalities signify that numbers x and y are proportional to a and b, numbers x and z are proportional to a and c, and numbers y and z are proportional to b and c, or briefly: x, y, z are proportional to a, b, c.

(b) Using the auxiliary function $f(X)$ we ascertain that equality $L = 0$ holds if and only if the discriminant of this function is equal to zero; then function $f(X)$ has a root which, by formula (2), satisfies the equations

$$aX-x = 0, \quad bX-y = 0, \quad cX-z = 0,$$

which is possible only if x, y, z are proportional to a, b, c.

REMARK 2. The theorem we have just proved can easily be generalized as follows:

If $a_1, a_2, \ldots, a_n$ and $x_1, x_2, \ldots, x_n$ are arbitrary numbers, then

$$(a_1^2+a_2^2+\ldots+a_n^2)(x_1^2+x_2^2+\ldots+x_n^2)-$$
$$-(a_1x_1+a_2x_2+\ldots+a_nx_n)^2 \geqslant 0,$$

or briefly:

$$\sum_{i=1}^{n} a_i^2 \times \sum_{i=1}^{n} x_i^2 - \left(\sum_{i=1}^{n} a_i x_i\right)^2 \geqslant 0, \tag{3}$$

equality occurring if and only if numbers $x_1, x_2, \ldots, x_n$ are proportional to numbers $a_1, a_2, \ldots, a_n$.

A simple proof is obtained by repeating the reasoning of method II. Method I can also be used.

Inequality (3) is sometimes called the *Schwarz*† *inequality*, although it was already known to Cauchy.

57. By the theorem of Pythagoras we have the relation

$$c^2 = a^2+b^2. \tag{1}$$

Let us multiply this equality by c^{n-2}:

$$c^n = a^2c^{n-2}+b^2c^{n-2}. \tag{2}$$

Since $c > a > 0$ and $n > 2$, we have $c^{n-2} > a^{n-2}$, whence

$$a^2c^{n-2} > a^n. \tag{3}$$

Analogously

$$b^2c^{n-2} > b^n. \tag{4}$$

Relations (2), (3) and (4) imply the required inequality:

$$c^n > a^n+b^n.$$

Remark. We show in the same way that if $n < 2$, then

$$c^n < a^n+b^n.$$

58. Inequality $m+4/m^2 \geqslant 3$ is equivalent to the inequality $m+4/m^2-3 \geqslant 0$ and thus also to the inequality $m^3+4-3m^2 \geqslant 0$ (since $m^2 > 0$). Now $m^3+4-3m^2 = m^3+1+3-3m^2 = (m^3+1)-$ $-3(m^2-1) = (m+1)(m^2-m+1)-3(m+1)(m-1) = (m+1)\times$ $\times(m^2-4m+4) = (m+1)(m-2)^2$. Since $m+1$ is positive for $m > 0$ and $(m-2)^2$ is non-negative, we have $m^3+4-3m^2 \geqslant 0$, which is what was to be proved.

It will be observed that the equality $m+4/m^2 = 3$ occurs only if $m-2 = 0$, i.e. if $m = 2$.

59. *Hint*: $(a-b)b \leqslant a^2/4$.

60. *Method I*. By dividing both sides of the equality $1 = a+b+c$ successively by a, b, c, we obtain

$$\frac{1}{a} = 1+\frac{b}{a}+\frac{c}{a}, \quad \frac{1}{b} = \frac{a}{b}+1+\frac{c}{b}, \quad \frac{1}{c} = \frac{a}{c}+\frac{b}{c}+1.$$

Consequently

$$\frac{1}{a}+\frac{1}{b}+\frac{1}{c} = 3+\left(\frac{a}{b}+\frac{b}{a}\right)+\left(\frac{b}{c}+\frac{c}{b}\right)+\left(\frac{c}{a}+\frac{a}{c}\right).$$

---

† H. A. Schwarz (1843–1921), a German mathematician.

Now

$$\frac{a}{b}+\frac{b}{a}=\frac{a^2+b^2}{ab}=\frac{(a-b)^2+2ab}{ab}=\frac{(a-b)^2}{ab}+2.$$

If follows immediately that

$$\frac{a}{b}+\frac{b}{a}\geqslant 2 \quad \text{and also} \quad \frac{b}{c}+\frac{c}{b}\geqslant 2, \quad \frac{c}{a}+\frac{a}{c}\geqslant 2.$$

Accordingly

$$\frac{1}{a}+\frac{1}{b}+\frac{1}{c}\geqslant 3+2+2+2,$$

i.e.

$$\frac{1}{a}+\frac{1}{b}+\frac{1}{c}\geqslant 9. \tag{1}$$

Method II. Let us calculate the difference

$$\frac{1}{a}+\frac{1}{b}+\frac{1}{c}-9=\frac{ab+bc+ca-9abc}{abc}. \tag{2}$$

Since the denominator of the expression on the right is positive, the proof of inequality (1) is reduced to showing that the numerator

$$L = ab+bc+ca-9abc$$

is non-negative.

Substituting in L the value $c = 1-a-b$, we obtain

$$L = ab+b(1-a-b)+a(1-a-b)-9ab(1-a-b).$$

We shall consider L as a quadratic function of variable a and we shall write it as

$$L = (9b-1)a^2+(9b-1)(b-1)a-b(b-1).$$

Let us find the discriminant Δ of this function:

$$\begin{aligned}\Delta &= (9b-1)^2(b-1)^2+4b(b-1)(9b-1)\\ &= (9b-1)(b-1)(9b^2-6b+1)\\ &= (9b-1)(b-1)(3b-1)^2.\end{aligned}$$

Observe that $b < 1$, and thus $b-1 < 0$, whereas $(3b-1)^2 \geqslant 0$; we shall therefore have to distinguish two cases:

(i) $9b-1 > 0$. Then $\Delta \leqslant 0$ and the quadratic function L has a non-negative value for any value of the variable a, and thus the right-hand side of (2) is non-negative, whence the conclusion that

$$\frac{1}{a}+\frac{1}{b}+\frac{1}{c}\geqslant 9.$$

(ii) $9b-1 \leqslant 0$. In this case we immediately conclude that

$$b \leqslant \frac{1}{9}, \quad \frac{1}{b} \geqslant 9,$$

and thus

$$\frac{1}{a}+\frac{1}{b}+\frac{1}{c} > 9.$$

Thus inequality (1) is true in both cases.

REMARK 1. It can be asked in what cases (under the given assumptions) we have the equality

$$\frac{1}{a}+\frac{1}{b}+\frac{1}{c} = 9, \tag{3}$$

i.e. for what positive values of a, b, c whose sum is 1 the sum of their reciprocals has its maximum value 9.

The answer to this question can easily be derived from either of the above proofs. Following method I we obtain equality (3) if and only if

$$\frac{a}{b}+\frac{b}{a} = \frac{b}{c}+\frac{c}{b} = \frac{c}{a}+\frac{a}{c} = 2,$$

i.e. if

$$a = b = c = \frac{1}{3}.$$

Following method II we obtain equality (3) only in case (1) if we postulate that $\Delta = 0$, and thus that $3b-1 = 0$, whence

$$b = \frac{1}{3}, \quad \text{and} \quad L = 2a^2-\frac{4}{3}a+\frac{2}{9} = 2\left(a-\frac{1}{3}\right)^2.$$

Thus the equality $L = 0$ holds if and only if

$$a = \frac{1}{3}, \quad b = \frac{1}{3}, \quad c = \frac{1}{3}.$$

REMARK 2. Let a, b, c be arbitrary positive numbers. Let us consider the numbers

$$m = \frac{a}{a+b+c}, \quad n = \frac{b}{a+b+c}, \quad p = \frac{c}{a+b+c}.$$

Since the sum of numbers m, n, p is equal to 1, by applying to them inequality (1) we obtain the inequality

$$\frac{a+b+c}{a}+\frac{a+b+c}{b}+\frac{a+b+c}{c} \geqslant 9,$$

i.e. the inequality

$$(a+b+c)\left(\frac{1}{a}+\frac{1}{b}+\frac{1}{c}\right) \geqslant 9,$$

which can be written as

$$\frac{a+b+c}{3} \geqslant \frac{3}{\frac{1}{a}+\frac{1}{b}+\frac{1}{c}}. \tag{4}$$

The left-hand side of inequality (4) is the *arithmetical mean* of the numbers a, b, c and its right-hand side is the reciprocal of the arithmetical mean of the numbers $1/a$, $1/b$, $1/c$, i.e. the *harmonic mean* of a, b, c.

Inequality (4) expresses the following theorem:

The arithmetical mean of three positive numbers is at least equal to the harmonic mean of those numbers.

These means are equal only if $a = b = c$ (see remark 1). We have deduced theorem (4) on the arithmetical mean and the harmonic mean from the theorem proved in problem 60. Conversely, the latter theorem is a conclusion from theorem (4) since if $a+b+$ $+c = 1$, then inequality (4) implies inequality (1). The two theorems are thus equivalent.

REMARK 3. Reasoning as in method I and in remark 1, we shall easily prove a more general theorem:

If the sum of positive numbers $a_1, a_2, \ldots, a_n$ is equal to 1, *then*

$$\frac{1}{a_1}+\frac{1}{a_2}+\ldots+\frac{1}{a_n} \geqslant n^2,$$

equality occurring only if $a_1 = a_2 = \ldots = a_n$.

We ascertain, as in remark 2, that the above theorem is equivalent to the following theorem:

If $a_1, a_2, \ldots, a_n$ are positive numbers, then their arithmetical mean is at least equal to their harmonic mean, i.e.

$$\frac{a_1+a_2+\ldots+a_n}{n} \geqslant n:\left(\frac{1}{a_1}+\frac{1}{a_2}+\ldots+\frac{1}{a_n}\right),$$

equality occurring only if $a_1 = a_2 = \ldots = a_n$.

61. *Method I.* We use the transformation

$$ab(a+b)+bc(b+c)+ca(c+a)-6abc$$
$$=a(b^2+c^2)-2abc+b(c^2+a^2)-2abc+c(a^2+b^2)-2abc$$
$$=a(b-c)^2+b(c-a)^2+c(a-b)^2.$$

We have obtained the sum of three non-negative numbers, i.e. a non-negative number; hence immediately follows the inequality

$$ab(a+b)+bc(b+c)+ca(c+a) \geqslant 6abc. \tag{1}$$

Method II. It is sufficient to prove that

$$\frac{ab(a+b)}{abc}+\frac{bc(b+c)}{abc}+\frac{ca(c+a)}{abc} \geqslant 6.$$

Now

$$\frac{ab(a+b)}{abc}+\frac{bc(b+c)}{abc}+\frac{ca(c+a)}{abc}$$
$$=\frac{a}{c}+\frac{b}{c}+\frac{b}{a}+\frac{c}{a}+\frac{c}{b}+\frac{a}{b}$$
$$=\left(\frac{a}{b}+\frac{b}{a}\right)+\left(\frac{b}{c}+\frac{c}{b}\right)+\left(\frac{c}{a}+\frac{a}{c}\right).$$

This number is not less than 6 because

$$\frac{a}{b}+\frac{b}{a} \geqslant 2, \quad \frac{b}{c}+\frac{c}{b} \geqslant 2, \quad \frac{c}{a}+\frac{a}{c} \geqslant 2.$$

Method III. The problem can be reduced to problem 60 in the following way:

Dividing both sides of (1) by the positive number abc we obtain an equivalent inequality,

$$\frac{a+b}{c}+\frac{b+c}{a}+\frac{c+a}{b} \geqslant 6,$$

which in turn can be replaced by the inequality

$$\frac{a+b+c}{c}+\frac{a+b+c}{a}+\frac{a+b+c}{b} \geqslant 9. \tag{2}$$

Numbers $a/(a+b+c)$, $b/(a+b+c)$, $c/(a+b+c)$ are three positive numbers whose sum is 1. According to problem 60, the sum of their reciprocals is not less than 9, which gives inequality (2).

We can also derive inequality (1) from the proposition that the harmonic mean of three positive numbers is not greater than their arithmetical mean because inequality (2) can be replaced by the inequality

$$\frac{a+b+c}{3} \geqslant 3:\left(\frac{1}{a}+\frac{1}{b}+\frac{1}{c}\right),$$

expressing this very proposition (see problem 60, remark 2).

Method IV. We shall apply the theorem stating that the arithmetical mean of two positive numbers is not less than their geometrical mean:

$$\frac{a+b}{2} \geqslant \sqrt{(ab)}, \quad \frac{b+c}{2} \geqslant \sqrt{(bc)}, \quad \frac{c+a}{2} \geqslant \sqrt{(ca)}.$$

Multiplying these inequalities and simplifying, we obtain the inequality

$$(a+b)(b+c)(c+a) \geqslant 8abc,$$

whence, performing the multiplication and applying an easy tranformation, we have

$$ab(a+b)+bc(b+c)+ca(c+a) \geqslant 6abc.$$

Remark 1. We have proved above (method III) that inequality (1) can be derived from the theorem given in problem 60. We shall ascertain that the converse is also true: the solution of problem 60 can be derived from inequality (1).

Suppose that $a > 0$, $b > 0$, $c > 0$ and $a+b+c = 1$. Inequality (1) gives in succession:

$$ab(a+b)+bc(b+c)+ca(c+a)-6abc \geqslant 0,$$

$$\frac{a+b}{c}+\frac{b+c}{a}+\frac{c+a}{b}-6 \geqslant 0,$$

$$\frac{a+b+c}{c}+\frac{a+b+c}{a}+\frac{a+b+c}{b}-9 \geqslant 0,$$

and since $a+b+c = 1$ we obtain

$$\frac{1}{c}+\frac{1}{a}+\frac{1}{b} \geqslant 9.$$

Remark 2. Inequality (1) can be generalized by showing that if $a_1, a_2, \ldots, a_n$ denote positive numbers, then

$$\begin{aligned} a_2a_3 \ldots a_n(a_2+a_3+ \ldots +a_n)+ \\ +a_3a_4 \ldots a_na_1(a_3+a_4+ \ldots +a_n+a_1)+ \ldots \\ +a_{n-1}a_1 \ldots a_{n-2}(a_{n-1}+a_1+ \ldots +a_{n-2}) \\ \geqslant n(n-1)a_1a_2 \ldots a_n. \end{aligned}$$

The proof of this inequality can easily be obtained by any of the methods I, II, III. Method IV is unsuitable for this purpose.

62. *Method I.* We shall factorize the polynomial $u^3+v^3+w^3-3uvw$.

It will be observed that if we substitute for u in this polynomial the expression $-v-w$, we obtain zero:

$$(-v-w)^3+v^3+w^3-3(-v-w)vw$$
$$= -v^3-3v^2w-3vw^2-w^3+v^3+w^3+3v^2w+3vw^2 = 0.$$

It follows that the polynomial in question is divisible by $u+v+w$†. This polynomial in the variables u, v, w is *homogeneous*, i.e. its terms are all of the same degree; moreover, it is *symmetrical*, i.e. it does not change with any permutation of the arguments u, v, w. The same two properties characterize the divisor $u+v+w$. Consequently the quotient of these polynomials must also be a homogeneous and symmetric polynomial in the variables u, v, w, whence

$$u^3+v^3+w^3-3uvw = (u+v+w)[a(u^2+v^2+w^2)+b(uv+uw+wu)].$$

Comparing the coefficients of the corresponding terms of the two sides, we obtain $a = 1$, $b = -1$. Since (see problem 55)

$$u^2+v^2+w^2-uv-vw-wu = \frac{1}{2}[(u-v)^2+(v-w)^2+(w-u)^2],$$

we finally obtain

$$u^3+v^3+w^3-3uvw = \frac{1}{2}(u+v+w)[(u-v)^2+(v-w)^2+(w-u)^2].$$

Since $u+v+w > 0$ and $(u-v)^2+(v-w)^2+(w-u)^2 \geqslant 0$, we have

$$u^3+v^3+w^3-3uvw \geqslant 0, \tag{1}$$

which is what was to be proved.

Method II. Let us introduce the notation

$$u^3 = a, \quad v^3 = b, \quad w^3 = c.$$

Inequality (1) assumes the form

$$\frac{a+b+c}{3} \geqslant \sqrt[3]{(abc)}. \tag{2}$$

† On the basis of the theorem stating that if a polynomial in a variable x becomes zero on substituting a for x, then the polynomial is divisible by $x-a$.

Inequality (2), which is to be proved, states that *the arithmetical mean of three positive numbers is at least equal to their geometrical mean.*

To begin with, it will be observed that an analogous theorem is true for two numbers, i.e. that

$$\frac{a+b}{2} \geqslant \sqrt{(ab)},$$

since

$$\frac{a+b}{2} - \sqrt{(ab)} = \frac{a+b-2\sqrt{(ab)}}{2} = \left(\frac{\sqrt{a}-\sqrt{b}}{2}\right)^2 \geqslant 0.$$

The theorem is also true for four numbers because

$$\frac{a+b+c+d}{4} = \frac{\dfrac{a+b}{2}+\dfrac{c+d}{2}}{2} \geqslant \frac{\sqrt{(ab)}+\sqrt{(cd)}}{2} \geqslant \sqrt[4]{(abcd)}. \quad (3)$$

In order to prove the theorem for three numbers, i.e. to prove inequality (2), we use inequality (3) taking $d = \frac{1}{3}(a+b+c)$; we obtain

$$\frac{a+b+c+\frac{1}{3}(a+b+c)}{4} \geqslant \sqrt[4]{\frac{abc(a+b+c)}{3}},$$

whence

$$\frac{1}{3}(a+b+c) \geqslant \sqrt[4]{\frac{abc(a+b+c)}{3}},$$

and consequently

$$\frac{1}{3^4}(a+b+c)^4 \geqslant \frac{abc(a+b+c)}{3}.$$

Dividing both sides of this inequality by the positive number $\frac{1}{3}(a+b+c)$ we obtain

$$\left(\frac{a+b+c}{3}\right)^3 \geqslant abc,$$

whence finally

$$\frac{a+b+c}{3} \geqslant \sqrt[3]{(abc)}.$$

Remark 1. The reasoning followed in method II is based on an idea of a famous French mathematician, A. L. Cauchy (1789–1857), who proved in this way the general theorem stating that *the arithmetical mean of n positive numbers* $a_1, a_2, \ldots, a_n$ *is at least equal to the geometrical mean of these numbers,* i.e. that

$$\frac{a_1+a_2+\ldots+a_n}{n} \geqslant \sqrt[n]{(a_1 a_2 \ldots a_n)}. \qquad (\alpha)$$

We shall reproduce this proof here. To begin with, we ascertain that

(1) Inequality (α) is true if $n = 2$ (as before in method II).

(2) If inequality (α) is true for $n = k \geqslant 2$, then it is also true for $n = 2k$. Indeed

$$\frac{a_1+a_2+\ldots+a_k+a_{k+1}+\ldots+a_{2k-1}+a_{2k}}{2k}$$

$$= \frac{\frac{a_1+a_2}{2}+\frac{a_3+a_4}{2}+\ldots+\frac{a_{2k-1}+a_{2k}}{2}}{k}$$

$$\geqslant \frac{\sqrt{(a_1a_2)}+\sqrt{(a_3a_4)}+\ldots+\sqrt{(a_{2k-1}a_{2k})}}{k} \geqslant \sqrt[2k]{(a_1 a_2 \ldots a_{2k})}.$$

Applying the principle of induction we conclude that inequality (α) is true if n is any natural power of number 2, i.e. if $n = 2^m$, where m is a natural number. Finally, let n be a natural number greater than 1 and not equal to a natural power of number 2. In that case there exists a natural number m such that

$$2^{m-1} < n < 2^m.$$

Let us denote the natural number $2^m - n$ by p; thus $n+p = 2^m$. If besides the n positive numbers $a_1, a_2, \ldots, a_n$ we take p more positive numbers, $a_{n+1}, a_{n+2}, \ldots, a_{n+p}$, we shall obtain $n+p = 2^m$ numbers for which, as has been shown before, we have the inequality

$$\frac{a_1+a_2+\ldots+a_n+a_{n+1}+\ldots+a_{n+p}}{n+p}$$

$$\geqslant \sqrt[n+p]{(a_1 a_2 \ldots a_n a_{n+1} \ldots a_{n+p})}. \qquad (\beta)$$

Let us choose numbers $a_{n+1}, a_{n+2}, \ldots, a_{n+p}$ in such a way that

$$a_{n+1} = a_{n+2} = \ldots = a_{n+p} = \frac{a_1+a_2+\ldots+a_n}{n}.$$

We then obtain on the left side of (β) the expression

$$\frac{a_1+a_2+\ldots+a_n+p\,\frac{a_1+a_2+\ldots+a_n}{n}}{n+p} = \frac{a_1+a_2+\ldots+a_n}{n}$$

and on the right-hand side of (β) the expression

$$\sqrt[n+p]{\left[a_1 a_2 \dots a_n \left(\frac{a_1+a_2+\dots+a_n}{n}\right)^p\right]}.$$

Inequality (β) will thus assume the form

$$\frac{a_1+a_2+\dots+a_n}{n} \geqslant \sqrt[n+p]{\left[a_1 a_2 \dots a_n \left(\frac{a_1+a_2+\dots+a_n}{n}\right)^p\right]}.$$

Let us raise both sides of this inequality to the power $n+p$:

$$\left(\frac{a_1+a_2+\dots+a_n}{n}\right)^{n+p} \geqslant a_1 a_2 \dots a_n \left(\frac{a_1+a_2+\dots+a_n}{n}\right)^p.$$

If we divide both sides of this inequality by the positive number $(a_1+a_2+\dots+a_n)^p/n$ and then extract from them the root of degree n, we shall obtain

$$\frac{a_1+a_2+\dots+a_n}{n} \geqslant \sqrt[n]{(a_1 a_2 \dots a_n)}.$$

Remark 2. If the numbers $a_1, a_2, \dots, a_n$ are all equal, then

$$\frac{a_1+a_2+\dots+a_n}{n} = \sqrt[n]{(a_1 a_2 \dots a_n)}. \qquad (\gamma)$$

The inverse theorem, which is a supplement of theorem (α), is also true: the equality of the two sides of inequality (α), i.e. equality (γ), holds only if $a_1 = a_2 = \dots = a_n$.

In order to prove that, we can use the argument given in remark 1. Namely, if the numbers $a_1, a_2, \dots, a_n$ satisfy equality (γ) then, performing the earlier transformations in the inverse order, we shall obtain, instead of inequality (β), the equality

$$\frac{a_1+a_2+\dots+a_n+a_{n+1}+\dots+a_{n+p}}{n+p} = \sqrt[n+p]{(a_1 a_2 \dots a_n a_{n+1} \dots a_{n+p})}.$$

On the basis of the inequalities given in (2) we conclude that

$$\frac{a_1+a_2}{2}+\frac{a_3+a_4}{2}+\dots+\frac{a_{n+p-1}+a_{n+p}}{2} = \sqrt{(a_1 a_2)}+\sqrt{(a_3 a_4)}+\dots+\sqrt{(a_{n+p-1} a_{n+p})}.$$

Since each component of the left-hand side of this equality is at least equal to the corresponding component of the right-hand side, the equality of the sums implies the equality of the correspon-ding components, i.e.

$$\frac{a_1+a_2}{2} = \sqrt{(a_1 a_2)},$$

whence $(\sqrt{a_1}-\sqrt{a_2})^2 = 0$ and $a_1 = a_2$. Since the successive order of the numbers $a_1, a_2, \ldots, a_n$ is arbitrary, it follows that these numbers are all equal.

We shall now give a simpler and more elegant proof of the inverse theorem in question.

Suppose that equality (γ) holds and that some of the numbers $a_1, a_2, \ldots, a_n$ are not equal, say $a_1 \neq a_2$. Then

$$\frac{a_1+a_2}{2} > \sqrt{(a_1 a_2)},$$

i.e.

$$\left(\frac{a_1+a_2}{2}\right)^2 > a_1 a_2.$$

In that case

$$\frac{a_1+a_2+\ldots+a_n}{n} = \frac{\frac{a_1+a_2}{2}+\frac{a_1+a_2}{2}+a_3+\ldots+a_n}{n}$$

$$\geqslant \sqrt[n]{\left(\frac{a_1+a_2}{2}\times\frac{a_1+a_2}{2}\times a_3 \ldots a_n\right)} = \sqrt[n]{\left[\left(\frac{a_1+a_2}{n}\right)^2\times a_3 \ldots a_n\right]}$$

$$> \sqrt[n]{(a_1 a_2 a_3 \ldots a_n)}.$$

We have found that the supposition of $a_1 \neq a_2$ leads to a contradiction of assumption (γ); that supposition is thus false, whence $a_1 = a_2 = \ldots = a_n$.

Remark 3. Denote the sum of positive numbers $a_1, a_2, \ldots, a_n$ by S and their product by P. Writing inequality (α) as

$$P \leqslant \left(\frac{S}{n}\right)^n \quad \text{or} \quad S \geqslant n\sqrt[n]{P},$$

we can state the following theorems:

(a) *The greatest value of the product of n positive numbers whose sum S is given is*

$$\left(\frac{S}{n}\right)^n.$$

This value is assumed if and only if the factors are all equal.

(b) *The least value of the sum of n positive numbers whose product P is given is*

$$n\sqrt[n]{P},$$

which is assumed by that sum if and only if the components are all equal.

Using these theorems we can solve a great many problems consisting in finding the greatest or the least value of a magnitude. Here are a few examples to serve as exercises:

1. Which of the triangles with a given perimeter has the greatest area?

2. Which of the parallelepipeds with a given volume has (a) the least sum of the edges, (b) the least surface area?

3. Which of the cylinders inscribed in a given sphere has the greatest volume?

4. We use a paper square of given side a to make a box in the shape of a parallelepiped with a square base by cutting off at the corners four equal squares of side x and then folding the paper in a suitable way. For what value of x will the volume of the box be greatest?

REMARK 4. Applying the theorem on the arithmetical mean and the geometrical mean to the numbers $1/a_1, 1/a_2, \ldots, 1/a_n$, we obtain the inequality

$$\frac{1}{n}\left(\frac{1}{a_1}+\frac{1}{a_2}+\ldots+\frac{1}{a_n}\right) \geqslant \sqrt[n]{\left(\frac{1}{a_1}\times\frac{1}{a_2}\times\frac{1}{a_n}\right)};$$

it follows that

$$\frac{n}{\dfrac{1}{a_1}+\dfrac{1}{a_2}+\ldots+\dfrac{1}{a_n}} \leqslant \sqrt[n]{(a_1 a_2 \ldots a_n)}.$$

Consequently: the harmonic mean of n positive numbers is at most equal to the geometrical mean of those numbers, the equality of the two means occurring only if the given numbers are all equal. Cf. problem 60.

63. *Method I.* We shall transform the left-hand side of the inequality

$$\left(1+\frac{1}{x}\right)\left(1+\frac{1}{y}\right) \geqslant 9 \tag{1}$$

considering that $x+y = 1$:

$$\left(1+\frac{1}{x}\right)\left(1+\frac{1}{y}\right) = \frac{x+1}{x}\times\frac{y+1}{y} = \frac{xy+2}{xy} = 1+\frac{2}{xy}.$$

As we know, the product of positive numbers x and y whose sum is constant is greatest if $x = y$; consequently if $x+y = 1$, then $xy \leqslant \frac{1}{4}$. It thus follows from the preceding equality that

$$\left(1+\frac{1}{x}\right)\left(1+\frac{1}{y}\right) \geqslant 1+2 : \frac{1}{4} = 9.$$

REMARK. Inequality (1) changes into an equality if and only if $x = y = \frac{1}{2}$.

Method II. We shall prove inequality (1) by showing that the difference between its left side and its right side is non-negative. Since $y = 1-x$, we have

$$\left(1+\frac{1}{x}\right)\left(1+\frac{1}{y}\right)-9 = \left(1+\frac{1}{x}\right)\left(1+\frac{1}{1-x}\right)-9$$

$$= \frac{x+1}{x}\times\frac{2-x}{1-x}-9 = \frac{8x^2-8x+2}{x(1-x)} = \frac{2(2x-1)^2}{x(1-x)}.$$

The numerator of the fraction obtained is non-negative, and the denominator is positive because $x > 0$ and $1-x > 0$; consequently the fraction is non-negative.

Incidentally, it will be observed that this fraction is equal to zero if and only if $x = \frac{1}{2}$.

64. *Method I.* To prove inequality

$$(a+b+c+d)^2 > 8(ac+bd) \tag{1}$$

we shall introduce in the calculation, instead of numbers b, c, d, the successive differences between the given numbers. Let

$$b-a = r, \quad c-b = s, \quad d-c = t;$$

then

$$b = a+r, \quad c = a+r+s, \quad d = a+r+s+t.$$

We substitute these values in the expression

$$m = (a+b+c+d)^2-8(ac+bd) \tag{2}$$

and obtain

$$m = (4a+3r+2s+t)^2 - 8[a(a+r+s)+(a+r)(a+r+s+t)],$$

whence, on performing the operations, we have

$$m = r^2+4rs+4s^2+4st-2rt+t^2$$

and finally

$$m = (r-t)^2+4rs+4s^2+4st. \tag{3}$$

If $a < b, b < c, c < d$ then $r > 0$, $s > 0, t > 0$; it then follows from (3) that $m > 0$, which means that inequality (1) holds.

REMARK. It can be shown that the inequality $m > 0$, and thus also the inequality given in the problem, are true under a much weaker assumption. To show this we transform formula (3) as follows:

$$m = (r-t+2s)^2+8st. \tag{4}$$

Equality (4) implies that if $st > 0$, i.e. if the given numbers satisfy the condition

$$b < c < d \quad \text{or} \quad d < c < b, \tag{5}$$

signifying that number c is to be contained between the numbers b and d, then $m > 0$.

It will be observed that if we interchange the letters a and c in (2), the value of the expression m will not change. Thus in inequalities (5) we can write a instead of c; as a new condition sufficient for having $m > 0$, we obtain the condition $b < a < d$ or $d < a < b$, i.e. the condition that number a should be contained between numbers b and c.

But expression m retains its value also if we write in it b and d instead of a and c, and *vice versa*. Thus for the inequality $m > 0$ to hold it is also sufficient that b or d should be contained between a and c. We have obtained the following result:

The inequality $(a+b+c+d)^2 > 8(ac+bd)$ is true if either of the numbers a, c is contained between the numbers b and d or if either of the numbers b, d is contained between the numbers a, c.

If none of the above cases occurs, number m can be positive as well as negative or equal to zero. For example, if $a = 0$, $b = 2$, $c = 1$, $d = 3$, then $m = -12$; if $a = 0$, $b = 2$, $c = 1$, $d = 10$, then $m = 9$; if $a = 0$, $b = 2$, $c = 1$, $d = 9$, then $m = 0$.

Method II. Let us order the expression $m = (a+b+c+$ $+d)^2-8ac-8bd$ according to the powers of one of the letters, say a:

$$m = a^2+2(b-3c+d)a+[(b+c+d)^2-8bd],$$

and let us find the discriminant of the quadratic function obtained:

$$\begin{aligned}\tfrac{1}{4}\Delta &= (b-3c+d)^2-(b+c+d)^2+8bd = (2b-2c+2d)(-4c)+8bd\\ &= 8(bd-bc+c^2-cd) = 8(b-c)(d-c).\end{aligned}$$

If $b < c < d$ or $d < c < b$, then $\Delta < 0$ and consequently $m > 0$; we have obtained the same result as the one contained in the remark to method I.

65. *Method I.* Number $\frac{1}{2}n(n-1)$ is equal to the sum of the first $n-1$ natural numbers. Consequently

$$2^{\frac{1}{2}n(n-1)} = 2^{1+2+3+\dots+(n-1)} = 2\times 2^2\times 2^3\times \dots \times 2^{n-1}.$$

If $m \geqslant 2$, then $2^m > m+1$, and thus

$$2\times 2^2\times 2^3\times \dots \times 2^{n-1} > 2\times 3\times 4\times \dots \times n = n!$$

and finally

$$2^{\frac{1}{2}n(n-1)} > n!.$$

The proof that for $m \geqslant 2$ we have the inequality $2^m > m+1$ is easily obtained by induction.

Method II. Induction can be applied directly to the proof of the inequality

$$2^{\frac{1}{2}n(n-1)} > n! \tag{1}$$

for any integer $n > 2$.

(a) If $n = 3$, inequality (1) is true because in that case

$$2^{\frac{1}{2}n(n-1)} = 2^3 = 8, \quad \text{and} \quad n! = 6.$$

(b) If inequality (1) is true for a certain $k \geqslant 3$, i.e. if

$$2^{\frac{1}{2}k(k-1)} > k!,$$

then

$$2^{\frac{1}{2}(k+1)k} = 2^{\frac{1}{2}k(k-1)+k} = 2^{\frac{1}{2}k(k-1)} \times 2^k > k!\times 2^k$$
$$> k!(k+1) = (k+1)!,$$

and thus

$$2^{\frac{1}{2}(k+1)k} > (k+1)!.$$

(a) and (b) imply that inequality (1) holds for any natural $n \geqslant 3$.

PART TWO

Geometry and Trigonometry

PROBLEMS

§ 5. Proving Theorems

66. Given that S is the mid-point of a segment AB of a straight line, and M is any point on the straight line, prove that

$$MA^2+MB^2 = 2\times SA^2+2\times SM^2.$$

67. A, B, C, D are points in a plane, no three of them being collinear; $AB = CD$ and $AD = BC$. Prove that if the segments AB and CD intersect, then the segments AC and BD are parallel.

68. Prove that if two medians of a triangle are equal, the triangle is isosceles.

69. Prove that if the sides of a triangle are equal, respectively, to the medians of another triangle, then each of the medians of the first triangle is equal to $\frac{3}{4}$ of the corresponding side of the other triangle.

70. Prove that the ratio of the sum of the medians of every triangle to its perimeter is contained in the interval $(\frac{3}{4}, 1)$ and is not contained in any smaller interval.

71. Prove that the altitudes of an acute-angled triangle are the bisectors of the angles of a triangle whose vertices are the feet of those altitudes.

72. Prove that if one of the sides a, b, c of a triangle is equal to the arithmetic mean of the remaining sides, then one of the altitudes of that triangle is 3 times as great as the radius of the circle inscribed in that triangle.

73. Triangle ABC is right-angled at C. Draw the altitude CD, and inscribe a circle in each of the triangles ABC, ACD and BCD. Prove that the sum of the radii of those circles equals CD.

74. Prove that the altitudes h_1, h_2, h_3 of a triangle and the radii r_1, r_2, r_3 of the escribed circles satisfy the equality

$$\frac{1}{h_1}+\frac{1}{h_2}+\frac{1}{h_3} = \frac{1}{r_1}+\frac{1}{r_2}+\frac{1}{r_3}.$$

75. A vertical pole of altitude a illuminated by the rays of the sun threw a shadow upon a horizontal plane; the length of the shadow was a at one moment, $2a$ at another moment and $3a$ at a third moment. Prove that the sum of the angles of incidence of the rays at those three moments equals a right angle.

76. On the successive sides of a square, choose points M, N, P, Q in such a way that all the sides of the quadrilateral $MNPQ$ are equal. Prove that the quadrilateral $MNPQ$ is a square.

77. The centres of four identical balls lie on a circle and the centre of gravity of the system of those balls lies at the centre of the circle. Show that the centres of the balls are the vertices of a rectangle.

78. Show whether the following theorems are true:

(a) if the four vertices of a rectangle lie on the four sides of a rhombus, then the sides of the rectangle are parallel to the diagonals of the rhombus;

(b) if the four vertices of a square lie on the four sides of a rhombus which is not a square, then the sides of the square are parallel to the diagonals of the rhombus.

79. The diagonals of a trapezium $ABCD$ intersect at S. Draw a straight line through S parallel to the bases AB and CD of the trapezium and intersecting the sides AD and BC at points M and N. Prove that MN is equal to the harmonic mean of the bases.

80. Inside a triangle ABC there lies a point P such that

$$\sphericalangle PAB = \sphericalangle BPC = \sphericalangle PCA = \varphi.$$

Prove that

$$\frac{1}{\sin^2\varphi} = \frac{1}{\sin^2 A} + \frac{1}{\sin^2 B} + \frac{1}{\sin^2 C}.$$

81. In a circle we draw a chord AB and a diameter MN, and determine the projections M' and N' of points M and N upon the line AB. Prove that $M'A = BN'$.

82. We choose three points, A, B and C, on a circle. Prove that the feet of the perpendiculars drawn from an arbitrary point M of the circle to the lines AB, BC and CA are collinear.

83. Given two circles, we draw one of their internal common tangents and both their external common tangents. Prove that the segment of the internal tangent contained between the external tangents is equal to the segment of an external tangent contained between its points of contact.

84. Given a convex quadrilateral, prove that if there exists a circle tangent to its four sides produced then the differences of the opposite sides of that quadrilateral are equal.

85. In a circle we draw two equal chords AB and AC and an arbitrary chord AD; the straight line AD intersects the straight line BC at a point E. Prove that the product $AE \times AD$ is independent of the position of point D on the circle, i.e. that $AE \times AD = AC^2$.

86. We are given a circle C and points A and B lying at different distances from its centre. Prove that the common chords of circle C and the circles passing through points A and B lie on straight lines having one point in common.

87. The diagonals of a quadrilateral inscribed in a circle intersect at a point K. Points M, N, P, Q are the projections of K on the sides of that quadrilateral. Prove that the straight lines KM, KN, KP and KQ bisect the angles of the quadrilateral $MNPQ$.

88. A quadrilateral $ABCD$ is inscribed in a circle. The straight lines AB and CD intersect at E, and the straight lines AD and BC intersect at F. The bisector of $\sphericalangle AEC$ intersects the side BC at M and the side AD at N; the bisector of $\sphericalangle BFD$ intersects the side AB at P and the side CD at Q. Prove that the quadrilateral $MPNQ$ is a rhombus.

89. Prove that in an isosceles trapezium circumscribed about a circle the segments joining the points of contact of the opposite sides with the circle pass through the point of intersection of the diagonals.

90. Prove that if a plane figure has two and only two axes of symmetry, those axes are perpendicular.

91. Prove that if two altitudes of a tetrahedron intersect, then the remaining two altitudes also intersect.

92. Prove that if the opposite edges of a tetrahedron $ABCD$ are equal, i.e. $AB = CD$, $AC = BD$, $AD = BC$, then the straight lines joining the mid-points of the opposite edges are perpendicular and constitute the axes of symmetry of the tetrahedron.

93. Prove that a plane which passes (a) through the mid-points of two opposite edges of a tetrahedron and (b) through the mid-point of one of the remaining edges divides the tetrahedron into two parts of equal volumes. Does this assertion remain true if we reject assumption (b)?

94. Given two skew straight lines m and n, mark off a segment AB of given length a on m and a segment CD of given length b on n. Prove that the volume of the tetrahedron $ABCD$ is independent of the position of the segments AB and CD on m and n.

§ 6. Finding Geometrical Magnitudes

95. In a parallelogram of given area S each vertex has been connected with the mid-points of the opposite two sides. In this manner the parallelogram has been cut into parts, one of them being an octagon. Find the area of that octagon.

96. In a triangle ABC we choose points D, E, F on sides BC, CA and AB respectively in such a way that

$$BD : DC = CE : EA = AF : FB = k,$$

where k is a given positive number. Given the area S of the triangle ABC, find the area of the triangle DEF.

97. In a triangle ABC a point M lies on the side BC, a point N lies on the side AC, and the segments AM and BN intersect at a point P. Given the ratios

$$BM : MC = m \quad \text{and} \quad AN : NC = n,$$

find the ratios $AP : PM$ and $BP : PN$.

98. On the sides BC, CA and AB of a triangle ABC we choose points M, N, P, respectively, in such a way that:

$$\frac{BM}{MC} = \frac{CN}{NA} = \frac{AP}{PB} = k,$$

where k denotes a given number greater than 1; we then draw the segments AM, BN and CP. Given the area S of the triangle ABC, find the area of the triangle determined by the lines AM, BN and CP.

99. In a triangle ABC the angle A is given. Is it possible to find the angles B and C if we know that a certain straight line passing through the point A divides the triangle into two isosceles triangles?

100. A circle is circumscribed about a triangle ABC. Knowing the radius R of that circle, find the radius of a circle passing through the centres of the three circles escribed to the triangle ABC.

101. Prove that the area S of a quadrilateral inscribed in a circle and having sides a, b, c, d is expressed by the formula

$$S = \sqrt{[(p-a)(p-b)(p-c)(p-d)]},$$

where $2p = a+b+c+d$.

102. Prove that if a, b, c, d are the sides of a quadrilateral which has a circle circumscribed about it and a circle inscribed in it, then the area S of the quadrilateral is expressed by the formula

$$S = \sqrt{(abcd)}.$$

103. Points A, B, C, D are the consecutive vertices of a regular polygon, and the following relation holds:

$$\frac{1}{AB} = \frac{1}{AC} + \frac{1}{AD}.$$

How many sides has this polygon?

104. The angle at which a vertical mast mounted on a tower is visible from a point on the ground has its greatest value α when the distance from that point to the axis of the mast is a. Find the height of the tower and the height of the mast.

105. The cross-section of a ball-bearing consists of two concentric circles, C and C_1, between which there are n small circles $K_1, K_2, \ldots, K_n$, each of them tangent to the neighbouring two and to the circles C and C_1. Given the radius r of the inside circle C and the natural number n, find the radius x of the circle C_2 passing through the points of contact of the circles $K_1, K_2, \ldots, K_n$ and the sum s of the lengths of those arcs of the circles $K_1, K_2, \ldots, K_n$ which lie outside the circle C_2.

106. A beam of length a is suspended horizontally by its ends by means of two parallel ropes of lengths b. We twist the beam through an angle φ about the vertical axis passing through the centre of the beam. How far will the beam rise?

107. A homogeneous circular disc is suspended in a horizontal position by means of a rope attached at its centre O. At three different points A, B, C of the edge on the disc we place weights p_1, p_2, p_3 without disturbing the equilibrium of the disc. Find the angles AOB, BOC, and COA.

108. Given the mutual distances of four points A, B, C, D in space, find the distance between the mid-point of the segment AB and the mid-point of the segment CD.

109. Find the volume V of a tetrahedron $ABCD$ given the length d of the edge AB and the area S of the projection of the tetrahedron upon a plane perpendicular to the line AB.

110. Through each vertex of a tetrahedron of a given volume V we draw a plane parallel to the opposite face of the tetrahedron. Find the volume of the tetrahedron formed by those planes.

§ 7. Loci

111. Given two intersecting straight lines on a plane, find on that plane the locus of a point for which the sum of the distances from the given lines is equal to a given segment a.

112. Find the locus of the centre of a rectangle whose vertices lie on the perimeter of a given triangle.

113. A triangle ABC has been inscribed in a given circle. Find the locus of the centre M of the circle inscribed in the triangle ABC if the vertices A and B of the triangle are stationary and the vertex C runs over the circumference of the given circle.

114. Given two concentric circles K_1 and K_2, find the locus of the vertex of an angle equal to a given angle α and having one arm tangent to the circle K_1 and the other arm tangent to the circle K_2.

115. Given two intersecting straight lines a and b, find the locus of a point M having the following property: the distance between the orthogonal projections of the point M upon the straight lines a and b is constant and equal to a given segment d.

116. We are given on a plane two parallel straight lines a and b and a point P lying on neither of those lines. A variable straight line m passes through point P and intersects the lines a and b at points A_1 and B_1 respectively; another variable line passes through the point P and intersects the lines a and b at points A_2 and B_2 respectively. Find the locus of the point of intersection of the lines A_1B_2 and A_2B_1.

117. A disc of radius r rolls without sliding along the inside of a rim of radius $2r$. What line is traced by a point chosen arbitrarily on the edge of the disc?

118. Given two intersecting planes A and B and a straight line m intersecting those planes, find the locus of the centres of segments, parallel to m, whose end-points lie on the planes A and B.

119. Find the locus of the mid-points of segments of a given length a whose end-points lie on two perpendicular (intersecting or skew) straight lines.

§ 8. Constructions

120. Given a straight line, a circle and a point A lying neither on the line nor on the circle, draw through the point A a segment having one end-point on the circle and the other on the straight line, the point A being the mid-point of that segment.

121. We are given on a plane a straight line p and points A and B lying on one side of that line.

(1) Place on the line p a segment MN of length d in such a way as to make the path $AM+MN+NB$ the shortest possible.

(2) Place on the line p a segment MN of length d so that $AM = NB$.

122. Construct a quadrilateral $ABCD$ given the lengths of the sides AB and CD and the angles of the quadrilateral.

123. Draw a pentagon in which the mid-points of the sides are the vertices of a given pentagon. Find generalizations of this problem.

124. Construct a rectangle by putting together nine squares with sides equal to 1, 4, 7, 8, 9, 10, 14, 15 and 18.

125. In a given square inscribe a square in which one side or one side produced passes through a given point K.

126. Draw three pairwise tangent circles whose centres are the vertices of a given triangle.

127. In a circle equal chords AB and AC have been drawn. Draw a third chord which will be divided by the chords AB and AC into 3 equal parts.

128. Given two points A and B of a circle, find a point C of that circle for which the sum $AC+CB$ is equal to a given segment a.

129. Given two intersecting circles of radii r and R, draw through one of the points of intersection a straight line which intersects the circles at two further points, P and Q, so that the segment PQ is of a given length d.

130. Given a circle K and straight lines MP and MQ tangent to that circle at points P and Q, draw another tangent to that circle, such that its segment included in the angle PMQ is of a given length l.

131. Given points M and N on the diameter of a circle, inscribe in that circle an isosceles triangle with one of the equal sides passing through the point M and the other passing through the point N.

132. Construct an equilateral triangle whose vertices lie on three given parallel straight lines.

133. Given two concentric circles and a point A, draw through the point A a straight line whose segment contained in the larger circle is divided by the smaller circle into three equal parts.

134. Given two concentric circles, draw a square with two vertices lying on one circle and the other two vertices on the other circle.

135. Given a point A, a straight line p and a circle k, construct a triangle ABC, such that B lies on p, C lies on k, $\sphericalangle A = 60°$ and $\sphericalangle B = 90°$.

136. Given points A and B and a circle k, draw a circle passing through A and B and having with the circle k a common chord of a given length d.

137. Given a circle and a line segment MN, find a point C on the circle such that if the points of intersection of MC, NC with the circle are A, B respectively, then $\triangle ABC$ is similar to $\triangle MNC$.

138. Through a point M given on the line AB outside triangle ABC draw a straight line intersecting the sides AC and BC at points N and P in such a way that the segments AN and BP are equal.

139. (a) Given non-collinear points A, B, C, determine on the plane ABC three parallel straight lines passing through the points A, B, C respectively, the distances between neighbouring parallels being equal. (b) Given non-coplanar points A, B, C, D, determine four parallel planes passing through the points A, B, C, D respectively, the distances between neighbouring planes being equal.

140. Given non-coplanar points A, B, C, D, draw a plane through the point A such that the orthogonal projection of the quadrilateral $ABCD$ upon that plane is a parallelogram.

§ 9. Maxima and Minima

141. Houses A and B stand on the opposite sides of a river whose banks at this place are rectilinear and parallel. Find where a foot-bridge, perpendicular to the banks, should be constructed in order to provide the shortest route from A to B.

142. Through a point M given inside an acute angle draw a straight line which cuts off from that angle a triangle of the least area.

143. Show that if a point M lies in a square $ABCD$ with side 1, then at most one of the distances MA, MB, MC and MD is greater than $\frac{1}{2}\sqrt{5}$, at most two are greater than 1, and at most three are greater than $\frac{1}{2}\sqrt{2}$.

144. A triangular piece of sheet metal weighs 900 g. Prove that if we cut that piece along a straight line passing through the centre of gravity of the triangle, it is impossible to cut off a piece weighing less than 400 g.

145. Given a straight line p and points A and B lying on opposite sides of the line, pass a circle through the points A and B in such a way that the chord of that circle lying on line p be the shortest possible.

146. Given a straight line m and points A and B not lying in the same plane, determine on line m such a point C that the sum of the segments AC and CB be the least possible.

147. Given a straight line p and points A and B within the same plane, find a point M on p for which the sum AM^2+BM^2 is the least possible.

148. Given four spheres with the same radius r, each touching the other three, find the radius of the smallest sphere that comprises them all.

149. Prove that of all triangles with a given base and a given area the isosceles triangle has the smallest perimeter.

150. Prove that for every triangle the radius R of the circumcircle and the radius r of the inscribed circle satisfy the inequality

$$R \geqslant 2r.$$

151. Find the radius of the smallest circle which can enclose any triangle with sides not longer than a given segment a.

152. We want to unscrew a square nut of side a by using a spanner with the aperture in the shape of a regular hexagon of side b. What condition should be satisfied by sides a and b to make that possible?

153. A cyclist sets off from point O and rides with constant velocity v along a rectilinear highway. A messenger, who is at a distance a from point O and at a distance b from the highway, wants to deliver a letter to the cyclist. What is the minimum velocity with which the messenger should run in order to attain his objective?

154. In a circular tower whose interior diameter is 2 m there is a winding staircase 6 m high. The height of each step is 0·15 m. In a horizontal projection the steps form contiguous circular sectors of 18°. The narrower ends of the steps are fixed in a circular column with diameter 0·64 m whose axis coincides with the axis of the tower. Find the greatest length of a rectilinear bar which could be carried up the stairs from the foot of the tower to its top. (The thickness of the bar and the thickness of the boards of which the steps are made are to be disregarded.)

§ 10. Trigonometrical Transformations

155. Prove that if the angles A, B, C of a triangle satisfy the equation

$$\cos 3A + \cos 3B + \cos 3C = 1,$$

then one of those angles is equal to 120°.

156. Show that if none of the angles of a convex quadrilateral $ABCD$ is a right angle, then the following equality holds:

$$\frac{\tan A + \tan B + \tan C + \tan D}{\tan A \tan B \tan C \tan D} = \cot A + \cot B + \cot C + \cot D.$$

157. Prove that if $A+B+C$, $A+B-C$, $A-B+C$ or $A-B-C$ is equal to an odd multiple of two right angles, then $\cos^2 A + \cos^2 B + \cos^2 C + 2\cos A \cos B \cos C = 1$ and that the converse is also true.

158. What algebraic relation holds between A, B and C if

$$\cot A + \frac{\cos B}{\sin A \cos C} = \cot B + \frac{\cos A}{\sin B \cos C}\,?$$

159. What algebraic relation holds between α, β and γ if the following equality holds:

$$\tan \alpha + \tan \beta + \tan \gamma = \tan \alpha \tan \beta \tan \gamma\,?$$

160. Prove that if $x_1, x_2, \ldots, x_n$ are angles between 0° and 180° and n is any natural number greater than 1, then

$$\sin(x_1 + x_2 + \ldots + x_n) < \sin x_1 + \sin x_2 + \ldots + \sin x_n.$$

SOLUTIONS

§ 5. Proving Theorems

66. In order to prove the equality

$$MA^2+MB^2 = 2\times SA^2+2\times SM^2 \tag{1}$$

we shall express the lengths MA and MB in terms of the lengths SM and SA.

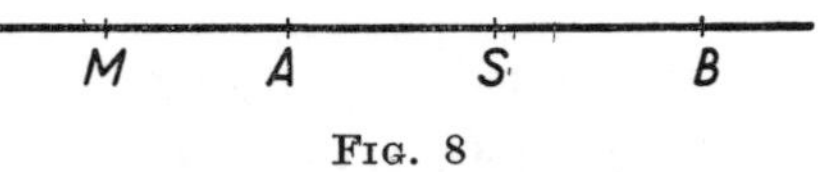

FIG. 8

If the point M lies as shown in Fig. 8, i.e. if it lies on BA produced, then

$$MA = MS-AS,$$

$$MB = MS+SB = MS+AS$$

and consequently

$$MA^2+MB^2 = (MS-AS)^2+(MS+AS)^2 = 2MS^2+2AS^2,$$

which gives the required equality (1).

In order to complete the proof we must investigate the remaining cases; namely if point M lies: (a) at point A, (b) between points A and S, (c) at point S, (d) between points S and B, (e) at point B, (f) on AB produced. In each of these cases the proof is easy and is left to the reader.

It is possible, however, to avoid the tiresome investigation of the above-mentioned cases by carrying out a general proof, i.e. a proof that is applicable to each of those cases. For this purpose we introduce, instead of the lengths of segments of a straight line, the *relative measures of directed segments*, i.e. *vectors* on a directed line (an axis).

We choose on a given line the positive direction, i.e. the direction from A to B. If P and Q are points of the line, then the *relative measure* PQ of the vector with the initial point P and end-point Q is: (a) the number equal to the length of the segment PQ if the direction from P to Q is positive; (b) the number opposite to the length PQ if the direction from P to Q is negative; (c) number zero if point Q coincides with point P.

By this agreement we have

$$PQ+QP = 0, \quad \text{whence} \quad QP = -PQ.$$

If P, Q, R are points of a line, then, regardless of their relative position, we always have the equality

$$PQ+QR = PR.$$

It will be observed that the square of the relative measure PQ of a segment is equal to the square of the length of that segment; we denote both by the same symbol PQ^2.

Using relative measures, we can write in our problem, without introducing a drawing,

$$MA = MS+SA, \quad MB = MS+SB,$$

whence

$$MA^2+MB^2 = (MS+SA)^2+(MS+SB)^2$$
$$= 2\times MS^2+2\times MS\times(SA+SB)+SA^2+SB^2.$$

Since S is the centre of the segment AB, we have

$$SB = AS = -SA, \quad \text{whence} \quad SA+SB = 0,$$

and finally

$$MA^2+MB^2 = 2\times MS^2+2\times SA^2,$$

which is what we wanted to prove.

Remark. Formula (1) is a particular case of the following theorem:

If point S is the centre of gravity of a system of n points $A_1, A_2, \ldots, A_n$ in space (we assume the points to have equal masses) and M is an arbitrary point in space, then

$$MA_1^2+MA_2^2+\ldots+MA_n^2 = n\times MS^2+SA_1^2+SA_2^2+\ldots+SA_n^2.$$

We shall prove this theorem for $n = 3$. The proof will be understood only by those readers who are acquainted with the notion and properties of the sum of vectors and the scalar product of vectors in space.

The symbol $\overrightarrow{PQ}$ will denote a vector with initial point P and end-point Q, and PQ will denote the length of that vector.

Let S (Fig. 9) be the centre of gravity of a triangle $A_1A_2A_3$ and M—any point in space.

Let us write the equalities:

$$\overrightarrow{MA_1} = \overrightarrow{MS}+\overrightarrow{SA_1}, \quad \overrightarrow{MA_2} = \overrightarrow{MS}+\overrightarrow{SA_2}, \quad \overrightarrow{MA_3} = \overrightarrow{MS}+\overrightarrow{SA_3}.$$

Squaring these equalities and then adding them, we obtain

$$MA_1^2+MA_2^2+MA_3^2 = 3\times MS^2+2\times\overrightarrow{MS}(\overrightarrow{SA}_1+\overrightarrow{SA}_2+\overrightarrow{SA}_3)+$$
$$+SA_1^2+SA_2^2+SA_3^2.$$

Now $\overrightarrow{SA}_1+\overrightarrow{SA}_2+\overrightarrow{SA}_3 = 0$ because the vector $\overrightarrow{SK}$, opposite to the vector $\overrightarrow{SA}_3$, is equal to the sum of the vectors $\overrightarrow{SA}_1+\overrightarrow{SA}_2$ (the quadrilateral SA_1KA_2 being a parallelogram); hence

$$MA_1^2+MA_2^2+MA_3^2 = 3\times MS^2+SA_1^2+SA_2^2+SA_3^2.$$

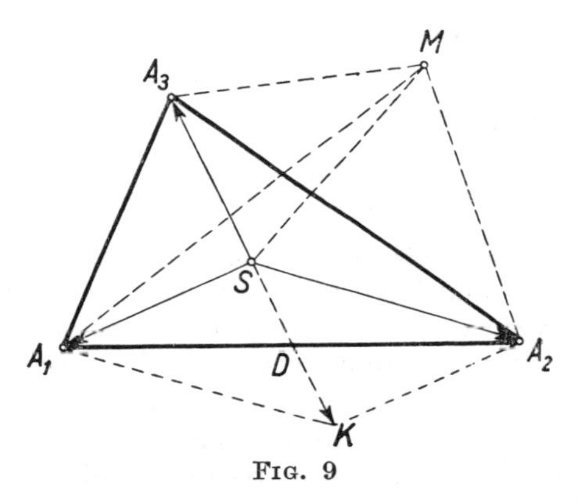

FIG. 9

If $n = 2$, the proof of our theorem, i.e. the proof of formula (1) if M is any point in space, can easily be carried out without the use of vector algebra.

67. *Method I.* Let M denote the intersection point of the segments AB and CD (Fig. 10). The triangles ABC and CDA with the common side AC are congruent because $AB = CD$, $BC = AD$; consequently $\sphericalangle BAC = \sphericalangle DCA$. It follows that AMC

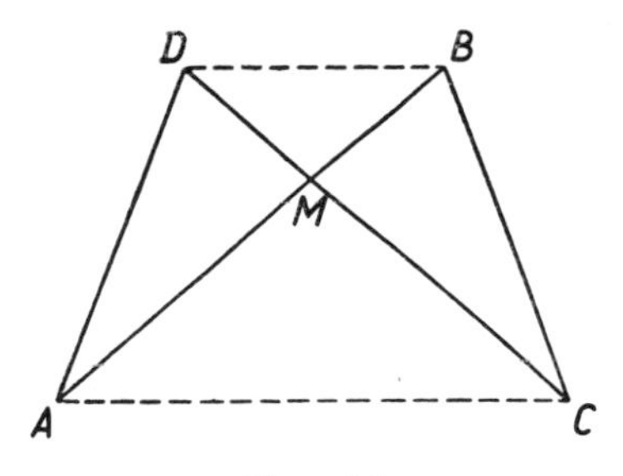

FIG. 10

between the sides a, b, c and the opposite angles α, β, γ of a triangle. For given a, b, c and γ, this formula defines uniquely the acute angle $(\alpha-\beta)/2$, and thus also the two angles α and β, since $\alpha+\beta = 180°-\gamma$.

68. *Method I.* Let AD and BE be medians of the triangle ABC (Fig. 13) with S as their intersection point.

If $AD = BE$, then $AS = BS$ since $AS = \frac{2}{3}AD$ and $BS = \frac{2}{3}BE$. Thus in the triangle ASB we have $\sphericalangle BAS = \sphericalangle ABS$, whence the triangles ABE and ABD are congruent, and this implies that $\sphericalangle BAC = \sphericalangle ABC$.

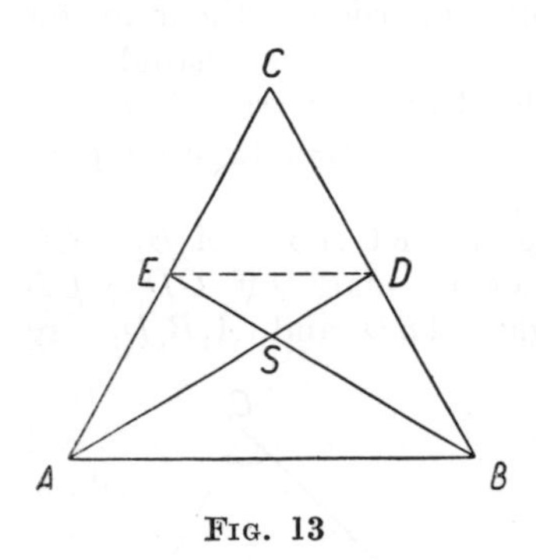

Fig. 13

Remark. The theorem can be generalized: if AD and BE are such segments in a triangle ABC (Fig. 13) that the points D and E divide the sides BC and AC in the same ratio and if $AD = BE$, then the triangle is isosceles.

Indeed, if $BD/BC = AE/AC$, then by the inverse of Thales' theorem the lines AB and DE are parallel; by Thales' theorem we thus have $AS/AD = BS/BE$, and, since $AD = BE$, we have $AS = BS$. This implies, as before, the equality of the angles A and B of triangle ABC.

Method II. Let a, b, c denote the lengths of the sides of a triangle and m_a and m_b the lengths of the medians drawn to the sides a and b. By a well-known formula for the length of a median of a triangle we have

$$m_a^2 = \tfrac{1}{2}b^2+\tfrac{1}{2}c^2-\tfrac{1}{4}a^2,$$

$$m_b^2 = \tfrac{1}{2}c^2+\tfrac{1}{2}a^2-\tfrac{1}{4}b^2.$$

Subtracting these equalities we obtain

$$m_a^2-m_b^2 = \tfrac{3}{4}b^2-\tfrac{3}{4}a^2.$$

If $m_a = m_b$, then the above equality gives $b^2 = a^2$, whence $a = b$.

Remark. The formula used above for the length of a median of a triangle is a particular case of a more general formula, very useful in calculations. Let us draw in triangle ABC a segment $AD = d_a$ (Fig. 14), write $BD = m$ and find d_a in terms of the sides a, b, c of the triangle and of m.

By the Cosine Rule applied to the triangles ABD and ABC† we have:

$$d_a^2 = c^2+m^2-2cm\cos B,$$

$$b^2 = c^2+a^2-2ca\cos B.$$

We multiply the first of these equations by a and the second by m, and then subtract the second from the first

$$ad_a^2-mb^2 = ac^2-mc^2+m^2a-ma^2,$$

whence

$$ad_a^2 = mb^2+(a-m)c^2-ma(a-m)$$

and

$$d_a^2 = \frac{m}{a}b^2+\frac{a-m}{a}c^2-m(a-m).$$

Fig. 14

The formula we have obtained is known as the Stewart‡ theorem.

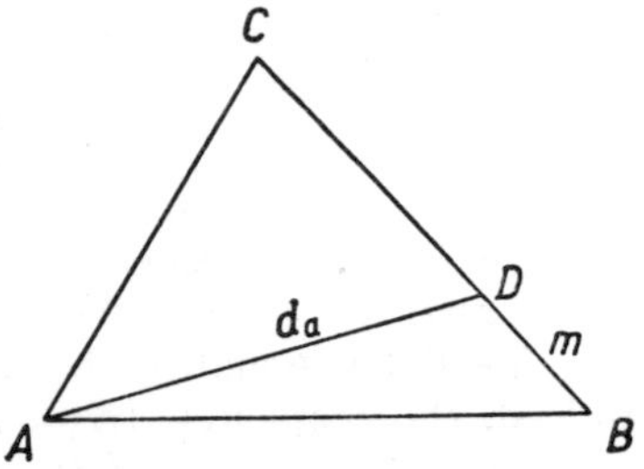

If $m = a/2$, the Stewart formula gives the formula which we have used before for the square of a median. If $(a-m)/m = b/c$, i.e. if $m = ac/(b+c)$, d_a is the bisector of $\sphericalangle A$; from the Stewart formula we easily obtain

$$d_a^2 = \frac{bc(a+b+c)(b+c-a)}{(b+c)^2};$$

we have found the length of the bisector d_a if the sides of the triangle are given.

We suggest to the reader that he should deduce the following theorems from Stewart's theorem:

(1) a theorem stating that, if two bisectors of a triangle are equal, then the triangle is an isosceles one;

(2) the theorem formulated in the remark on method I.

69. *Method I.* Let AD, BE and CF be the medians of the triangle ABC (Fig. 15).

---

† The theorem on cosines can of course be replaced here by the "generalized theorem of Pythagoras" and thus the use of trigonometry can be dispensed with.

‡ Stewart (1717–1785)—a Scottish mathematician, professor of the University of Edinburgh.

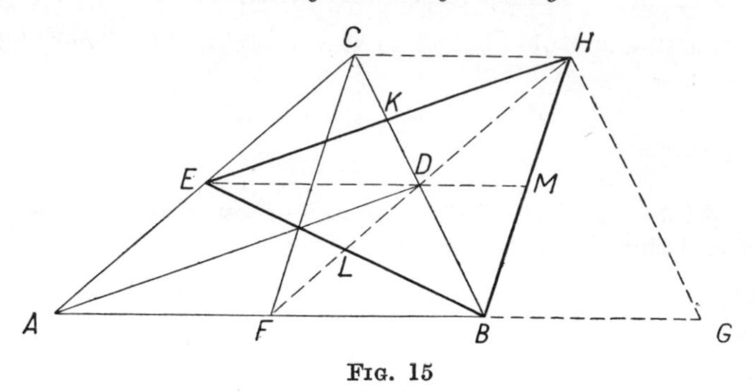

FIG. 15

Let us translate triangle ABC to the position FGH and consider triangle BHE. The sides BE and BH of this triangle are equal to the medians BE and FC of triangle ABC; we shall show that the side EH is equal to the third median, AD, of triangle ABC. Indeed, the segments AE and DH lie on parallel lines AC and FH, and $AE = FD = DH$, whence the quadrilateral $ADHE$ is a parallelogram and $EH = AD$.

The medians of the triangle BHE are the segments BK, HL, EM; each of them equals $\frac{3}{4}$ of the corresponding side of the triangle ABC since

$$BK = BD+DK = BD+\tfrac{1}{2}DC = \tfrac{3}{4}BC,$$

$$HL = LD+DH = \tfrac{1}{2}FD+DH = \tfrac{3}{4}FH = \tfrac{3}{4}AC,$$

$$EM = ED+DM = AF+\tfrac{1}{2}FB = \tfrac{3}{4}AB.$$

Method II. Let a, b, c denote the sides of the triangle and m_a, m_b, m_c the corresponding medians. The lengths of the medians can be found from the formulas:

$$m_a^2 = \tfrac{1}{4}(2b^2+2c^2-a^2),$$

$$m_b^2 = \tfrac{1}{4}(2c^2+2a^2-b^2),$$

$$m_c^2 = \tfrac{1}{4}(2a^2+2b^2-c^2).$$

Now let t_a denote the median in a triangle with sides m_a, m_b, m_c drawn to the side m_a. By a formula analogous to the preceding ones we have

$$t_a^2 = \tfrac{1}{4}(2m_b^2+2m_c^2-m_a^2).$$

Substituting the above values of m_a, m_b, m_c, we obtain

$$t_a^2 = \tfrac{1}{4}[\tfrac{1}{2}(2c^2+2a^2-b^2)+\tfrac{1}{2}(2a^2+2b^2-c^2)-\tfrac{1}{4}(2b^2+2c^2-a^2)]$$
$$= \tfrac{1}{4}\times\tfrac{9}{4}a^2 = \tfrac{9}{16}a^2.$$

Consequently

$$t_a = \tfrac{3}{4}a.$$

REMARK. Method I has a certain superiority over method II because it includes a proof that we can always construct a triangle whose sides are equal to the medians of any given triangle, i.e. that the sum of two medians of a triangle is always greater than the third median. It would be rather cumbersome to draw this conclusion from the formulas of method II.

70. Denote the sides of the triangle by $BC = a$, $CA = b$, $AB = c$, and the medians by m_a, m_b, m_c. Let S denote the intersection point of the medians, i.e. the centre of gravity of the triangle ABC (Fig. 16).

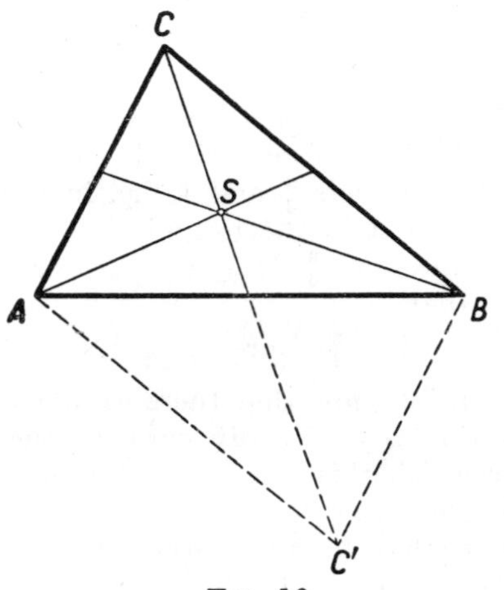

FIG. 16

We have

$$AS = \tfrac{2}{3}m_a, \quad BS = \tfrac{2}{3}m_b, \quad CS = \tfrac{2}{3}m_c.$$

In triangle BSC we have the inequality $BS+SC > BC$, i.e. $\frac{2}{3}m_b+\frac{2}{3}m_c > a$; hence

$$m_b+m_c > \tfrac{3}{2}a,$$

and analogously

$$m_c+m_a > \tfrac{3}{2}b, \quad m_a+m_b > \tfrac{3}{2}c.$$

Adding these inequalities and reducing, we obtain

$$m_a+m_b+m_c > \tfrac{3}{4}(a+b+c),$$

whence

$$\frac{m_a+m_b+m_c}{a+b+c} > \frac{3}{4}. \tag{1}$$

Let C' be the point symmetric to the vertex C with respect to the mid-point of the side AB. We have $AC' = BC = a$, $BC' = AC = b$. From the triangle $C'AC$ we obtain the inequality $C'A+AC > C'C$, i.e.

$$a+b > 2m_c$$

and analogously

$$b+c > 2m_a, \quad c+a > 2m_b.$$

Adding these inequalities and reducing, we obtain

$$a+b+c > m_a+m_b+m_c,$$

whence

$$\frac{m_a+m_b+m_c}{a+b+c} < 1. \tag{2}$$

Inequalities (1) and (2) express the proposition that the ratio of the sum of the medians of any triangle to its perimeter is contained between the numbers $\frac{3}{4}$ and 1, i.e. in the interval $(\frac{3}{4}, 1)$. It remains to prove that in the formulation of this theorem we cannot replace the interval $(\frac{3}{4}, 1)$ by a narrower interval, i.e. such an interval (α, β) that

$$\alpha > \tfrac{3}{4} \quad \text{or} \quad \beta < 1.$$

Accordingly we must show that there exist triangles for which the ratio in question is arbitrarily close to the number $\frac{3}{4}$, and also triangles for which this ratio is arbitrarily close to 1. This fact can easily be demonstrated.

Namely, observe that in an isosceles triangle in which the angle at the vertex is close to 180° the perimeter differs very little from the base doubled, each of the equal medians differs very little from $\frac{3}{4}$ of the base and the third median is small, and consequently the sum of the medians of the triangle differs little from $\frac{3}{4}$ of its perimeter. And if the angle at the vertex of an isosceles triangle is close to 0°, then the perimeter of the triangle differs little from a doubled arm and each of the equal medians differs little from $\frac{1}{2}$ of an arm, whereas the third median differs little from a whole arm; consequently the sum of the medians differs little from the perimeter of the triangle.

The above explanation is not a mathematical proof but it gives a hint how to conduct the proof.

Suppose that in a triangle ABC in which $AC = BC$ (Fig. 17) points M and P are the mid-points of the sides BC and AB and the segment MH is perpendicular to the side AB. In the triangle AMH we have the inequality

$$AM < AH+HM,$$

whence

$$m_a < \tfrac{3}{4}c + \tfrac{1}{2}h,$$

where $h = CP$ denotes the altitude of the isosceles triangle ABC (and also the median drawn to its base). Since $m_b = m_a$ and $m_c = h$, we obtain from the above inequality

$$m_a + m_b + m_c < \tfrac{3}{2}c + 2h. \tag{3}$$

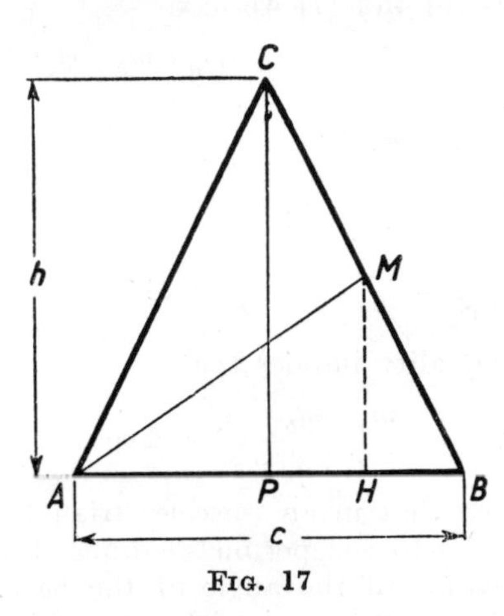

FIG. 17

Since $c < a+b$, we have $2c < a+b+c$ and $\tfrac{3}{2}c < \tfrac{3}{4}(a+b+c)$, and consequently inequality (3) gives

$$m_a + m_b + m_c < \tfrac{3}{4}(a+b+c) + 2h,$$

whence

$$\frac{m_a + m_b + m_c}{a+b+c} - \frac{3}{4} < \frac{2h}{a+b+c}. \tag{4}$$

Now

$$\frac{2h}{a+b+c} = \frac{2h}{2a+c} < \frac{h}{a} = \sin A;$$

thus by the preceding inequality we have

$$\frac{m_a + m_b + m_c}{a+b+c} - \frac{3}{4} < \sin A. \tag{5}$$

We have proved that in an isosceles triangle the ratio of the sum of the medians to the perimeter differs from $\frac{3}{4}$ by less than the sine of the angle at the base. Consequently this difference is arbitrarily small when $\measuredangle A$ is sufficiently close to $0°$.

Next it will be observed that in triangle BPC we have the inequality $BC < BP+PC$, i.e. $a < \frac{1}{2}c+h$, whence we obtain

$$a+b+c < 2c+2h. \tag{6}$$

But $MH < AM$ and thus $\frac{1}{2}h < m_a$ and analogously $\frac{1}{2}h < m_b$; since $h = m_c$ we have

$$2h < m_a+m_b+m_c. \tag{7}$$

By inequalities (6) and (7) we have

$$a+b+c < 2c+(m_a+m_b+m_c),$$

whence

$$1-\frac{m_a+m_b+m_c}{a+b+c} < \frac{2c}{a+b+c}.$$

Since

$$\frac{2c}{a+b+c} = \frac{2c}{2a+c} < \frac{c}{a} = 2\times\frac{c}{2a} = 2\cos A,$$

the preceding inequality implies that

$$1-\frac{m_a+m_b+m_c}{a+b+c} < 2\cos A. \tag{8}$$

We have proved that in an isosceles triangle the ratio of the sum of the medians to the perimeter differs from unity by less than twice the cosine of the angle at the base. This difference is thus arbitrarily small if the angle at the base is sufficiently close to 90°.

71. Let S be the intersection of the altitudes AM, BN and CP of the acute-angled triangle ABC (Fig. 18).

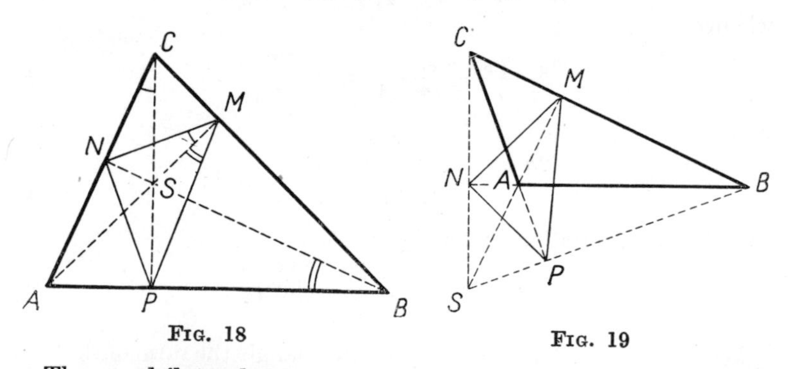

FIG. 18 FIG. 19

The quadrilateral $SMCN$, in which the angles at the vertices M and N are right angles, is inscribed in a circle with diameter SC; consequently

$$\sphericalangle SMN = \sphericalangle SCN$$

(angles inscribed in the same arc are equal).

Similarly in the quadrilateral $SMBP$

$$\sphericalangle SMP = \sphericalangle SBP.$$

But $\sphericalangle SCN = \sphericalangle SBP$ since each of these angles equals $90° - \sphericalangle BAC$; consequently $\sphericalangle SMN = \sphericalangle SMP$.

The altitude AM is thus indeed the bisector of the angle NMP.

Remark. In an obtuse-angled triangle (Fig. 19) the altitude AM drawn from the vertex of the obtuse angle A is the bisector of the angle NMP and the remaining two altitudes are the bisectors of the exterior angles of the triangle MNP; the proof is analogous to the preceding one.

72. Suppose, for example, that $a = \frac{1}{2}(b+c)$. In the formulation of the theorem it is not indicated which altitude is to be proved to be three times as long as the radius ϱ of the inscribed circle. However, we easily conclude that under the above assumption it can only be the altitude h_a drawn to the side a.

Indeed, if the assumption $a = \frac{1}{2}(b+c)$ implied that, for instance, $h_b = 3\varrho$, then an exchange of letters b and c would lead to the conclusion that also $h_c = 3\varrho$, whence $h_b = h_c$, and consequently $b = c$. We should thus obtain the proposition: "If $a = \frac{1}{2}(b+c)$ then $b = c$", which is obviously false.

We are thus to prove that $h_a = 3\varrho$. We shall find the required proof if we investigate the relations between the radius of the inscribed circle and the other lengths in a triangle.

We know, for instance, that between the area $P = \frac{1}{2}ah_a$ the perimeter $2p = a+b+c$ and the radius ϱ of the inscribed circle, we have the relation $P = p\varrho$; consequently

$$ah_a = \varrho(a+b+c). \tag{1}$$

By our assumption we have $b+c = 2a$; thus $a+b+c = 3a$ and consequently (1) gives

$$ah_a = 3a\varrho, \qquad \text{whence} \qquad h_a = 3\varrho.$$

We could also base our reasoning on well-known formulas for the segments determined on the sides of a triangle by the points of contact of the inscribed circle. We leave it to the reader to develop this idea.

Remark. The converse theorem is also true:

If $h_a = 3\varrho$, then $a = \frac{1}{2}(b+c)$.

In order to prove this, it is sufficient to substitute $h_a = 3\varrho$ in equality (1).

73. We adopt the notation indicated in Fig. 20; the radii of the circles inscribed in triangles ABC, BCD, ACD will be denoted by r, r_1 and r_2 respectively.

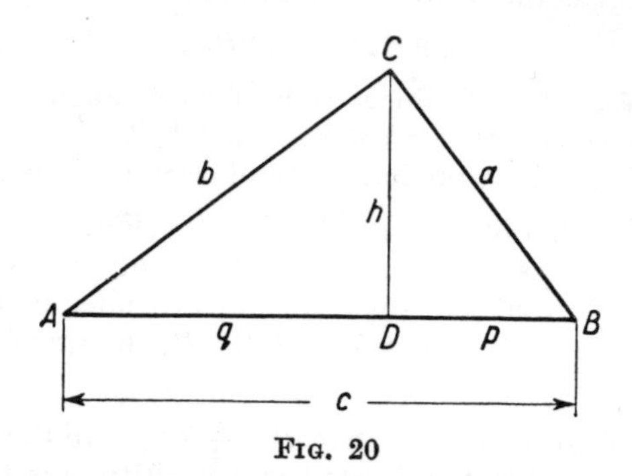

Fig. 20

Method I. Since the right-angled triangles ABC, CBD and ACD are similar, the radii of the circles inscribed in these triangles are proportional to their hypotenuses:

$$r_1 : r_2 : r = a : b : c.$$

Hence

$$\frac{r_1+r_2+r}{a+b+c} = \frac{r}{c} \quad \text{and} \quad r_1+r_2+r = \frac{(a+b+c)r}{c}.$$

The product of the perimeter of a triangle by the radius of the circle inscribed in that triangle equals twice the area of the triangle, i.e.

$$(a+b+c)r = ch.$$

Substituting this in the preceding formula, we obtain

$$r_1+r_2+r = h,$$

which is what was to be proved.

Method II. Let us apply the formula expressing the radius of the circle inscribed in a right-angled triangle in terms of the sides of that triangle.

As we know, the segment CK (Fig. 21) joining the vertex C of any triangle ABC with the point of contact K of the inscribed circle has the length

$$CK = \frac{a+b-c}{2}.$$

Since in the case in question we have $C = 90°$, the quadrilateral $OKCL$ is a square and we obtain the formula

$$r = OK = CK = \frac{a+b-c}{2}. \tag{1}$$

Analogously, for the triangles BCD and ACD (Fig. 20) we have

$$r_1 = \frac{h+p-a}{2}, \tag{2}$$

$$r_2 = \frac{h+q-b}{2}. \tag{3}$$

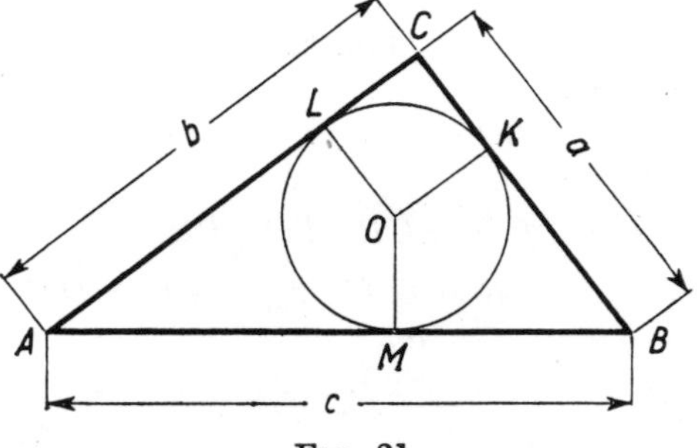

FIG. 21

Adding equalities (1), (2) and (3) and considering that $p+q = c$, we obtain the required equality,

$$r+r_1+r_2 = h.$$

Method III. We shall show that the altitude CD of triangle ABC can be divided into three parts equal to the radii r, r_1 and r_2 respectively.

To begin with, we ascertain, as in method I, that the radii r_1, r_2, r are proportional to the sides a, b, c. The radii r_1, r_2, r are thus equal to the sides of a triangle similar to triangle ABC. That triangle is right-angled; the side equal to r is its hypotenuse.

Let point O be the centre of the circle inscribed in the triangle ABC, and let the segments OK, OL, OM (Fig. 22) be the radii of that circle which are perpendicular to the sides of the triangle. The quadrilateral $OKCL$ is a square of side r.

Let us draw the perpendiculars: $LP \perp CD$, $OQ \perp LP$, $OR \perp CD$. The triangles CLP and LOQ are right-angled triangles with hypotenuses equal to r; these triangles are similar to triangle ABC, which we easily verify by considering the corresponding angles of the three triangles. According to the remark made at the beginning, the other sides of the triangles CLP and LOQ are equal to r_1 and r_2, respectively:

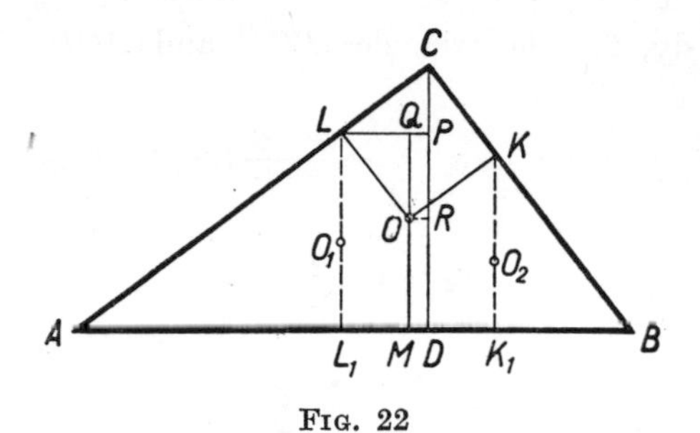

FIG. 22

$$LP = OQ = r_2, \quad CP = LQ = r_1.$$

Since

$$CD = CP+PR+RD,$$

$$CP = r_1, \quad PR = QO = r_2, \quad RD = OM = r,$$

we have

$$CD = r_1+r_2+r.$$

REMARK. Let us draw LL_1 and KK_1 perpendicular to AB. Since

$$L_1D = LP = r_2,$$

in order to determine the centre O of the circle inscribed in triangle ADC we mark off on LL_1 a segment L_1O_1 equal to r_2.

Considering that $L_1M = LQ = r_1$, we can see that the triangle MO_1L_1 is congruent to the triangle CLP, whence $MO_1 = r$.

Similarly, the centre O_2 of the circle inscribed in triangle BDC lies on K_1K, and $MO_2 = r$.

Thus the three centres, O_1, O_2, O, of the circles inscribed in the triangles ADC, BDC and ABC lie on a circle with centre M.

74. *Hint.* Multiply by the area of the triangle both sides of the equality to be proved.

75. The angles mentioned in the problem are the angles ACB, ADB, AEB in Fig. 23, in which $AB = a$, $BC = a$, $BD = 2a$, $BE = 3a$, $\measuredangle B = 90°$. Since $\measuredangle ACB = 45°$, it is sufficient to prove that

$$\measuredangle ADB + \measuredangle AEB = 45°.$$

For this purpose let us rotate triangle ABD through 90° about A to the position AFG and let us consider triangle ADG. It is

right-angled and isosceles because $\measuredangle DAG = 90°$, $AD = AG$.
Consequently

$$\measuredangle AGD = \measuredangle AGF + \measuredangle CGD = 45°.$$

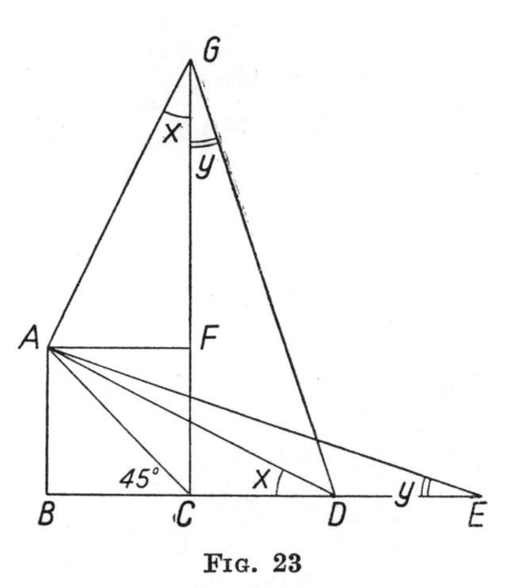

FIG. 23

But $\measuredangle AGF = \measuredangle ADB$ and $\measuredangle CGD = \measuredangle AEB$ (since $\triangle DGC = \triangle AEB$). Therefore

$$\measuredangle ADB + \measuredangle AEB = 45°.$$

Figures 24 and 25 represent two modifications of the above proof. The main idea is the construction of a right-angled isosceles

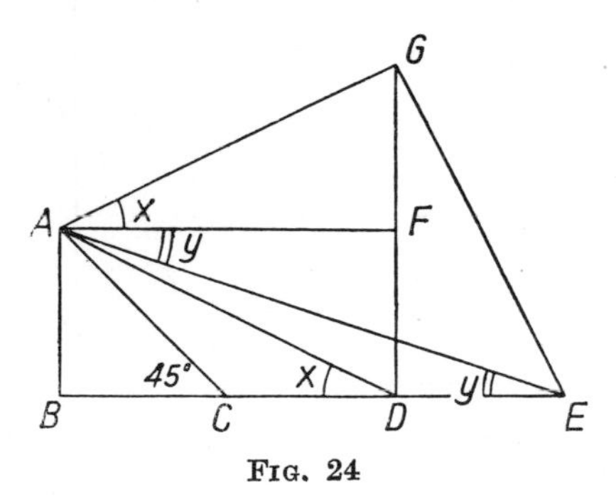

FIG. 24

triangle with the acute angle equal to the sum of the angles $x = \measuredangle ADB$ and $y = \measuredangle AEB$. In Fig. 24 this triangle is $\triangle AEG$, and in Fig. 25 $\triangle ADG$.

REMARK. Using trigonometry we can make the proof short:

$$\tan x = \frac{AB}{BD} = \frac{1}{2}, \qquad \tan y = \frac{AB}{BE} = \frac{1}{3};$$

consequently

$$\tan(x+y) = \frac{\tan x + \tan y}{1 - \tan x \tan y} = \frac{\frac{1}{2}+\frac{1}{3}}{1-\frac{1}{2}\times\frac{1}{3}} = 1,$$

whence $x+y = 45°$.

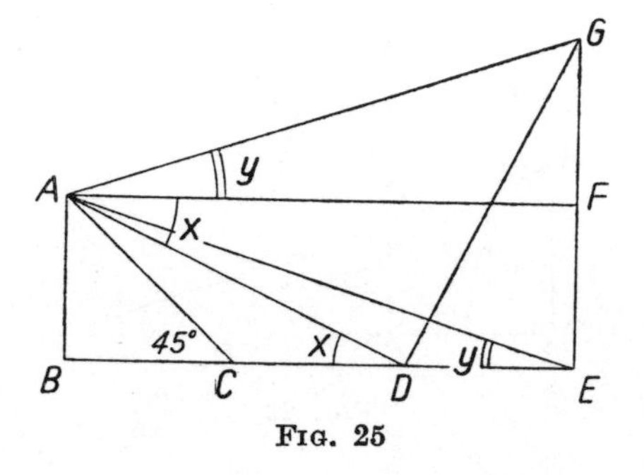

FIG. 25

76. *Method I.* By our assumption, the quadrilateral $MNPQ$ (Fig. 26) is a rhombus; the diagonals MP and NQ are thus perpendicular and bisect each other. In order to prove that the

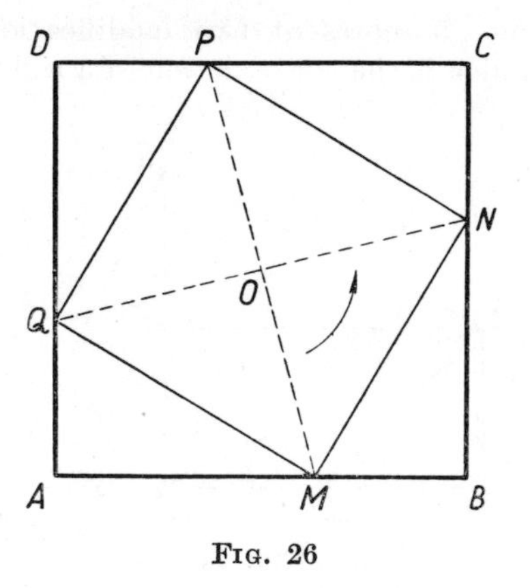

FIG. 26

rhombus $MNPQ$ is a square it is sufficient to show that $MP = NQ$.

We shall prove first that the intersection point O of the diagonals MP and NQ is the centre of the square $ABCD$.

Indeed, point O is the centre of symmetry of the rhombus $MNPQ$; consequently the lines AB and CD are symmetrical with respect to point O since they are parallel and pass through the symmetric points M and P. The lines AD and CB are also symmetrical. Thus point O is the centre of symmetry of the whole figure: it is therefore the centre of the square $ABCD$.

Let us rotate the figure through 90° about O. The line OM, perpendicular to the line ON, will then coincide with ON, whereas AB will coincide with BC; point M will thus assume the position of N. It follows that $OM = ON$, whence $MP = NQ$, which is what we were to prove.

Method II. In order to prove that the quadrilateral $MNPQ$ with equal sides is a square it is sufficient to show that all its angles are equal.

To begin with, it will be observed that if a segment of given length with end-points M and N lying on the arms of angle K

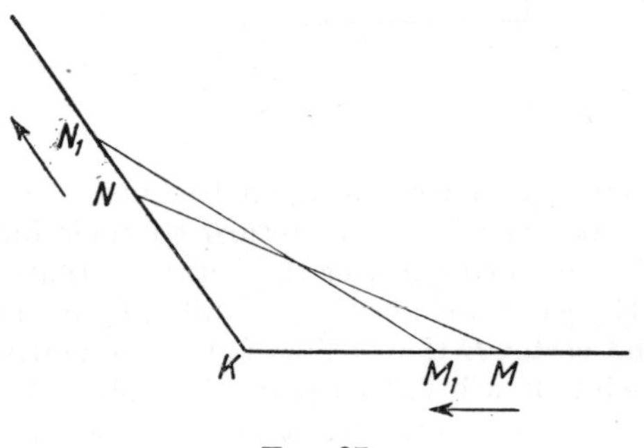

Fig. 27

(Fig. 27) is shifted in such a way that the end-point M approaches the vertex K, then the end-point N will get farther away from K, and *vice versa*.

Let us rotate the whole figure (Fig. 28) through 90° in the direction of the arrow about the centre O of the square $ABCD$. The vertices A, B, C, D, will fall on points B, C, D, A respectively. We shall show that the points M, N, P, Q will fall on points N, P, Q, M respectively.

Indeed, if the point M fell not on point N but, for instance, on point M_1, lying closer to the vertex C than point N, then, according to the above remark, points N, P, Q would coincide with points N_1, P_1, Q_1 lying nearer the vertices D, A, B, respectively, than P, Q, M, as shown in Fig. 28. The new figure is congruent to the preceding one (but it is turned about). If we rotated it again through 90° in the same direction, the vertices of the interior quadrilateral would again approach the respective vertices of the square $ABCD$—and this would be repeated if we rotated the figure through 90° for the third and the fourth time, i.e. if we rotated it through 360° from the initial position.

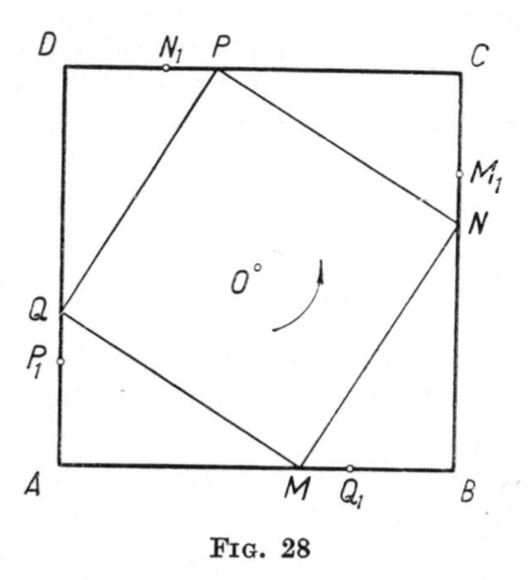

FIG. 28

We have obtained a contradiction because a rotation of 360° makes all the points of a figure return to their initial positions. Thus point M_1 cannot lie nearer the vertex C than point N.

Analogously, we conclude that point M_1 cannot lie farther away from the vertex C than point N. Consequently, after a rotation of 90° point M will fall on point N; points N, P, Q will fall on points P, Q, M respectively. It follows that the angles of the quadrilateral $MNPQ$ are equal.

REMARK I. In method II instead of rotating the figure four times through 90°, we could make one 90° rotation only and reason as follows:

If points M_1, N_1, P_1, Q_1 lay as shown in Fig. 28, we would have the relations:

$$\begin{aligned} MB &= M_1C < NC, \\ NC &= N_1D < PD, \\ PD &= P_1A < QA, \\ QA &= Q_1B < MB, \end{aligned}$$

and we would obtain the contradiction $MB < MB$.

REMARK 2. Method II is superior to method I in making it possible to prove the following, more general theorem:

If we choose points $P_1, P_2, \ldots, P_n$ on the successive sides of a regular polygon **of** *n* **sides** *in such a way that all the sides of the polygon $P_1, P_2, \ldots, P_n$ are equal, the polygon is regular.*

In the proof the figure must of course be turned through the angle $360°/n$.

77. Let A, B, C, D denote the successive vertices of the quadrilaterel formed by the centres of the balls, and M, N the centres of two opposite sides AB and CD of that quadrilateral.

Let us replace the balls with centres A and B by a ball twice as heavy with centre M, and the balls with centres C and D—by a ball twice as heavy with centre N. The centre of gravity of the new pair of balls is the same as the centre of gravity of the previous four, because the forces of gravity acting at points A and B have been replaced by their resultant force, acting at point M, and the forces of gravity acting at points C and D have been replaced by their resultant force, acting at point N. Consequently, the centre of gravity of the pair of balls placed at points M and N, i.e. the mid-point of the segment MN, lies at the centre of the circle under consideration.

The chords AB and CD, whose mid-points M and N lie on a diameter of the circle, are perpendicular to that diameter; they are thus parallel. The remaining two sides, BC and AD, of the quadrilateral $ABCD$ are also parallel. Consequently, the quadrangle is a parallelogram inscribed in a circle, i.e. it is a rectangle.

78. (a) The first proposition is not true; in order to demonstrate this, it is sufficient to give a *counterexample*, i.e. to show a figure contradicting the theorem.

From the centre O of the rhombus $ABCD$ (Fig. 29) let us draw a circle with a radius greater than the distance from the centre

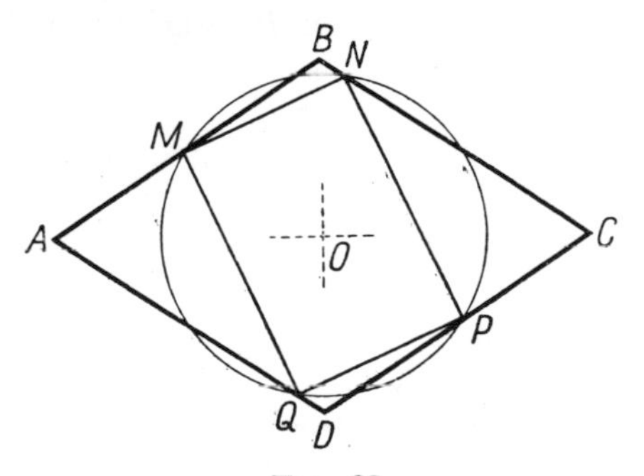

FIG. 29

of the rhombus to its side and smaller than half the shorter diagonal of the rhombus.

The whole figure is symmetrical with respect to each of the lines AC and BD, and also with respect to point O. The circle intersects each side of the rhombus at two points. Let M be one of the points of intersection of the side AB with the circle, and let N be that point of intersection of the side BC with the circle which is not symmetrical to point M with respect to BD. Next, let P and Q be points symmetrical to points M and N with respect to point O. The segments MP and NQ are thus diameters of the circle; in view of the symmetry of the figure with respect to O, points P and Q are points of intersection of the sides CD and DA with the circle.

The quadrilateral $MNPQ$ is a rectangle since each of its angles is an angle inscribed in a semicircle (e.g. the diameter MP subtends $\sphericalangle MNP$). The side MN of this rectangle is not perpendicular to BD since MN does not pass through a point symmetric to point M with respect to BD; consequently, the side MN is not parallel to the diagonal AC. The side MN is not parallel to the diagonal BD either, since it joins the points M and N, lying on opposite sides of BD. The vertices M, N, P, Q of the rectangle $MNPQ$ lie on the sides AB, BC, CD, DA, respectively, of the rhombus, but the sides of the rectangle are not parallel to the diagonals of the rhombus. The rectangle $MNPQ$ provides thus a counterexample disproving proposition (a).

(b) We shall show that the second proposition is true.

Let $MNPQ$ be a square inscribed in the rhombus $ABCD$, the vertices M, N, P, Q of the square lying, respectively, on the sides AB, BC, CD, DA of the rhombus.

To begin with, we shall prove that the centre O of the square is also the centre of the rhombus. Let us rotate the figure through 180° about point O. The vertex M of the square will then take the place of the opposite vertex P (Fig. 30) and the line AB

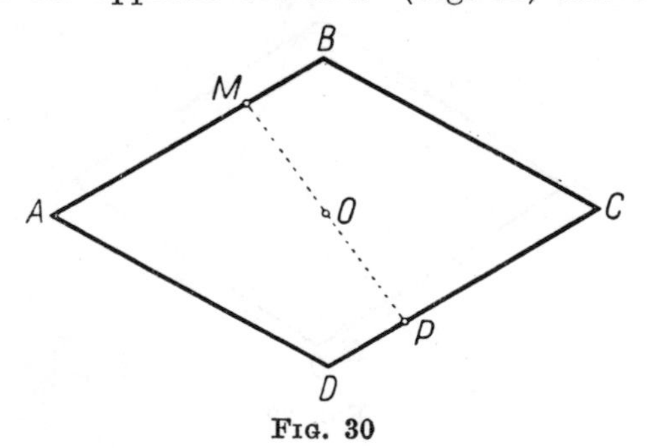

Fig. 30

will pass through point P running parallel to AB, i.e. it will coincide with CD. Since the same argument is applicable to any side of the rhombus, it follows that after the rotation the rhombus will coincide with itself, which means that the centre of rotation O is the centre of symmetry of the rhombus.

Let us now rotate the whole figure about point O through 90° so as to make OA coincide with OB (in Fig. 31 the direction of

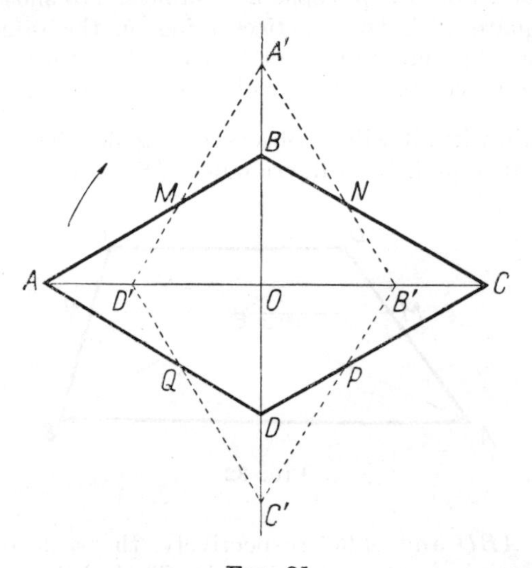

Fig. 31

rotation is indicated by an arrow). After the rotation the square $MNPQ$ will become the square $NPQM$, i.e it will coincide with itself; the rhombus $ABCD$ will become the rhombus $A'B'C'D'$, which does not coincide with rhombus $ABCD$, since by hypothesis $ABCD$ is not a square. For example if $OA > OB$, then $OA' > OB$ and $OD' = OD < OA$; consequently the segments $D'A'$ and AB intersect. The intersection point of these segments is the vertex M of the square, since after the rotation it is on point M that point Q of segment AD will fall. Analogously, the segments BC and $A'B'$ intersect at the vertex N of the square.

The figure consisting of the two rhombi is symmetric with respect to the line BD; in this symmetry point M corresponds to point N. Thus MN is perpendicular to the axis of symmetry BD, i.e. it is parallel to AC, which is what was to be proved.

Remark. If we rejected the assumption that the rhombus $ABCD$ is not a square, the theorem would not be true, because infinitely many squares can be inscribed in a square, only one of them having sides parallel to the diagonals of the given square.

Theorem (b), which we have proved, implies that a rhombus which is not a square can have only one square inscribed in it, one vertex of the square lying on each side of the rhombus. A stronger theorem can be proved: there exists only one square whose vertices lie on the boundary of a rhombus $ABCD$ which is not a square. For this purpose it is sufficient to show that there exists no square with two vertices lying on the same side of a rhombus and the remaining vertices on the other sides of the rhombus. We leave this to the reader as an exercise.

79. To begin with, it will be observed (Fig. 32) that the segments MS and SN are equal. Indeed, triangles MSD and SNC are similar

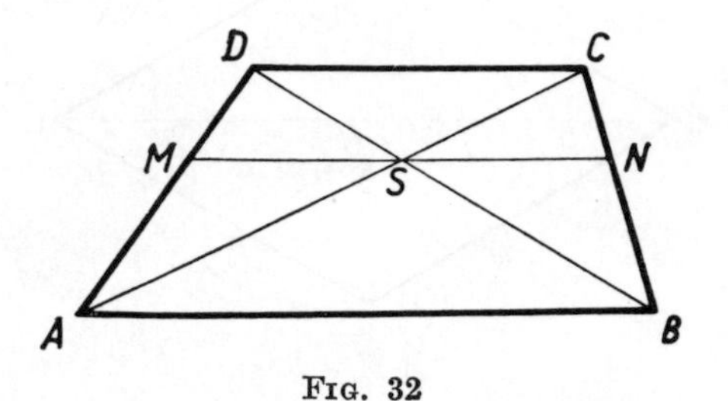

Fig. 32

to triangles ABD and ABC respectively, the scale of similarity being the same in both cases since by Thales' theorem we have the equality $MD/AD = NC/BC$. Consequently each of the segments MS and SN is in the same ratio to the segment AB, which means that $MS = SN$.

The relation between the length of the segment MN and the lengths of the bases AB and CD of the trapezium will be obtained by considering the similar triangles ABD and MSD, in which

$$\frac{MS}{AB} = \frac{SD}{BD},$$

and the similar triangles DBC and SBN, in which

$$\frac{SN}{DC} = \frac{BS}{BD}.$$

Adding these equalities, we obtain

$$\frac{MS}{AB} + \frac{SN}{DC} = \frac{BS+SD}{BD} = 1,$$

and since $MS = SN = \frac{1}{2}MN$, the last equality gives

$$\frac{2}{MN} = \frac{1}{AB} + \frac{1}{DC},$$

which means exactly that the segment MN is the harmonic mean of the bases AB and DC of the trapezium.

REMARK 1. The notion of the harmonic mean of two segments is connected with the notion of the harmonic quadruple of points.

We shall consider *directed segments*, i.e. *vectors*, on a straight line p (Fig. 33) denoting by the symbols AB, AC, BC etc. the

A C B D p

FIG. 33

relative measures of the corresponding vectors (see problem 60). We say that the four points A, B, C, D lying on p form a harmonic quadruple (A, B, C, D) or that the pairs (A, B) and (C, D) divide each other harmonically if the following relation holds:

$$\frac{AC}{BC} = -\frac{AD}{BD}, \quad \text{and thus also} \quad \frac{CA}{DA} = -\frac{CB}{DB}. \tag{1}$$

Equations (1) signify that the ratios in which the end-points of one of the segments AB and CD divide the other segments are opposite numbers.

If (A, B, C, D) is a harmonic quadruple of points, then the segment AB is the harmonic mean of the segments AC and AD. That is because equality (1) implies, successively, the equalities:

$$AC \times BD = -BC \times AD,$$

$$AC \times (AD - AB) = -(AC - AB) \times AD,$$

$$2AC \times AD = AB \times AD + AB \times AC,$$

$$\frac{2}{AB} = \frac{1}{AC} + \frac{1}{AD}. \tag{2}$$

Performing the above transformations in the inverse order, we find that equality (2) implies equality (1). Thus if the segment AB is the harmonic mean of the segments AC and AD, then the points A, B, C, D form a harmonic quadruple.

REMARK 2. Let us consider the point of intersection T of the non-parallel sides AD and BC of the trapezium (Fig. 34) and points P and Q at which the straight line ST intersects the sides

AB and CD. We shall prove that the points P, Q, S, T form a harmonic quadruple.

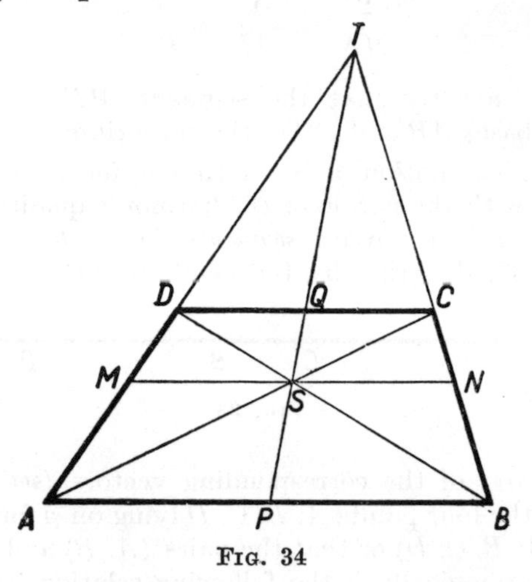

Fig. 34

Indeed, the segments AB and DC are homothetic with respect to the centre T; consequently

$$\frac{AB}{DC}=\frac{PT}{QT}.$$

The segments AB and CD are also homothetic with respect to the centre S, whence

$$\frac{AB}{DC}=-\frac{PS}{QS}.$$

We have put the minus sign on the right-hand side since the segments PS and QS run in opposite directions. The equalities obtained give

$$\frac{PT}{QT}=-\frac{PS}{QS},$$

which is what we were to prove.

The above theorem together with remark 1 leads to a very short proof of the theorem stating that in a trapezium (Fig. 34) the segment MN, parallel to the bases AB and CD, is the harmonic mean of the bases. Indeed, the segments AB, MN and DC, being homothetic with respect to the centre T, are proportional to the segments PT, ST, QT:

$$AB:MN:DC=PT:ST:QT.$$

Since P, Q, S, T form a harmonic quadruple of points, according to remark 1 the segments PT, ST, QT satisfy the relation

$$\frac{2}{ST} = \frac{1}{PT} + \frac{1}{QT}.$$

The same relation holds between the segments AB, MN and DC, which are proportional to PT, ST, QT. Namely

$$\frac{2}{MN} = \frac{1}{AB} + \frac{1}{DC}.$$

REMARK 3. The theorem given in remark 2 is a limiting case of an important theorem on the complete quadrilateral.

A complete quadrilateral is a figure formed by four straight lines a, b, c, d in a plane which intersect one another at six points A, B, C, D, E, F (Fig. 35). The four lines are termed the sides and the six

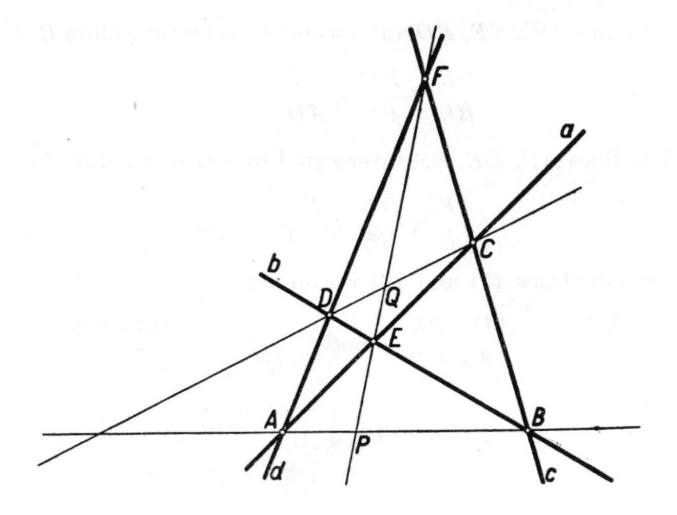

FIG. 35

points—the vertices of a complete quadrilateral. Two vertices which do not lie on the same side are called opposite vertices. A complete quadrilateral has three pairs of opposite vertices, each vertex belonging to one pair. In Fig. 35 A and B, C and D, E and F are opposite vertices.

Lines passing through opposite vertices are called *diagonals of the complete quadrilateral*. In Fig. 35 the diagonals are the lines AB, CD, EF. The following theorem holds:

On each diagonal of a complete quadrilateral the pair of vertices of the quadrilateral divides harmonically the pair of the points of intersection of that diagonal with the remaining two diagonals.

We shall prove, for example, that in Fig. 35 points E, F, P, Q constitute a harmonic quadruple.

The proof will be based on the theorem of Menelaus†:

If a line intersects lines AB, BC, CA at points M, N, and P respectively, then

$$\frac{AM}{MB}\times\frac{BN}{NC}\times\frac{CP}{PA}=-1.$$

We shall apply this theorem

(1) to lines AE, EF, FA, intersected by a line at points C, Q, D:

$$\frac{AC}{CE}\times\frac{EQ}{QF}\times\frac{FD}{DA}=-1;$$

(2) to lines DE, EF, FD, intersected by a line at points B, P, A:

$$\frac{DB}{BE}\times\frac{EP}{PF}\times\frac{FA}{AD}=-1;$$

(3) to lines AD, DE, EA intersected by a line at points F, B, C:

$$\frac{AF}{FD}\times\frac{DB}{BE}\times\frac{EC}{CA}=-1.$$

From equalities (1) and (2) we obtain

$$\frac{EP}{PF}=-\frac{AD\times BE}{FA\times DB}\quad\text{and}\quad\frac{EQ}{QF}=-\frac{DA\times CE}{FD\times AC},$$

whence

$$\frac{EP}{PF}:\frac{EQ}{QF}=-\frac{BE\times FD\times AC}{FA\times DB\times CE}=\frac{BE\times FD\times CA}{AF\times DB\times EC},$$

and, in view of equality (3), we have

$$\frac{EP}{PF}:\frac{EQ}{QF}=-1,$$

which means that points E, F, P, Q form a harmonic quadruple.

80. We shall derive the required relation from the equation

$$\text{area}\,\triangle ABC=\text{area}\,\triangle APB+\text{area}\,\triangle BPC+\text{area}\,\triangle CPA \quad (1)$$

---

† See problem 97 (we consider directed segments).

expressing the areas of the triangles in terms of the angles A, B, C, φ and of the radius R of the circle circumscribed on triangle ABC (Fig. 36).

(a) We know that

$$\text{area}\,\triangle ABC = 2R^2 \sin A \sin B \sin C. \tag{2}$$

(b) Observe that

$$\measuredangle APB = 180° - B, \quad \measuredangle BPC = 180° - C, \quad \measuredangle CPA = 180° - A,$$

since, for example,

$$\measuredangle APB = 180° - \varphi - \measuredangle PBA = 180° - \varphi - (B - \varphi) = 180° - B.$$

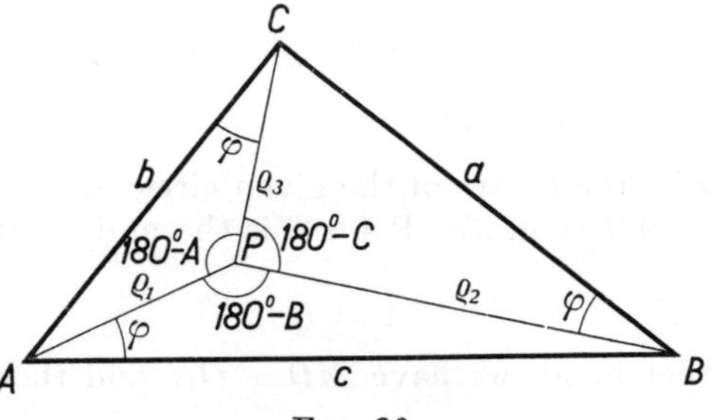

FIG. 36

Let us write $AP = \varrho_1$, $BP = \varrho_2$, $CP = \varrho_3$. These segments can be calculated by applying the Sine Rule to triangles APC, APB, BPC. We obtain

$$\varrho_1 = \frac{b}{\sin(180° - A)} \sin\varphi = \frac{2R\sin B}{\sin A} \sin\varphi$$

and similarly

$$\varrho_2 = \frac{2R\sin C}{\sin B} \sin\varphi, \quad \varrho_3 = \frac{2R\sin A}{\sin C} \sin\varphi.$$

Consequently

$$\text{area}\,\triangle APB = \frac{1}{2}\varrho_1 c \sin\varphi = \frac{1}{2} \times \frac{2R\sin B}{\sin A} \sin\varphi \times 2R \sin C \sin\varphi,$$

whence

$$\text{area}\,\triangle APB = 2R^2 \frac{\sin B \sin C}{\sin A} \sin^2\varphi. \tag{3}$$

Analogously

$$\text{area}\,\triangle BPC = 2R^2 \frac{\sin A \sin C}{\sin B} \sin^2\varphi, \tag{4}$$

$$\text{area}\,\triangle CPA = 2R^2\frac{\sin A \sin B}{\sin C}\sin^2\varphi. \tag{5}$$

Substituting expressions (2), (3), (4), (5) in (1), we obtain

$$2R^2 \sin A \sin B \sin C = 2R^2\frac{\sin B \sin C}{\sin A}\sin^2\varphi +$$

$$+2R^2\frac{\sin A \sin C}{\sin B}\sin^2\varphi + 2R^2\frac{\sin A \sin B}{\sin C}\sin^2\varphi;$$

dividing both sides of this equality by $2R^2 \sin A \sin B \sin C \sin^2\varphi$, we finally obtain

$$\frac{1}{\sin^2\varphi} = \frac{1}{\sin^2 A} + \frac{1}{\sin^2 B} + \frac{1}{\sin^2 C},$$

which is what was to be proved.

81. Let O be the centre of the given circle and S its projection upon the line AB (Fig. 37). Point S is the mid-point of the chord AB, i.e.

$$SA = BS. \tag{1}$$

On the other hand, we have $MO = ON$ and therefore by projection

$$M'S = SN'. \tag{2}$$

Adding equalities (1) and (2) we obtain

$$M'S+SA = BS+SN'. \tag{3}$$

Since

$$M'S+SA = M'A \quad \text{and} \quad BS+SN' = BN', \tag{4}$$

equality (3) gives

$$M'A = BN', \tag{5}$$

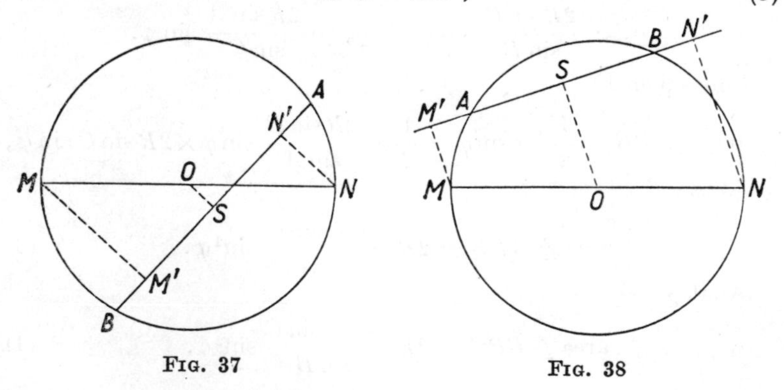

Fig. 37 Fig. 38

REMARK. In the above reasoning we used Fig. 37. But, according to the position of points A, B, M, N on the circle, the relative positions of points A, B, M', N' on AB may vary. If the proof is to be correct we should investigate all the possible cases. The proof will be similar in each of those cases. For example, in the case presented in Fig. 38 we should write, as before, equalities (1) and (2), but then we should have to subtract them from each other. We would obtain

$$M'S - SA = SN' - BS \tag{3a}$$

whence $MA' = BN'$.

Such breaking up of a proof into several cases is very inconvenient: the argument becomes long and tedious; moreover, one must be particularly careful not to omit any of the possibilities. In mathematics we give priority to general proofs, applicable to every case. In our problem we shall obtain a proof of this kind if we consider on AB, instead of segments in the usual sense, directed segments or *vectors*. (See problem 60.)

Then equalities (1)–(5) obtained before are true for any position of points A, B, M, N on the circle (cf. Figs. 37 and 38). The proof given above is general: it can even be carried out without the aid of a drawing. What is more, the result obtained, $M'A = BN'$, signifies not only that the segments $M'A$ and BN' are of the same length but that they have the same direction. As we say in mathematics, we have proved a stronger theorem than the preceding one.

82. The feet of the perpendiculars drawn from point M to the lines BC, CA, AB will be denoted by P, Q and R respectively. We are to prove that P, Q, R are collinear. The line in question is the so called *Simson*† *line* of triangle ABC with respect to point M.

If point M coincides with one of the points A, B, C, then two of the points P, Q, R also lie at that point; in this case the theorem is of course true.

Let point M lie inside one of the inscribed angles with vertices A, B, C, say in the angle BAC. The quadrilateral $ABMC$ is inscribed in a circle, and thus $\sphericalangle ABM + \sphericalangle ACM = 180°$.

If ABM and ACM are right angles, the theorem is true, since then point R lies at point B and point Q at point C; consequently points P, Q, R lie on BC.

It remains to investigate the case of one of those angles, say $\sphericalangle ABM$, being obtuse; consequently $\sphericalangle ACM$ is acute.

† Robert Simson (1687–1768), a Scottish mathematician.

Then point R lies on AB produced, whereas point Q may lie either on the segment CA or on CA produced. Each of these possibilities will be considered separately.

(a) Let Q lie on the segment AC (Fig. 39). In that case P lies on the segment BC. This follows from the fact that in triangle BMC the angles at the vertices B and C are acute. Namely

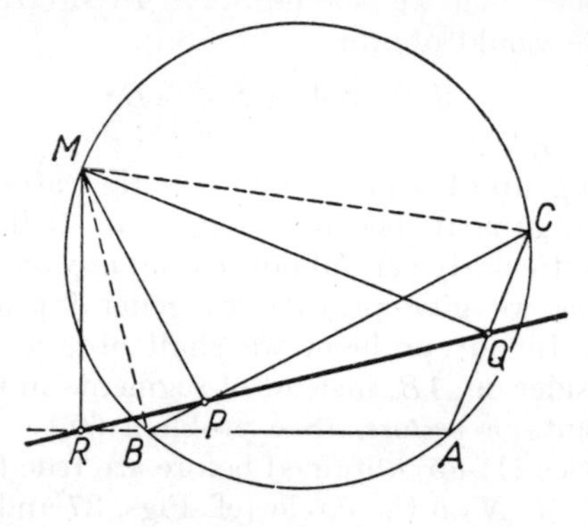

Fig. 39

$\sphericalangle MBC = \sphericalangle MAC$, since they are inscribed in the same arc, and $\sphericalangle MAC$ is an acute angle of a right-angled triangle MAQ; $\sphericalangle MCB$ is acute as part of the acute angle MCA. Thus the foot P of the altitude of triangle BMC drawn from vertex M lies on the base BC of this triangle.

In order to prove that points P, Q, R are collinear it is sufficient to show that the angles RPB and QPC are equal.

Now points M, P, B, R lie on a circle because the angles at points P and R are right angles; moreover, points P and M lie on the same side of the chord RB and consequently $\sphericalangle RPB = \sphericalangle RMB$.

Similarly points M, P, Q, C lie on a circle because the angles at points P and Q are right angles: points P and M lie on the same side of QC, whence $\sphericalangle QPC = \sphericalangle QMC$.

But the angles RMB and QMC are equal because

$$\sphericalangle RMB = 90° - \sphericalangle MBR$$
$$= 90° - (180° - \sphericalangle MBA) = \sphericalangle MBA - 90°,$$

$$\sphericalangle QMC = 90° - \sphericalangle MCQ = 90° - \sphericalangle MCA$$
$$= 90° - (180° - \sphericalangle MBA) = \sphericalangle MBA - 90°.$$

Thus $\sphericalangle RPB = \sphericalangle QPC$ and points P, Q, R are collinear.

(b) Let point Q lie on CA produced (Fig. 40). In that case point P lies on CB produced because $\sphericalangle MBC$ is obtuse: this follows from the fact that $\sphericalangle MBC = \sphericalangle MAC$, $\sphericalangle MAC$ being obtuse since it is adjacent to the acute angle MAQ. We prove the collinearity of points P, Q, R by showing that the angles RPB and QPC together add up to 180°.

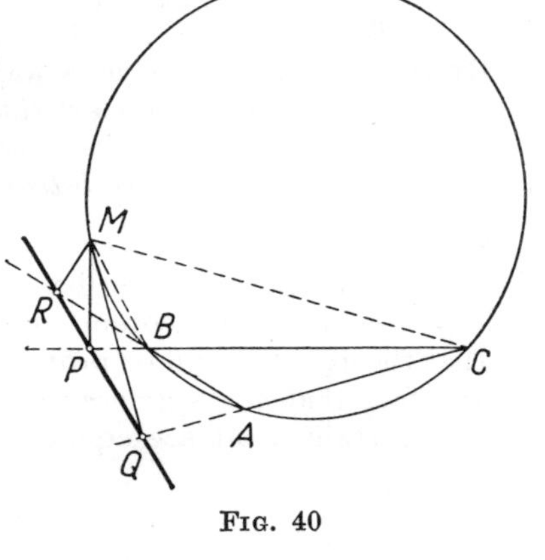

FIG. 40

Indeed, reasoning as in case (a), we find that $\sphericalangle RPB = 180° - \sphericalangle RMB$ since points M, R, P, B lie on a circle, points M and P lying of the opposite sides of the chord RB; further, we find that $\sphericalangle QPC = \sphericalangle QMC$ and finally that $\sphericalangle RMB = \sphericalangle QMC$.

Consequently $\sphericalangle RPB = 180° - \sphericalangle QPC$, whence we conclude that P, Q, R are collinear.

83. We shall obtain the proof of the theorem in question by using a theorem stating that the segments of the tangents drawn to a circle from a point are equal. For example in Fig. 41, repre-

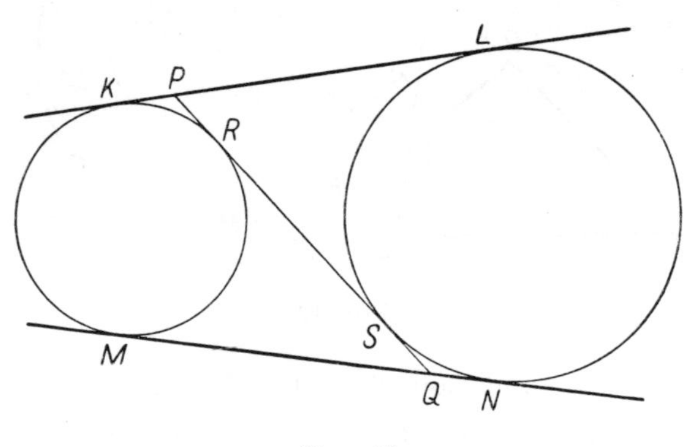

FIG. 41

senting the figure under consideration, $PK = PR$, $PL = PS$, $QM = QR$, $QN = QS$.

Considering that the symmetric segments KL and MN are equal, we can use a short argument:

$$2KL = KL+MN = KP+PL+MQ+QN$$
$$= PR+PS+RQ+SQ = (PR+RQ)+(PS+SQ) = 2PQ,$$

and therefore $KL = PQ$.

REMARK 1. The above proof is applicable also in the case of the given circles being tangent (externally); then points R and S coincide.

REMARK 2. In an analogous way the following theorem can be proved: *The segment of an exterior tangent of two circles which is contained between the interior tangents is equal to the segment of an interior tangent contained between its points of contact.*

84. From school geometry we know the theorem stating that in a quadrilateral circumscribed on a circle the sums of opposite sides are equal. The theorem which we are to prove is analogous to that theorem and can be proved in the same way on the grounds of the fact that the segments of the tangents to a circle drawn from a certain point are equal.

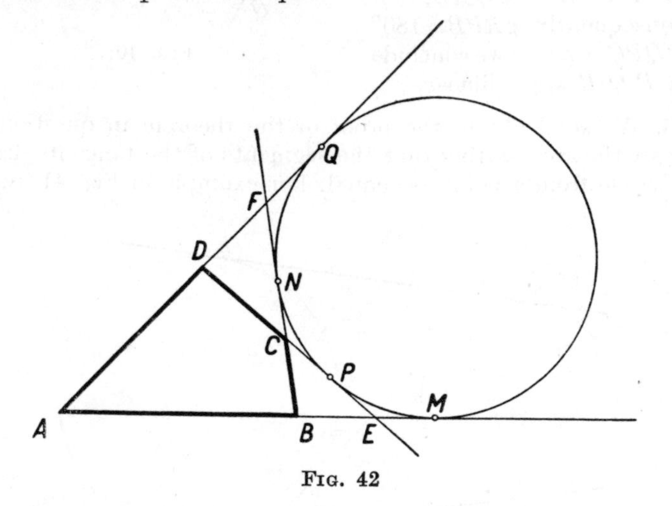

FIG. 42

According to the notation adopted in Fig. 42 we have

$$AB = AM-BM, \qquad AD = AQ-DQ,$$
$$DC = DP-CP, \qquad BC = BN-CN,$$

whence

$$AB-DC = AM-BM-DP+CP,$$
$$AD-BC = AQ-DQ-BN+CN;$$

since

$$AM = AQ, \quad BM = BN, \quad CN = CP, \quad DP = DQ,$$

we have

$$AB - DC = AD - BC.$$

REMARK. A circle tangent to the extensions of the sides of a quadrilateral $ABCD$, which we shall call, for brevity, a circle escribed to the quadrilateral, can exist only if the quadrilateral has no parallel sides. For, if $AB || CD$, then the lines AB and CD divide the plane into three domains I, II, III (Fig. 43) such that the required circle can lie in none of them, since in domain I there are no points of AB, in domain III there are no points of DC and in domain II there are no extensions of the sides AD and BC. Let us assume that the half-lines AB and DC intersect at point E and the half-lines AD and BC—at point F.

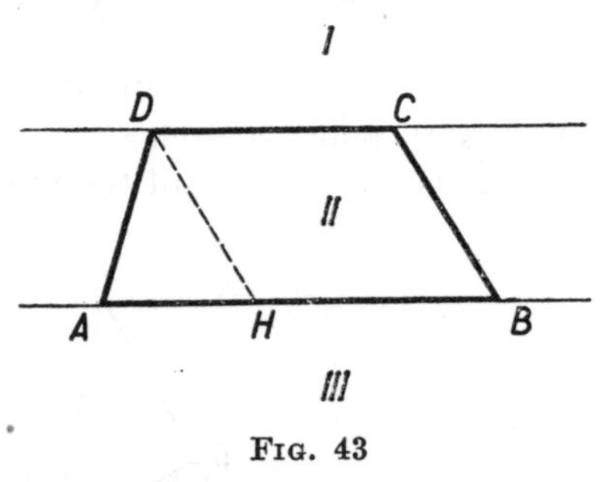

FIG. 43

A circle escribed to quadrilateral $ABCD$ (Fig. 42) is at the same time escribed to triangle ABF and to triangle ADE. From this observation we can derive a necessary and sufficient condition of the existence of that circle. Namely, each of the equal segments AM and AQ determined on the lines AB and AD by the circle escribed to triangles ABF and ADE is equal to half the perimeter of the triangle ABF and also equal to half the perimeter of the triangle ADE†. Consequently, the perimeters of triangles ABF and ADE are equal. Conversely, if triangles ABF and ADE have equal perimeters, then the circles escribed to these triangles and contained in the angle BAD are both tangent to the lines AB and AD at the same points M and Q, whose distance from point A equals half the common perimeter of the triangles; thus the two circles coincide and form a circle escribed to the quadrilateral $ABCD$. Consequently:

A necessary and sufficient condition of the existence of a circle escribed to the quadrilateral $ABCD$ is the equality of the perimeters of the triangles ABF and ADE.

---

† We remind the reader of the proof known from textbooks of geometry: $AM = AB + BM = AB + BN$, $AQ = AF + FQ = AF + FN$, whence $AM + AQ = AB + BN + AF + FN = AB + AF + BF$.

Using the above statement we can complete the preceding result and prove the inverse theorem:

If the differences of opposite sides in a convex quadrilateral ABCD which is not a parallelogram are equal, then there exists a circle escribed to that quadrilateral.

Let us assume that $AB-DC = AD-BC$ and that the sides AD and BC are not parallel. It is easy to ascertain that then the sides AB and DC are not parallel either. Indeed, if the sides AB and CD were parallel (Fig. 43), then by drawing in the trapezium $ABCD$ a segment $DH \parallel CB$ we would obtain $AB-DC = AB-HB = AH > AD-DH = AD-BC$, whence the trapezium would not satisfy our assumption that $AB-DC = AD-BC$.

Let us therefore consider a figure like the one represented in Fig. 44. On the grounds of the preceding argument it is sufficient

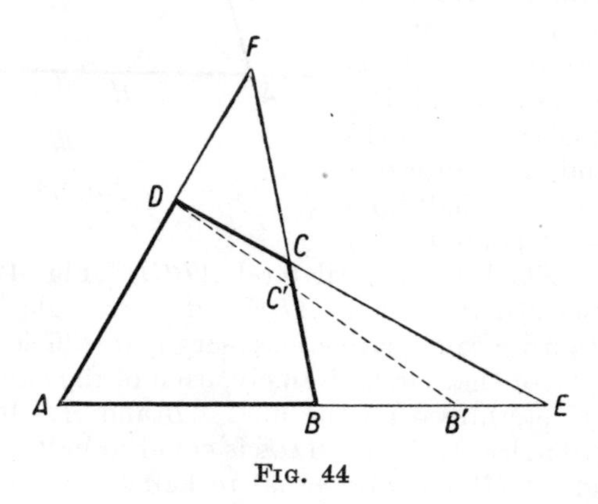

Fig. 44

to prove that the perimeters of triangles ADE and ABF are equal. Using the *reductio ad absurdum* method, suppose that, say, the perimeter of $\triangle ADE$ is greater than the perimeter of $\triangle ABF$. Let us determine on the line AE a point B' for which the perimeter of the triangle ADB' is equal to the perimeter of the triangle ABF†.

Point B' lies inside the segment BE because the triangle ADB' must be contained in the triangle ADE with a larger perimeter and must contain the triangle ABD with a smaller perimeter than the perimeter of the triangle ABF. Thus the segment DB' intersects the segment BC at a point C'. In view of the equality

† The point B' can be found by marking off on AE a segment AG equal to $AB+BF+FD$ and drawing the perpendicular bisector of segment DG.

of the perimeters of triangles ABF and ADB' the quadrilateral $ABC'D$ has an escribed circle, whence, as we already know,

$$AB-DC' = AD-BC'.$$

But

$$AB-DC = AD-BC,$$

by hypothesis. From these equalities we obtain by subtraction

$$DC-DC' = BC-BC' = CC',$$

which is impossible. The theorem is thus proved.

85. *Method I.* Writing the required equality as

$$\frac{AE}{AC} = \frac{AC}{AD},$$

we notice that it expresses the proportionality of two pairs of sides of triangles AEC and ACD with the common angle A.

The proof is thus reduced to showing that triangles AEC and ACD are similar.

To prove this it is sufficient to indicate one more pair of equal angles in these triangles.

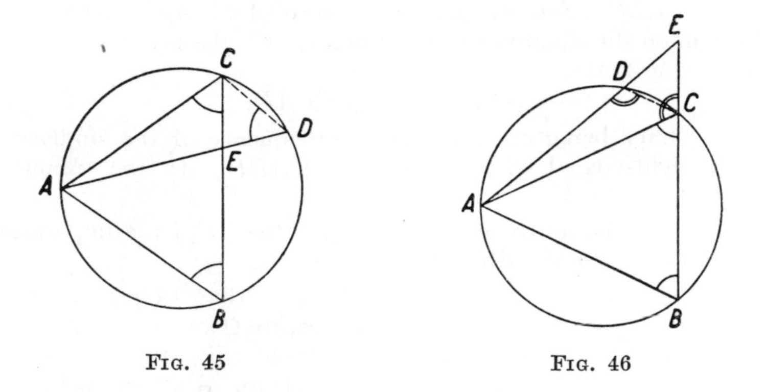

FIG. 45 FIG. 46

If point E lies between the points B and C (Fig. 45), then $\angle D = \angle B$ (since they are angles inscribed in the same arc) and $\angle B = \angle ACB$ (since $AC = AB$), whence $\angle D = \angle ACB$. If point E lies on BC produced (Fig. 46), then $\angle ADC + \angle B = 180°$, (these angles being opposite in the inscribed quadrilateral $ABCD$) and $\angle ACE + \angle ACB = 180°$; since $\angle B = \angle ACB$, we have $\angle ADC = \angle ACE$.

Method II. The fact that our figure is determined by two secants of the circle intersecting at point E suggests the well-known relation

$$AE \times ED = BE \times CE$$

and a suitable transformation of it.

Hint: Draw the perpendicular AH to BC.

Method III. Another, equally successful device is to take into consideration the chord AF lying on the axis of symmetry of

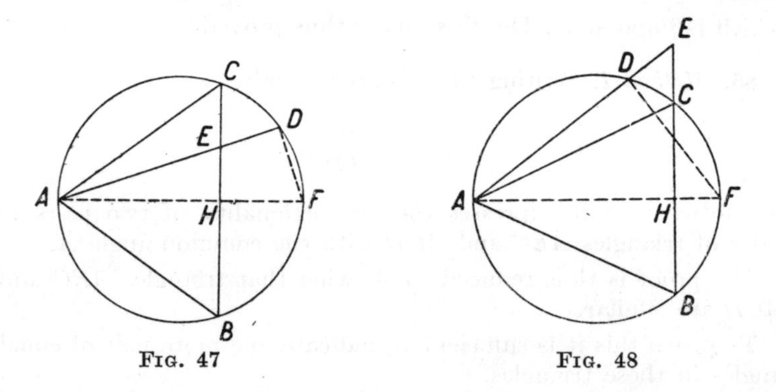

FIG. 47 FIG. 48

triangle ABC (Fig. 47 and 48). Since $AH \times AF = AC^2$ (the theorem on the square of a chord in a circle), the proof is reduced to showing that

$$AE \times AD = AH \times AF,$$

this equality being an immediate consequence of the similarity of the right-angled triangles AHE and ADF with the common angle A.

REMARK. The above problem is connected with the important notion of *inversion* with respect to a circle.

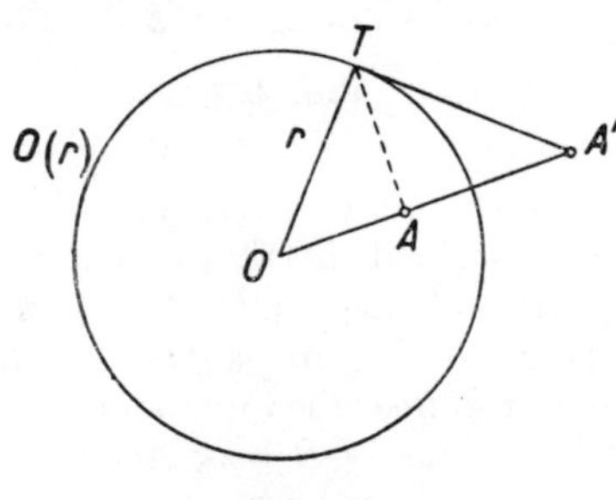

FIG. 49

Suppose that $O(r)$ is a circle with centre O and radius r given in a plane (Fig. 49).

If A is an arbitrary point of the plane different from O, then the point A' on OA defined by

$$OA' = \frac{r^2}{OA}$$

is called the *image* of point A in inversion with respect to the circle $O(r)$.

Since

$$OA = \frac{r^2}{OA'},$$

the point A is—*vice versa*—the image of point A'.

The relation between the segments OA and OA' is usually written down in the symmetric form

$$OA \times OA' = r^2.$$

If $OA = r$, then $OA' = r$ and the points A and A' coincide. If $OA < r$, then $OA' > r$, and *vice versa*.

Given one of the points A and A', it is easy to find the other point by construction, e.g. as shown in Fig. 49, in which the angles OAT and OTA' are right angles.

Every point of the plane except point O has a definite image on the plane and every point except point O is the image of a definite point of the plane. We say that *inversion with respect to the circle $O(r)$ is a transformation of the plane without point O onto itself.*

This transformation has many interesting properties. We shall confine ourselves here to the determination of the images of straight lines and circles in inversion, i.e. to the determination of figures into which straight lines and circles are transformed.

A circle with centre O is transformed into a circle with the same centre. In particular, the circle $O(r)$ coincides with its image in such a way that each point of the circle is its own image. Points lying inside the circle $O(r)$ have their images outside $O(r)$ and *vice versa.*

The image of every straight line drawn from the centre O is the same straight line, but every point of that line is transformed into another point except its point of intersection with $O(r)$: this point remains at the same place.

The image of a straight line m not passing through the centre O is a circle passing through O with the exception of point O itself; conversely, the image of such an "interrupted" circle is a straight line which does not pass through the centre O. The proof of this theorem is simple (Fig. 50).

Suppose that the images of points A and B in inversion with respect to the circle $O(r)$ are A' and B'. From the equality $OA \times OA' = OB \times OB'$, which gives the proportion

$$\frac{OA}{OB} = \frac{OB'}{OA'},$$

we infer that the triangles OAB and $OB'A'$ are similar and that $\sphericalangle OAB = \sphericalangle OB'A'$. Thus if point B runs over a straight line m perpendicular to OA, then point B' describes a circle m' with diameter OA'—and *vice versa.*

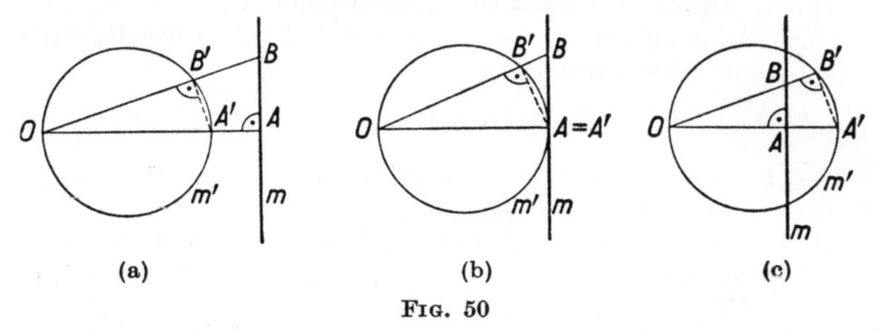

Fig. 50

In Figure 50a we have $OA > OA'$ and the line m lies outside the circle m'; in Fig. 50b we have $OA = OA'$, m and m' being tangent; finally, in Fig. 50c we have $OA < OA'$, the line m intersects the circle m' and we obtain the same figure as in problem 85. In fact, the essential point of that problem was to show that inversion with centre A and radius AC (Figs. 45–48) transforms the circle passing through points A, B, C into the straight line BC.

We shall prove in addition that *the image of a circle m which does not pass through the inversion centre O is also a circle which does not pass through the centre O.*

Let a point A' be the image of point A of circle m (Fig. 51) and let the straight line OA intersect the circle m at one more point, B. Then

$$OA \times OA' = r^2$$

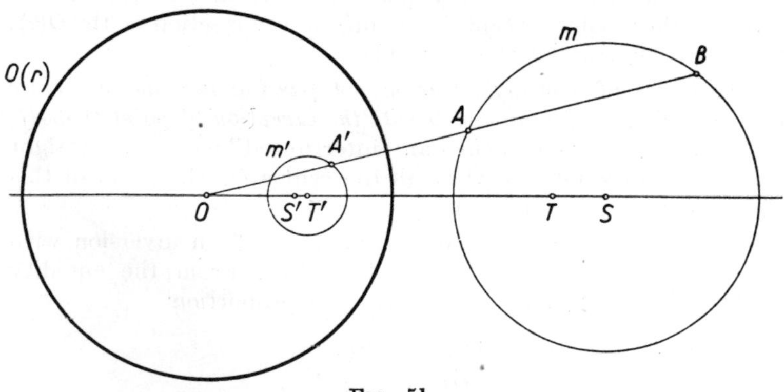

Fig. 51

and

$$OA \times OB = k^2$$

where k is the same number for every secant passing through O.

We obtain

$$OA' = \left(\frac{r}{k}\right)^2 \times OB.$$

Point A' thus corresponds to point B in a homothety with centre O; the homothety ratio is the number $(r/k)^2$.

If point A, and point B along with it, describe the circle m, point A' describes the circle m' homothetic to m with respect to point O. The circle m' is thus the image of the circle m both in a homothety with respect to point O and in inversion with respect to the circle $O(r)$. But each of these transformations associates the points of one circle with the points of the other circle in a different manner. For example point A' corresponds to point A in the inversion, and to point B in the homothety. The centre S of circle m (Fig. 51) corresponds in the inversion not to the centre of circle m' but to another point S'; similarly, the centre T' of circle m' corresponds to point T, other than the centre of the circle m. In Figs. 51, 53 and 54 the centre O lies outside, and in Fig. 52 inside the circle m.

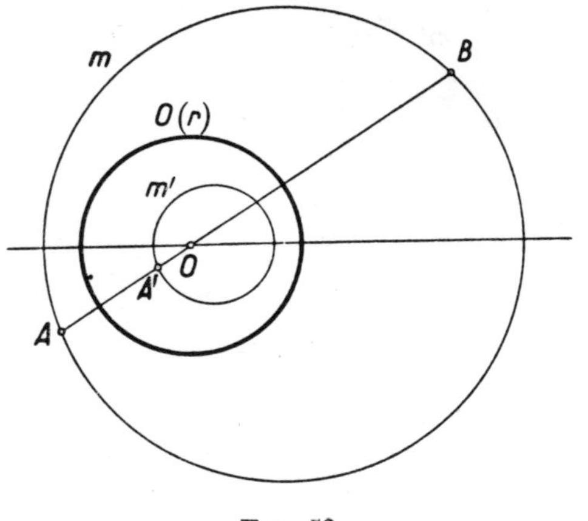

FIG. 52

If the circle m intersects the inversion circle $O(r)$, then its image m' intersects the circle $O(r)$ at the same points.

EXERCISE. When do the circles m and m' coincide?

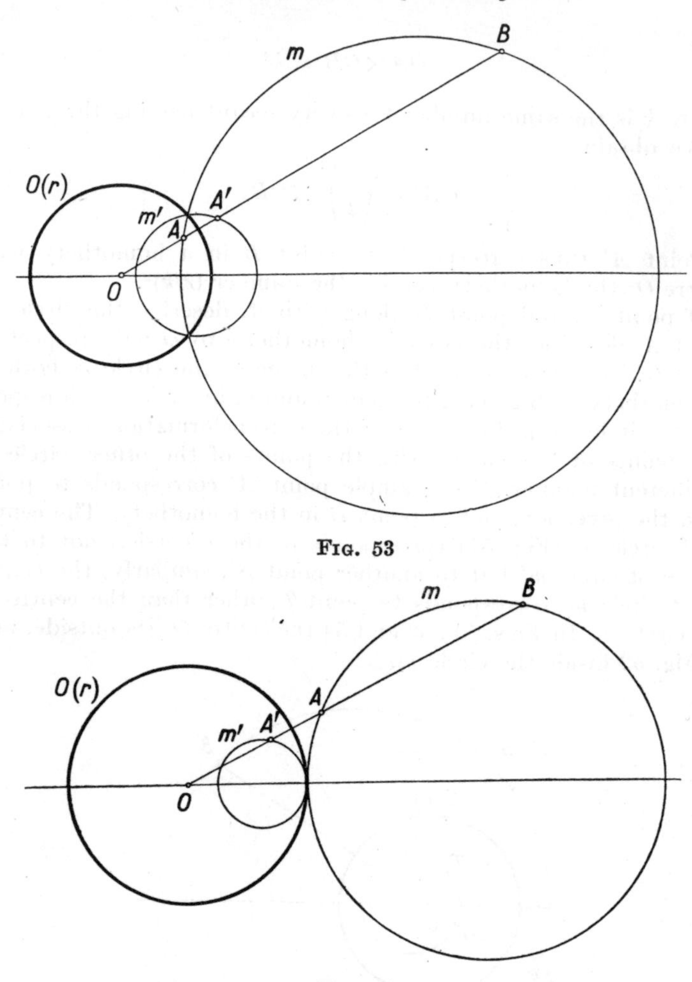

Fig. 53

Fig. 54

86. If one of the given points, say point A, lies on the circle C, the theorem is obvious because then all the chords mentioned in the theorem have the common end-point A.

We shall carry out the proof for the case where neither of the points A and B lies on the given circle C.

Let us consider two circles, K_1 and K_2, passing through the given points A and B and intersecting the circle C, the first at points M and N and the second at points P and Q (Figs. 55 and 56).

The lines MN and PQ intersect. Indeed, if the lines MN and PQ were parallel, then the segments MN and PQ, as parallel chords of circle C, would have a common axis of symmetry, passing through the centre of circle C. That axis of symmetry

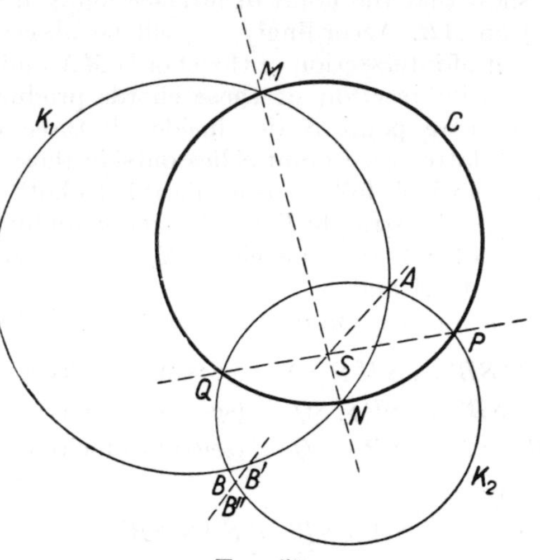

FIG. 55

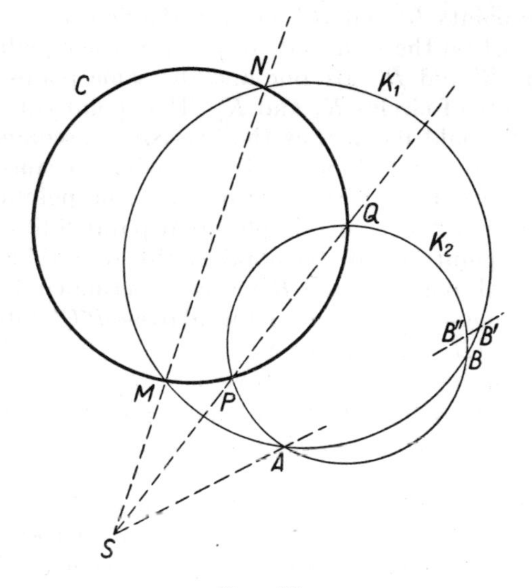

FIG. 56

would also pass through the centres of the circles K_1 and K_2 and thus it would also be the axis of symmetry of the segment AB. Thus the centre of circle C would be equally distant from A and from B, which is contrary to our assumption.

We shall show that the point of intersection S of the lines MN and PQ lies on AB. Accordingly, it will be observed that S is either the point of intersection of the chords MN and PQ (Fig. 55) or the point of intersection of these chords produced (Fig. 56). In the former case point S lies inside all three circles, C, K_1 and K_2, in the latter case point S lies outside those three circles. The argument which follows is applicable to both cases.

Let us consider the straight line SA and denote by B' the second common point of SA and the circle K_1 and by B'' the second common point of SA and K_2.†

By a well-known theorem on the secant of a circle we can write

$$SA\times SB' = SM\times SN \quad \text{(secants of circle } K_1),$$
$$SA\times SB'' = SP\times SQ \quad \text{(secants of circle } K_2),$$
$$SM\times SN = SP\times SQ \quad \text{(secants of circle } C).$$

Consequently

$$SA\times SB' = SA\times SB'',$$

whence

$$SB' = SB''.$$

Since the points B' and B'', both in the first and in the second case, lie on SA on the same side of point S, the equality obtained implies that B' and B'' are one and the same point—one of the common points of circles K_1 and K_2. That point cannot be point A since that would mean that the line SA is a common tangent of K_1 and K_2 at point A, which is impossible because the circles K_1 and K_2 are not tangent. Consequently, the points B' and B'' coincide with point B, which implies that point S lies on AB.

The same point S will be obtained by describing any circle passing through points A and B and intersecting circle C at points P' and Q', since—according to the above—$P'Q'$ intersects MN at a point of AB, i.e. at point S.

The theorem has thus been proved.

Remark 1. If points A, B are equidistant from the centre of circle C, then the common chords of C and the circles passing

† We do not know beforehand whether or not either of the points B' and B'' coincides with point A, which of course could happen only in a case like that presented in Fig. 56. Our subsequent reasoning will prove this to be impossible.

through points A, B are parallel to AB, the figure being symmetric with respect to the perpendicular bisector of the segment AB.

REMARK 2. The argument applied above to the circles K_1 and K_2 can also be used if one or even both of those circles are tangent to circle C provided that, instead of the secant MN or PQ, we consider the common tangent of the circles C and K_1 or C and K_2. Those common tangents pass through the point S of AB which has been determined above.

REMARK 3. The theorem which has been proved above by using the theorem on the secants of a circle can be deduced as a simple consequence of the properties of the *radical axis* of two circles. We shall explain this briefly.

Let $O(r)$ be a circle with centre O and radius r, and let P be a point lying in the plane of the circle at a distance d from O. The number

$$d^2-r^2$$

is termed *the power of point P for the circle O(r)*. This number is positive, negative or equal to zero according to whether point P lies outside the circle, inside it or on its circumference.

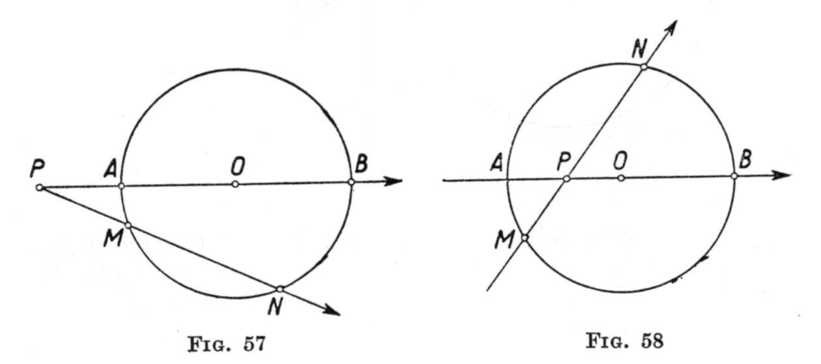

FIG. 57 FIG. 58

Using the relative measures of directed segments on the axis PO (Figs. 57 and 58) we have

$$PO = d, \quad -OA = OB = r.$$

Consequently

$$d^2-r^2 = (d-r)(d+r) = (PO+OA)(PO+OB) = PA \times PB.$$

Let us draw through point P a straight line intersecting the circle at points M and N. By a well-known theorem we have

$$PM \times PN = PA \times PB.$$

We thus have the equality

$$d^2 - r^2 = PM \times PN.$$

This equality expresses the following theorem:

The power of a point P for a circle is equal to the product of the relative measures of the segments of an arbitrary secant passing through P which are directed from point P towards the points of intersection of the secant with the circle.

In the wording of this theorem we can replace "relative measures of directed segments" by "lengths of segments", at the same time changing the word "power" to the words "absolute value of the power".

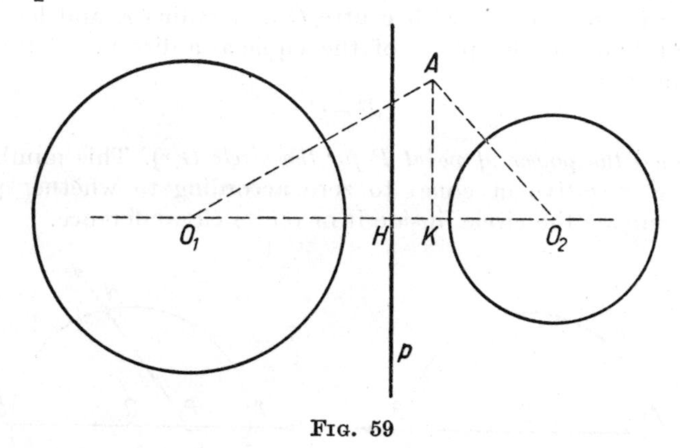

Fig. 59

Let us consider two non-concentric circles, $O_1(r_1)$ and $O_2(r_2)$, on a plane; let $O_1O_2 = a$ (Fig. 59). We shall show that on the line O_1O_2 there exists one and only one point H whose powers for the given circles are equal, i.e. that

$$O_1H^2 - r_1^2 = HO_2^2 - r_2^2.$$

Let us write the above equality as

$$O_1H^2 - HO_2^2 = r_1^2 - r_2^2. \tag{1}$$

If O_1H and HO_2 denote the relative measures of the segments on the axis O_1O_2, then

$$O_1H + HO_2 = a. \tag{2}$$

Dividing (1) by (2), we obtain

$$O_1H - HO_2 = \frac{r_1^2 - r_2^2}{a}. \tag{3}$$

From equalities (2) and (3) we can find O_1H and HO_2:

$$O_1H = \frac{1}{2}\left(a+\frac{r_1^2-r_2^2}{a}\right), \qquad HO_2 = \frac{1}{2}\left(a-\frac{r_1^2-r_2^2}{a}\right).$$

These values satisfy condition (1) and uniquely determine point H. This immediately implies the following theorem:

The geometrical locus of the points of a plane whose powers for two non-concentric circles are equal is a straight line perpendicular to the line joining their centres and passing through the point H, defined above, of that line.

Indeed, if A is any point of the plane and K the projection of point A upon the line O_1O_2 (Fig. 59), then

$$AO_1^2-r_1^2 = AK^2+O_1K^2-r_1^2,$$

$$AO_2^2-r_2^2 = AK^2+KO_2^2-r_2^2,$$

whence we can see that the equality of the powers

$$AO_1^2-r_1^2 = AO_2^2-r_2^2$$

holds if and only if we have the equality of the powers

$$O_1K^2-r_1^2 = KO_2^2-r_2^2,$$

i.e. if point K coincides with point H determined before. The required locus is thus the line p drawn through H and perpendicular to O_1O_2.

The straight line p is termed the ***radical axis*** of the circles $O_1(r_1)$ and $O_2(r_2)$.

A common point of two circles belongs to the radical axis of those circles because its power for either of them is 0. Thus the power **radical** of intersecting circles passes through their points of intersection, and the radical axis of tangent circles is their common tangent at their point of contact; the radical axis of circles having no points in common lies outside those circles.

If the centres of three circles are not collinear, then the three radical axes of three pairs of those circles pass through one point, termed the radical centre of the three circles in question.

Indeed, the three radical axes mentioned, being perpendicular to the three sides of the triangle formed by the centres of the circles, intersect pairwise; the point of intersection of two radical axes has equal powers for all three circles and thus it lies on the third radical axis as well.

The theorem on the radical centre makes it easy to draw the radical axis of two circles having no points in common.

We describe an auxiliary circle intersecting the given circles and we determine the radical centre of the three circles; it lies on the required power axis.

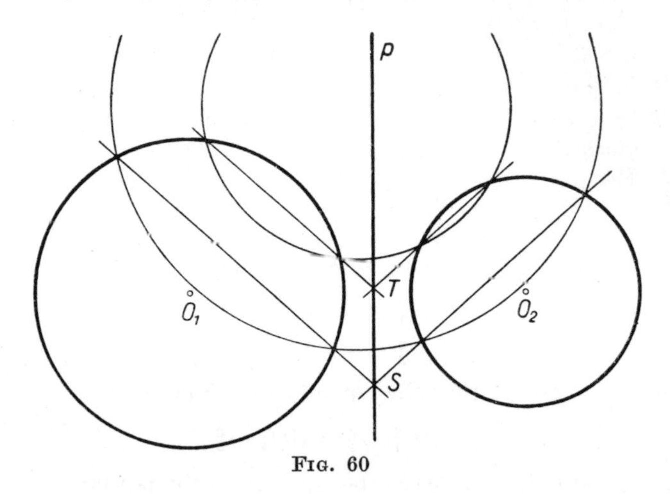

FIG. 60

Figure 60 shows two points, S and T, of the required radical axis p which have been determined by the use of two auxiliary circles.

If the centres of three circles are different collinear points, then the three radical axes of three pairs of those circles either are three parallel lines or coincide and form one straight line.

Let us return to the theorem of problem 86 and to Figs. 55 and 56. The proof of that theorem can now be put very briefly.

All circles K_1, K_2 passing through points A and B have a common radical axis—the straight line AB. The radical axis of the circles K_1 and C is the straight line MN. The point of intersection S of the lines AB and MN is the radical centre of the circles C, K_1 and an arbitrary circle K_2 passing through the points A and B. If the circle K_2 intersects the circle C at points P and Q, then PQ, as the radical axis of the circles C and K_2, passes through point S, which is what we were to prove.

The theorem which we have proved enables us to solve by a simple method the following construction problem.

Draw a circle tangent to a given circle C and passing through two given points A and B.

If points A and B are not equidistant from the centre of circle C, we describe an auxiliary circle passing through points A and B and intersecting circle C at points M and N. We determine the point of intersection S of AB and MN. From point S we draw a tangent to circle C. If the point of contact T of that tangent with C does not lie on AB, then the circle passing through points A, B, T is the required circle.

If points A and B are equidistant from the centre of circle C, then point T is obtained at the intersection of the perpendicular bisector of the segment AB with C.

The problem can have two solutions, one solution or no solutions.

A detailed discussion of these possibilities (which is left to the reader as an exercise) gives the following result:

(1) If points A and B both lie inside circle C or both lie outside circle C while the straight line AB is not tangent to the circle, the problem has two solutions.

(2) If points A and B lie outside circle C and AB is tangent to it, the problem has one solution.

(3) If one of the points A and B lies on the circle and the other point lies either inside the circle or outside it but in such a way that AB is not tangent to the circle, the problem has one solution.

(4) In all the remaining cases, i.e. if one of the points A, B lies on circle C and the other point lies on the tangent to the circle at the first point, and also if one point lies inside circle C and the other outside it, and finally if the points A and B both lie on the given circle C, the problem has no solution.

87. Since $\sphericalangle AQK = \sphericalangle AMK = 90°$ (Fig. 61), the points A, M, K, Q lie on the circle with diameter AK; consequently

$$\sphericalangle KMQ = \sphericalangle KAQ = \sphericalangle CAD$$

(angles inscribed in the same arc).

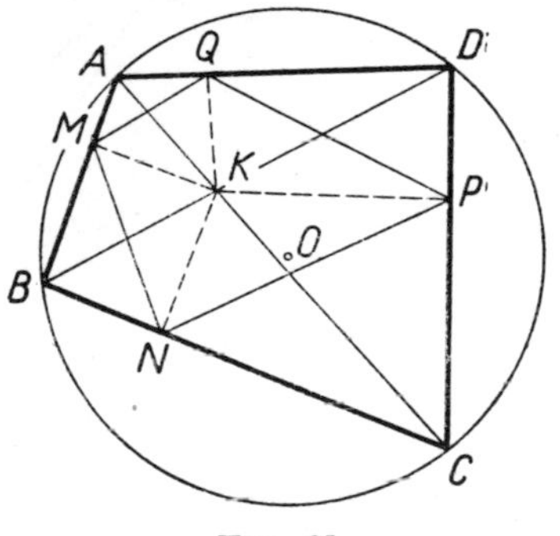

FIG. 61

Analogously

$$\sphericalangle KMN = \sphericalangle KBN = \sphericalangle DBC.$$

But $\sphericalangle CAD = \sphericalangle DBC$ (angles contained in the same arc), and consequently $\sphericalangle KMQ = \sphericalangle KMN$, which means that MK is the bisector of the angle QMN. For the other angles of quadrilateral $MNPQ$ the proof follows the same lines.

In Figure 61 the centre O of the circle lies inside quadrilateral $ABCD$. If point O lies outside quadrilateral $ABCD$, two of the projections of point K upon the sides of the quadrilateral lie on those sides produced (Fig. 62). The quadrilateral $MNPQ$ is then concave.

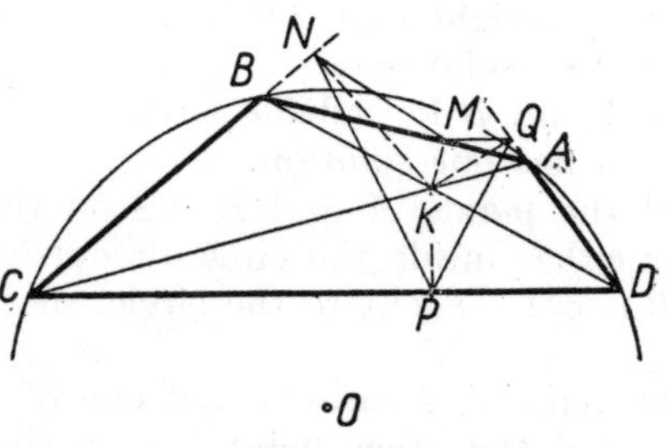

FIG. 62

The proof is essentially the same; it undergoes a slight modification with regard to the concave angle QMN:

$$\sphericalangle KMQ = 180° - \sphericalangle KAQ = \sphericalangle CAD,$$
$$\sphericalangle KMN = 180° - \sphericalangle KBN = \sphericalangle DBC,$$

whence, as before, we obtain $\sphericalangle KMQ = \sphericalangle KMN$.

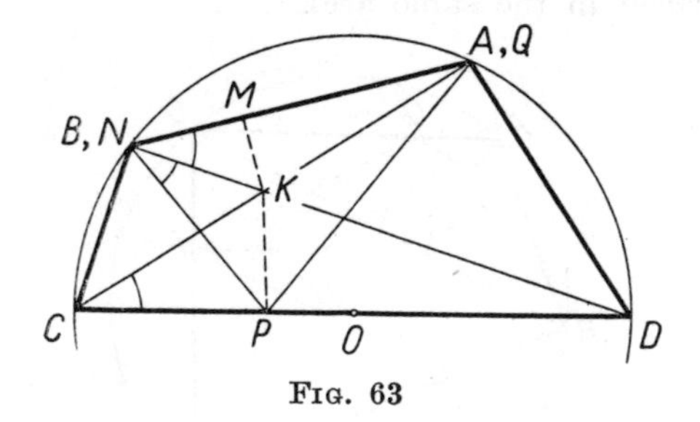

FIG. 63

If the centre O of the circle lies on the perimeter of quadrilateral $ABCD$, e.g. on the side CD (Fig. 63), then point N coincides with point B and point Q coincides with point A. In this "limiting case" we obtain a deformed quadrilateral $MNPQ$ in which the vertices M, N, Q are collinear and the angle at the vertex M is equal to 180°.

Also, in this case the perpendiculars KM, KN, KP, KQ are the bisectors of the angles of the "quadrilateral" $MNPQ$. For KM this is obvious; for KP the proof is the same as before. For KN (and similarly for KQ) the argument is as follows:

Points N, C, P, K lie on the circle with diameter CK (since $\sphericalangle CNK = \sphericalangle CPK = 90°$) and consequently

$$\sphericalangle KNP = \sphericalangle KCP = \sphericalangle DCA, \qquad \sphericalangle KNM = \sphericalangle DBA,$$

and since $\sphericalangle DCA = \sphericalangle DBA$, we have $\sphericalangle KNP = \sphericalangle KNM$, which is what was to be proved.

88. Disregarding for the time being the condition that the points A, B, C, D lie on a circle, let us consider an arbitrary convex quadrilateral $ABCD$ in which the sides AB and CD produced intersect at point E and the sides AD and BC produced —at point F.

Let us draw the bisectors EO and FO of the angles E and F and the segment EF; denote the angles as shown in Fig. 64 and consider the triangles EAF, ECF, EOF with the common base EF.

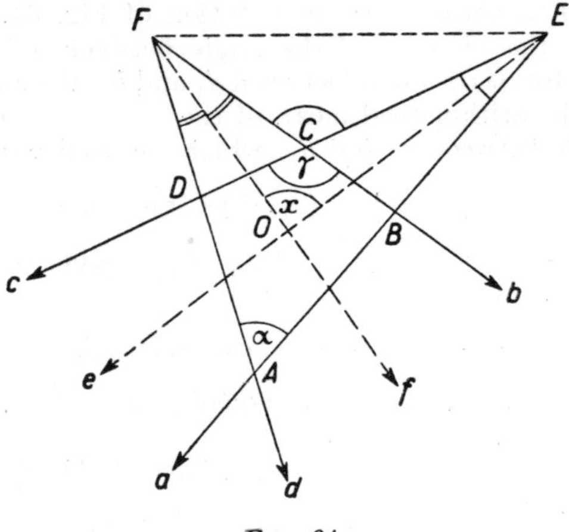

Fig. 64

Each of the angles at the base EF in the triangle EOF is the arithmetical mean of the angles of triangles EAF and ECF at the same vertex; this implies that the third angle x of triangle EOF

is the arithmetical mean of the remaining angles, α and γ, of the triangles EAF and ECF:

$$x = \frac{\alpha+\gamma}{2}.$$

The same conclusion can be reached in the following way:

From an arbitrary point M (Fig. 65) draw half-lines $a_1, b_1, c_1,$

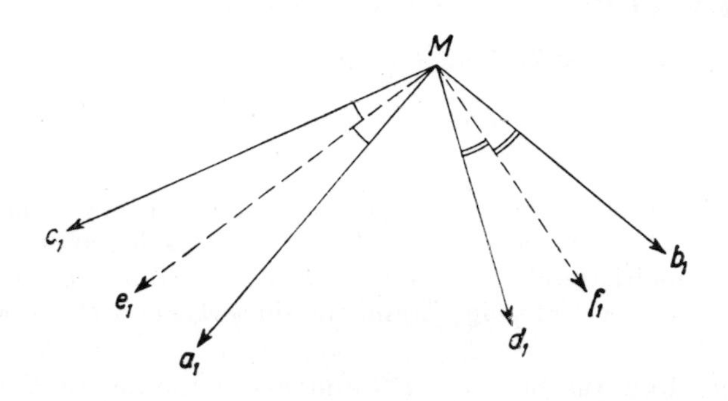

FIG. 65

d_1, e_1, f_1 having, respectively, the directions of the half-lines a, b, c, d, e, f, according to the notation of Fig. 64.

Since e_1 is the bisector of the angle between a_1 and c_1, and f_1 is the bisector of the angle between d_1 and b_1, the angle between e_1 and f_1 is the arithmetical mean of the angle between a_1 and d_1 and the angle between c_1 and b_1, which we shall write down as:

$$(e_1, f_1) = \tfrac{1}{2}[\measuredangle(a_1, d_1)+\measuredangle(c_1, b_1)].$$

But $\measuredangle(e_1, f_1) = \measuredangle(e, f)$, $\measuredangle(a_1, d_1) = \measuredangle(a, d)$, $\measuredangle(c_1, b_1) = \measuredangle(c, b)$, whence

$$\measuredangle(e, f) = \tfrac{1}{2}[\measuredangle(a, d)+\measuredangle(c, b)].$$

We have obtained the same equality as before, which can easily be verified.

Let us now assume that the quadrilateral $ABCD$ is inscribed in a circle (Fig. 66).

Accordingly, $\alpha+\gamma = 180°$ and the preceding equality gives

$$x = \tfrac{1}{2}(\alpha+\gamma) = 90°.$$

This means that the diagonals of the quadrilateral $MNPQ$ are perpendicular.

It will be observed that, consequently, the bisector EO of angle E in triangle PEQ is perpendicular to the side PQ, whence triangle PEQ is isosceles and point O is the mid-point of the segment PQ. Analogously, point O is the mid-point of the segment MN.

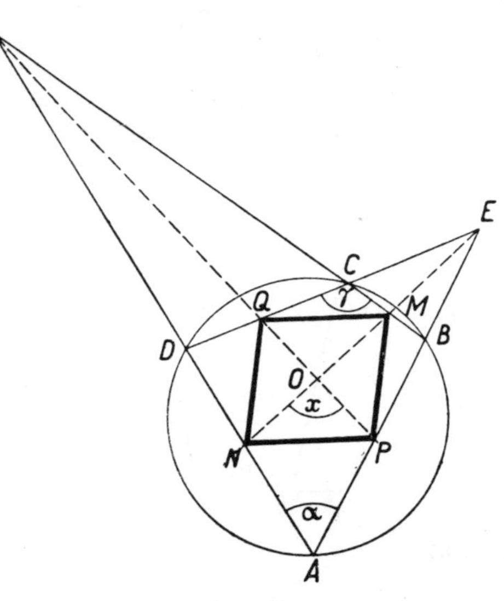

FIG. 66

The quadrilateral $MPNQ$, whose diagonals bisect each other and are perpendicular, is therefore a rhombus.

89. Let E, F, G, H denote, respectively, the points of contact of the sides AB, BC, CD, DA with the circle inscribed in the trapezium $ABCD$.

Method I. Let M be the point of intersection of the segments EG and HF (Fig. 67). Since the trapezium $ABCD$ is isosceles, the straight line EG is an axis of symmetry of the figure, point F being symmetric to point H and $HF \perp EG$. The parallel lines AB, HF and DC determine proportional segments on the lines EG and BC, whence

$$\frac{EM}{MG} = \frac{BF}{FC}.$$

Let N be the point of intersection of the diagonals AC and BD of the trapezium. Since these diagonals are symmetric with respect to the line EG, point N lies on the segment EG (Fig. 68). Triangles AEN and CGN are homothetic with respect to point N, whence

$$\frac{EN}{NG} = \frac{AE}{CG}.$$

But $AE = EB = BF$ and $FC = CG$, whence $AE/CG = BF/FC$; the above proportions imply that

$$\frac{EM}{MG} = \frac{EN}{NG}.$$

Points M and N divide the segment EG in the same ratio and consequently they coincide, which is what we were to prove.

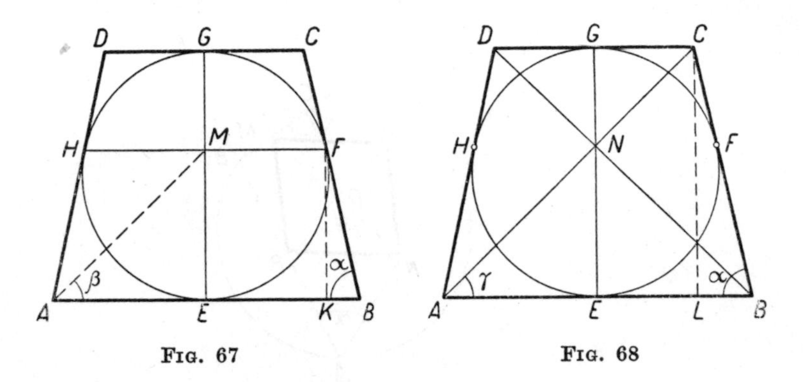

FIG. 67 FIG. 68

Method II. Let N denote, as in method I, the point of intersection of the diagonals AC and BD; it lies on the segment EG. We shall prove that HN is the bisector of the angle AND (Fig. 69).

Indeed, the triangles AEN and CGN are homothetic with respect to point N, and thus

$$\frac{AN}{CN} = \frac{AE}{CG}.$$

But $CN = DN$, $AE = AH$, $CG = GD = DH$ and thus the above proportion gives

$$\frac{AN}{DN} = \frac{AH}{DH}.$$

By the theorem on the bisector of an angle in a triangle, this implies that HN is the bisector of angle AND.

Similarly, the straight line NF is the bisector of the angle BNC. The bisectors of the vertically opposed angles AND and BNC are collinear, and consequently the line HF passes through point N.

Method III. Let M denote the point of intersection of the segments EG and HF (Fig. 70). Let us join point M with points A and C. Then

$$\tan \sphericalangle AME = \frac{AE}{EM}, \quad \tan \sphericalangle CMG = \frac{CG}{GM}.$$

Since $AE = EB = BF$, $CG = CF$, we have

$$\tan \sphericalangle AME = \frac{BF}{EM}, \quad \tan \sphericalangle CMG = \frac{CF}{GM}.$$

But (see method I) $BF/EM = CF/GM$ whence

$$\tan \sphericalangle AME = \tan \sphericalangle CMG.$$

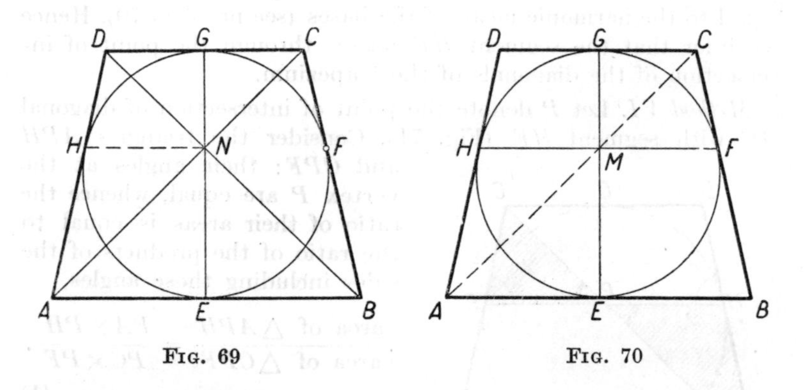

FIG. 69 FIG. 70

Since the angles AME and CMG are acute, we have $\sphericalangle AME = \sphericalangle CMG$, which implies that points A, M, C are collinear, which is what was to be proved.

Method IV. Let us return to the figures represented in Figs. 67 and 68; let us draw segments AM, $FK \perp AB$, $CL \perp AB$ and write $\sphericalangle ABC = \alpha$, $\sphericalangle MAB = \beta$, $\sphericalangle NAB = \gamma$. It will be observed that $AL = AE + EL = EB + GC = BF + FC = BC$. Now

$$\tan \beta = \frac{EM}{AE} = \frac{KF}{BF} = \sin \alpha,$$

$$\tan \gamma = \frac{CL}{AL} = \frac{CL}{BC} = \sin \alpha.$$

Consequently $\tan \beta = \tan \gamma$, whence $\beta = \gamma$, which implies that the points M and N coincide.

Method V. We shall give here only the essential idea of the proof, leaving its full development to the reader. Let $AB = a$, $CD = b$, $HF = x$. It is easy to find that

$$x = \frac{2ab}{a+b},$$

i.e. that the length x is the harmonic mean of the lengths of the bases a and b. For instance, we can draw in the trapezium the altitude from vertex D and consider two similar right-angled triangles whose hypotenuses are equal to $(a+b)/2$ and $b/2$ and sides lying opposite vertex D are equal to $(a-b)/2$ and $(x-b)/2$ respectively.

As we know, in a trapezium the segment passing through the point of intersection of its diagonals and parallel to its bases is equal to the harmonic mean of the bases (see problem 70). Hence we infer that the segment HF passes through the point of intersection of the diagonals of the trapezium.

Method VI. Let P denote the point of intersection of diagonal AC with segment HF (Fig. 71). Consider the triangles APH and CPF: their angles at the vertex P are equal, whence the ratio of their areas is equal to the ratio of the products of the sides including those angles,

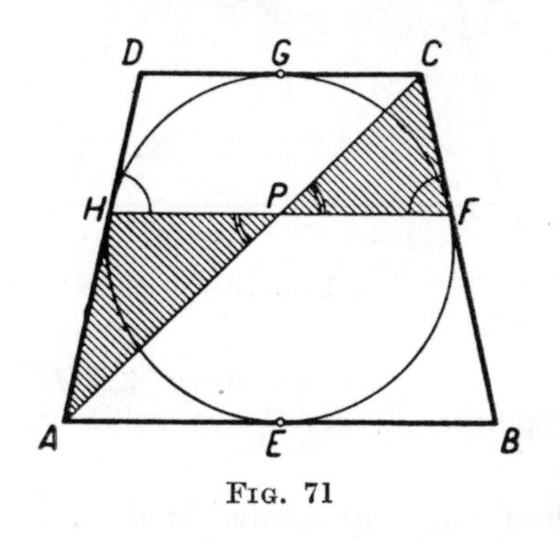

FIG. 71

$$\frac{\text{area of } \triangle APH}{\text{area of } \triangle CPF} = \frac{PA \times PH}{PC \times PF}. \tag{1}$$

Next, it will be observed that the angles (of those triangles) at the vertices H and F are supplementary because $\measuredangle AHP + \measuredangle PHD = 180°$, and $\measuredangle PHD = \measuredangle PFC$, being angles formed by the chord HF and the tangents of the circle at points H and F. Consequently $\sin \measuredangle AHP = \sin \measuredangle PFC$ and, since the area of a triangle equals half the product of two sides multiplied by the sine of the included angle, we have

$$\frac{\text{area of } \triangle APH}{\text{area of } \triangle CPF} = \frac{AH \times PH}{PF \times CF}. \tag{2}$$

Equalities (1) and (2) give

$$\frac{PA \times PH}{PC \times PF} = \frac{AH \times PH}{PF \times CF},$$

whence

$$\frac{PA}{PC} = \frac{AH}{CF}. \tag{3}$$

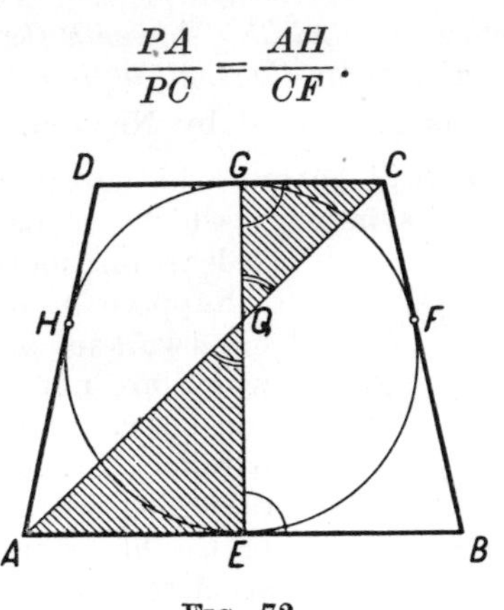

FIG. 72

Let Q denote the point of intersection of diagonal AC with segment EG (Fig. 72). Applying to the triangles AQE and CQG the argument used before with regard to the triangles APH and CPF, we obtain

$$\frac{QA}{QC} = \frac{AE}{CG}. \tag{4}$$

Since $AH = AE$ and $CF = CG$, we infer from (3) and (4) that

$$\frac{PA}{PC} = \frac{QA}{QC}. \tag{5}$$

Equality (5) denotes that points P and Q divide segment AC in the same ratio, i.e. they coincide. We have thus proved that the diagonal AC passes through the point of intersection of segments EG and HF. The same is of course true of the diagonal BD, and thus the theorem has been proved.

REMARK. We have got acquainted with six different proofs of the theorem formulated in the problem. If we reflect upon them, we shall notice that in each of the first five proofs we used the assumption of the circumscribed quadrilateral $ABCD$ being an isosceles trapezium while in the sixth proof we resorted neither to the fact of the lines AB and CD being parallel nor to the equality $AD = BC$ but based the proof solely on the fact that the quadrilateral $ABCD$ is circumscribed on a circle. Thus proofs 1–5 do not really matter because proof 6 shows that a much more general theorem is true. Namely:

In a quadrilateral circumscribed about a circle the segments joining the points of contact of opposite sides with the circle pass through the point of intersection of the diagonals of the quadrilateral.

This theorem was discovered by Newton.

90. The theorem will be proved if we show that a plane figure having two axes of symmetry which are not perpendicular has at least one more axis of symmetry.

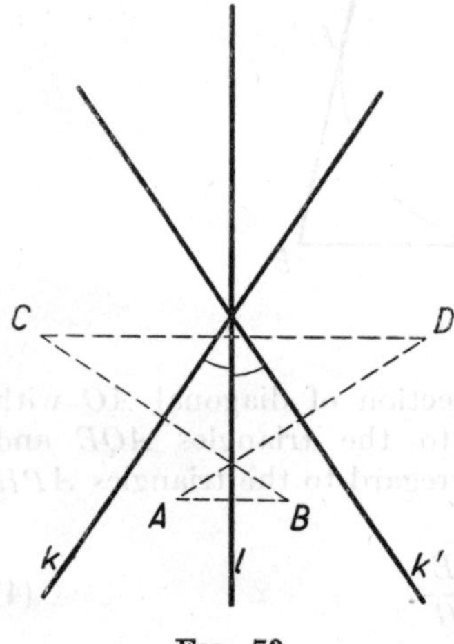

Fig. 73

Suppose that a plane figure F has two axes of symmetry, k and l, which are not perpendicular. They can either intersect (Fig. 73) or be parallel (Fig. 74). In both cases the reasoning will be the same. Let k' be the line symmetric to k with respect to l. The line k' is of course different from k and different from l. We shall prove that k' is an axis of symmetry of figure F.

If a point A belongs to figure F, then point B symmetric to A with respect to axis l also belongs to it, and the same holds for point C symmetric to B with respect to axis k and for point D symmetric to C with respect to l. It will be observed that the segment AD is symmetric to the segment BC with respect to the axis l because points A and D are symmetric to points B and C, respectively, with respect to the same axis. Consequently

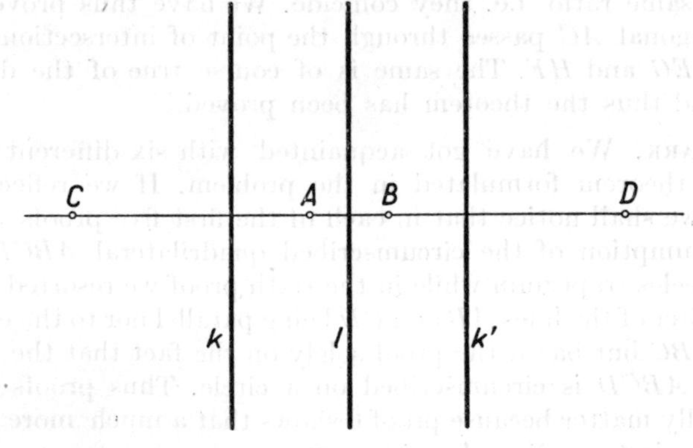

Fig. 74

the perpendicular bisector of segment AD is symmetric to the perpendicular bisector of segment BC (i.e. to k) with respect to axis l, i.e. the perpendicular bisector of AD is the line k'.

We have proved that, if a point A belongs to figure F, then point D symmetric to A with respect to k' also belongs to it, i.e. that k' is an axis of symmetry of figure F.

91. Let AM and BN be altitudes of the tetrahedron $ABCD$ (Fig. 75) intersecting at point S:

$$AS \perp \text{plane } BCD, \qquad BS \perp \text{plane } ACD.$$

The plane ABS passing through the lines AS and BS perpendicular to the planes BCD and ACD is perpendicular to those planes, and thus it is also perpendicular to the line of their intersection, i.e. to CD:

$$\text{plane } ABS \perp CD.$$

Consequently CD is perpendicular to every straight line lying in the plane ABS, and in particular

$$CD \perp AB.$$

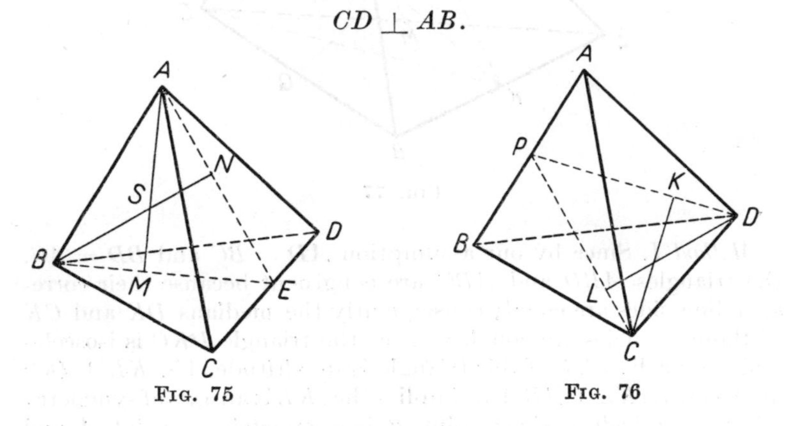

Fig. 75 Fig. 76

Thus through CD we can draw a plane perpendicular to AB. If CP is an altitude of triangle ABC (Fig. 76), then the relations

$$CP \perp AB, \qquad CD \perp AB$$

imply that

$$\text{plane } CDP \perp AB.$$

The altitudes CK and DL of triangle CDP are altitudes of the tetrahedron.

Indeed, $CK \perp AB$ (since CK lies in the plane CDP) and $CK \perp PD$, whence $CK \perp$ plane ABD. Similarly $DL \perp$ plane ABC.

Since the altitudes CK and DL of the tetrahedron are altitudes of a triangle, they intersect, which is what we were to prove.

92. Let K, L, M, N, P, Q denote the mid-points of the edges of the tetrahedron $ABCD$, as shown in Fig. 77. It is sufficient to prove that any of the lines KL, MN, PQ, say KL, is an axis of symmetry of the tetrahedron and that it is perpendicular to either of the remaining lines, say $KL \perp PQ$.

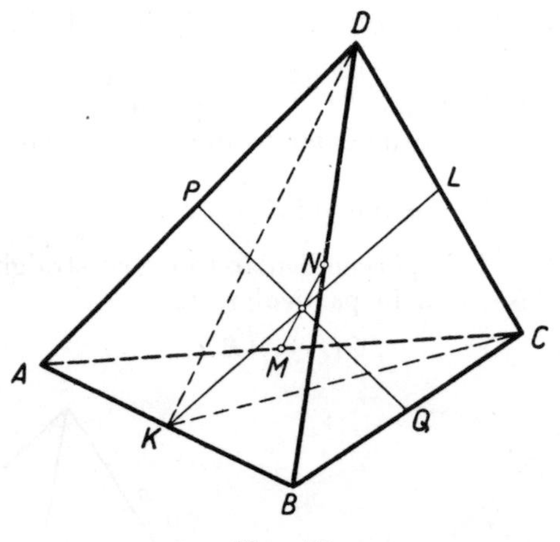

FIG. 77

Method I. Since by our assumption $AD = BC$ and $BD = AC$, the triangles ABD and ABC are congruent because their corresponding sides are equal; consequently the medians DK and CK of those triangles are equal, whence the triangle DKC is isosceles and the median KL of this triangle is its altitude, i.e. $KL \perp DC$; analogously, $KL \perp AB$. This implies that KL is an axis of symmetry of the tetrahedron since point B is symmetric to point A and point C is symmetric to point D with respect to KL. The segment BC is thus symmetric to the segment AD, whence the mid-point Q of segment BC is symmetric to the mid-point P of segment AD; the straight line PQ passing through points P and Q symmetric with respect to KL intersects KL and is perpendicular to it.

Method II. Since the segment KP (Fig. 78) joins the mid-points of the sides AB and AD of triangle ABD, we have $KP \parallel BD$ and $KP = \frac{1}{2}BD$; similarly $QL \parallel BD$ and $QL = \frac{1}{2}BD$, whence $KP \parallel QL$ and $KP = QL$; the quadrilateral $KPLQ$ is thus a parallelogram. From the assumption that $AC = BD$ it follows further that $KP = PL$ because $KP = \frac{1}{2}BD$ and $PL = \frac{1}{2}AC$; the parallelogram $KPLQ$ is thus a rhombus. Consequently, the segments KL and PQ bisect each other and are perpendicular.

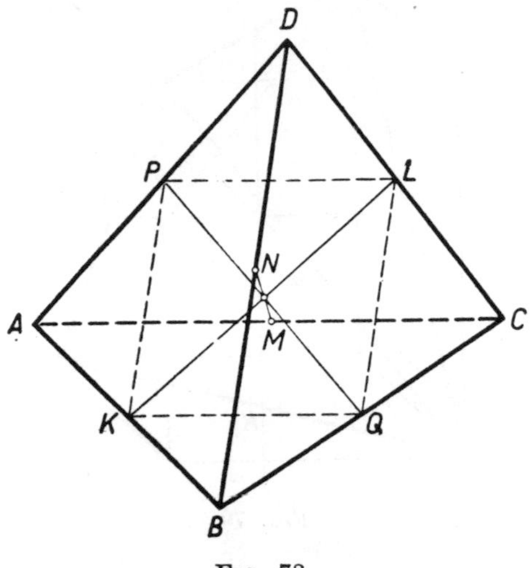

FIG. 78

Similarly, the segments KL and MN bisect each other and are perpendicular. It follows that if we rotate the figure about KL through 180° the points K, Q, M fall upon the points K, P, N respectively. Thus, if the figure is rotated in the above way, AC, which passes through point M and is parallel to KQ, will replace the straight line passing through point N and parallel to KP, i.e. the line BD; similarly, BC will replace AD, the triangle ABC will replace the triangle BAD and the tetrahedron $ABCD$ will pass into itself. The straight line KL is thus an axis of symmetry of the tetrahedron.

Method III. About every tetrahedron we can circumscribe a parallelepiped, i.e. construct a parallelepiped in which opposite faces pass through the opposite edges of the tetrahedron. If the opposite edges of the tetrahedron are equal, then the diagonals of each face of the circumscribed parallelepiped are equal, whence those faces are rectangles and the solid is a rectangular parallelepiped (Fig. 79).

As we know, a rectangular parallelepiped has three mutually perpendicular axes of symmetry, passing through the centres of the opposite faces of the solid. The same lines are axes of symmetry of the tetrahedron inscribed in it; they pass through the mid-points of the opposite edges of the tetrahedron.

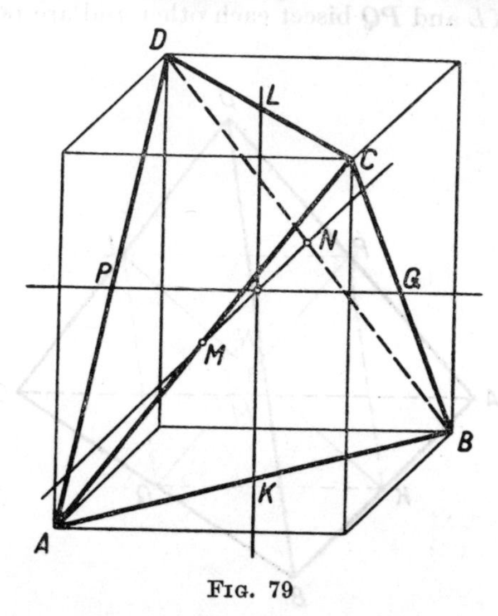

FIG. 79

93. The plane passing through the mid-points M and P of the opposite edges AB and CD of the tetrahedron $ABCD$ and through the mid-point N of the edge BC (Fig. 80) must also pass through the mid-point Q of the edge AD, opposite to the edge BC. Indeed, each of the straight lines NP and MQ is parallel to BD (by the

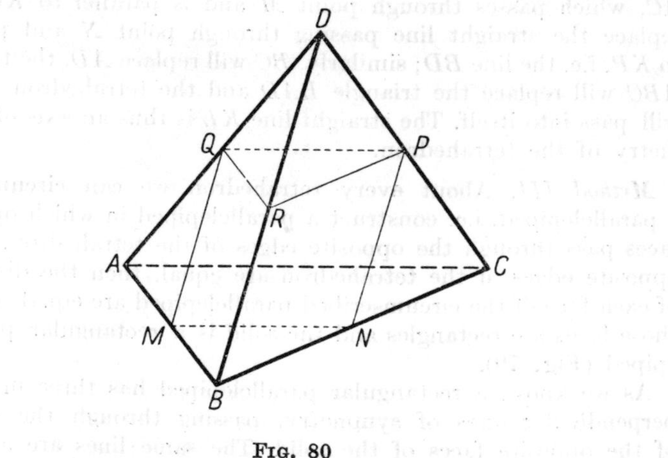

FIG. 80

theorem on the line joining the mid-points of two sides of a triangle), whence $MQ \parallel NP$; thus the plane MNP contains the line MQ.

The plane $MNPQ$ divides the tetrahedron into two parts; let us find the volume of the part lying on the same side of the plane $MNPQ$ as the edge BD, i.e. the volume of the pentahedron bounded by the parallelogram $MNPQ$, the triangle BMN, the trapeziums $BMQD$ and $BNPD$ and the triangle DPQ. Let us draw the plane PQR through the mid-point R of the edge BD. This plane divides the pentahedron into a tetrahedron, $DPQR$, and a triangular prism, $PQRBMN$. The tetrahedron $DPQR$ has a volume equal to $\frac{1}{8}$ the volume V of the tetrahedron $ABCD$ since it is similar to the given one, the ratio of similitude being $\frac{1}{2}$. The volume of the prism $PQRBMN$, with a base equal to $\frac{1}{4}$ the base of the tetrahedron $ABCD$ and an altitude equal to $\frac{1}{2}$ the altitude of that tetrahedron, is $\frac{1}{4}\times\frac{1}{2}\times 3V = \frac{3}{8}V$. Consequently, the volume of the pentahedron $DBMNPQ$ is $\frac{1}{8}V+\frac{3}{8}V = \frac{1}{2}V$, i.e. the plane $MNPQ$ divides the given tetrahedron into parts equal in volume.

In order to answer the last question posed in problem 93, let us draw through the mid-points M and P of the edges AB and CD an arbitrary plane α. If the plane α passes through the edge CD, it cuts the tetrahedron into two tetrahedrons, $AMCD$ and $BMCD$, whose volumes are equal because they have bases, AMC and BMC, of equal areas and a common altitude from the vertex D. If the plane α intersects CD, then points C and D lie on opposite sides of that plane, and consequently the plane α intersects one of the edges AD and AC. It suffices to consider the case where the plane α intersects AD at point Q' lying inside the segment QD (Fig. 81); we shall show that it then intersects BC at an interior point N' of the segment NC.

Indeed, the straight line $Q'P$, along which the planes α and ACD intersect, cuts AC produced at a point T; the straight line MT, which is the intersection of the planes α and ABC, cuts the segment NC, since segment MT joins points M and T lying on opposite sides of BC, and lies in the strip between the parallel lines AC and MN.

The plane α cuts the tetrahedron into two parts; let us find the volume of one of them, say the pentahedron $DBMN'PQ'$. Since, as follows from the above, points N' and Q' lie on opposite sides of the plane $MNPQ$, we have

$$\text{vol. } DBMN'PQ' = \text{vol. } DBMNPQ - \text{vol. } Q'MPQ + \text{vol. } N'MNP. \quad (1)$$

We shall prove that

$$\text{vol. } Q'MPQ = \text{vol. } N'MNP. \tag{2}$$

Since the bases MPQ and MNP of the tetrahedrons $Q'MPQ$ and $N'MNP$ are congruent triangles, being halves of the parallelogram $MNPQ$, it suffices to prove that the vertices N' and Q' are equidistant from the plane $MNPQ$. This amounts to showing that point S, at which the diagonals $N'Q'$ and PM of the plane quadrilateral $MN'PQ'$ intersect, bisects the segment $N'Q'$.

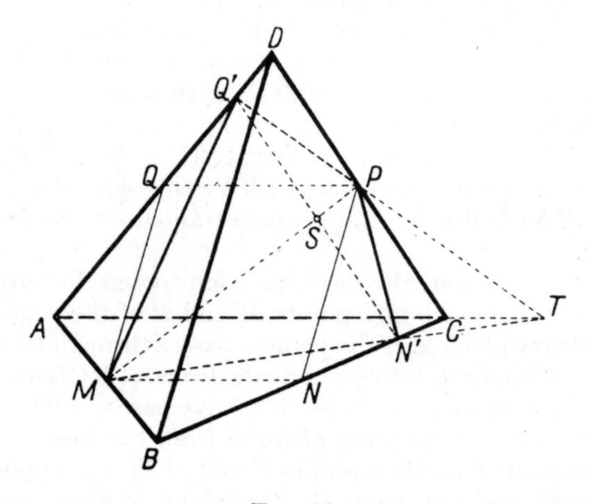

FIG. 81

Now the mid-points of all segments joining the points of AD with the points of BC lie in the same plane, parallel to the two skew lines AD and BC. This plane contains the points P and M, whence it also contains point S; consequently, point S is indeed the mid-point of the segment $N'Q'$.

Equalities (1) and (2) imply that

$$\text{vol. } DBMN'PQ' = \text{vol. } DBMNPQ = \tfrac{1}{2}V.$$

We have proved that every plane passing through the mid-points of two opposite edges of a tetrahedron divides it in two parts of equal volumes.

REMARK. The second part of this proof, i.e. the argument concerning the division of the tetrahedron $ABCD$ by an arbitrary plane α passing through points M and P, can be made considerably shorter.

Let us draw the orthogonal projection of the tetrahedron upon a plane perpendicular to MP (Fig. 82).

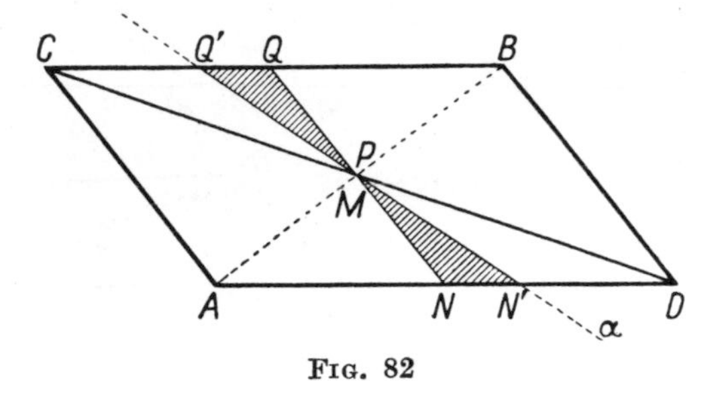

FIG. 82

The projections of the edges AB and CD are two lines bisecting each other at a point which is the common projection of points M and P; thus the projections of the remaining edges form the parallelogram $ADBC$, and the projection of the parallelogram $MNPQ$ coincides with the projection of the segment NQ. It can be seen at once that any plane α passing through MP intersects two opposite edges of the tetrahedron, e.g. AD at point N' and BC at point Q', the projections of points N' and Q' being symmetric with respect to the centre of the parallelepiped $ADBC$; this implies equality (1), and also equality (2) because the altitudes of the tetrahedrons $Q'MPQ$ and $N'MNP$ drawn from vertices Q' and N' are equal to the altitudes of triangles $N'MN$ and $Q'PQ$ drawn from the vertices Q' and N' in the figure in Fig. 82.

94. The figure formed by two skew lines m and n is most conveniently represented by means of projections on two perpendicular planes. As the horizontal plane of projection we shall take any plane parallel to both m and n (Fig. 83). The vertical projections of the given lines will then be parallel lines m'' and n'', the distance d between them being equal to the distance between the skew lines m and n. The horizontal projections m' and n' will be two intersecting lines forming an angle φ equal to the angle between the skew lines m and n. The horizontal projections of the segments AB and CD, parallel to the horizontal plane of projection, have lengths $A'B' = a$ and $C'D' = b$.

In order to find the volume of the tetrahedron $ABCD$, we shall consider, to begin with, a certain parallelepiped "circumscribed" about the tetrahedron, namely the parallelepiped for which the segments AB and CD are diagonals of two opposite faces. The

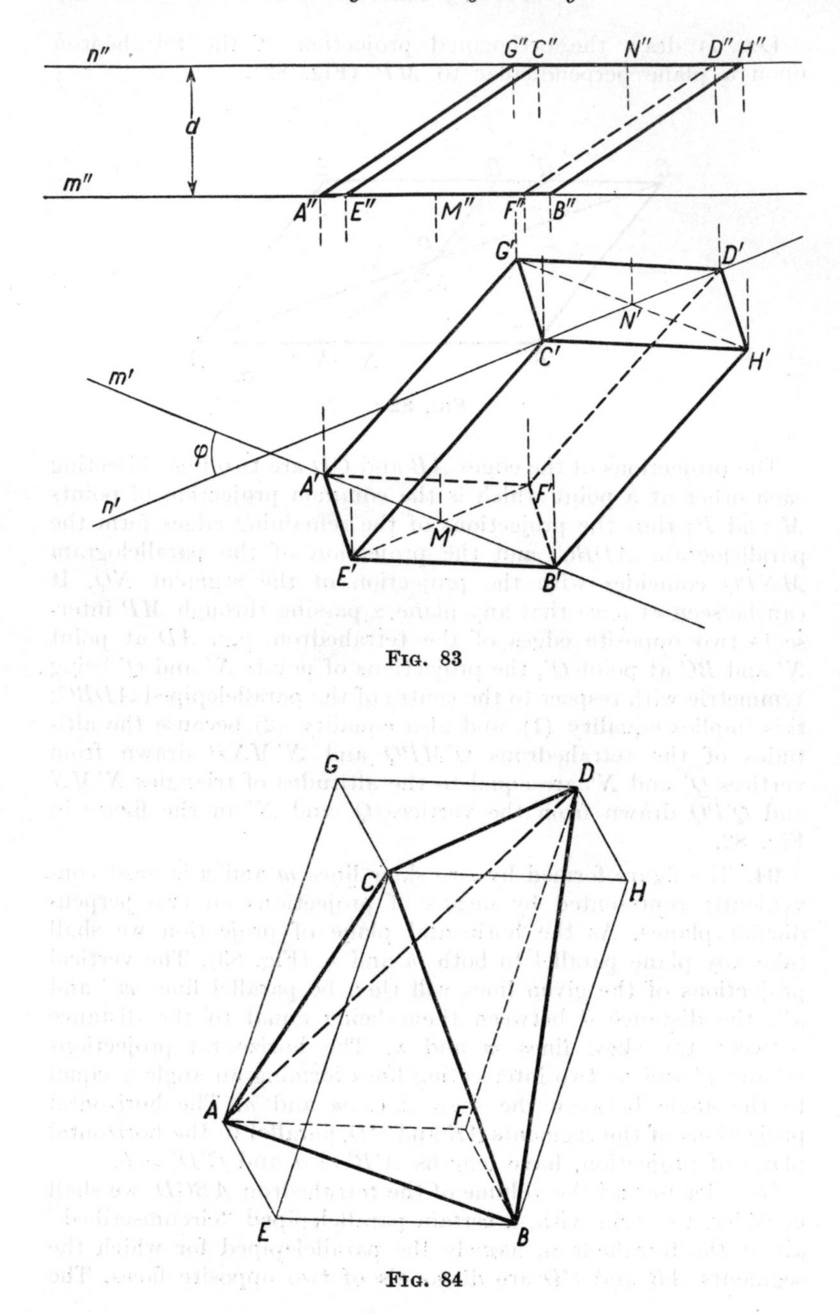

FIG. 83

FIG. 84

other two diagonals of the same faces are: the segment $EF = b$, parallel to segment CD and having a common mid-point M with segment AB, and the segment $GH = a$, parallel to segment AB and having a common mid-point N with segment CD.

The volume V of the parallelepiped constructed in this way is equal to the product of the area of face $AEBF$ and its distance from the opposite face, namely d.

But the area of the parallelogram $AEBF$, whose diagonals have lengths a and b and form an angle φ, is equal to $\frac{1}{2}ab\sin\varphi$, whence $V = \frac{1}{2}abd\sin\varphi$.

The tetrahedron $ABCD$ is formed from the parallelepiped by cutting off the four corner tetrahedrons, $EABC$, $FABD$, $GACD$, $HBCD$ (Fig. 84). The volume of each of those tetrahedrons, with bases equal to $\frac{1}{2}$ the bases of the parallelepiped and altitudes equal to d, is $\frac{1}{6}V$.

Consequently, the volume of the tetrahedron $ABCD$ is $\frac{1}{3}V = \frac{1}{6}abd\sin\varphi$, whence it depends only on the lengths a, b, d and angle φ, and is independent of the position of segments AB and CD on lines m and n.

§ 6. Finding Geometrical Magnitudes

95. Denote the mid-points of the sides of the parallelogram $ABCD$ by K, L, M, N, as shown in Fig. 85, and the centre of the parallelogram by O. The segment joining vertex A with

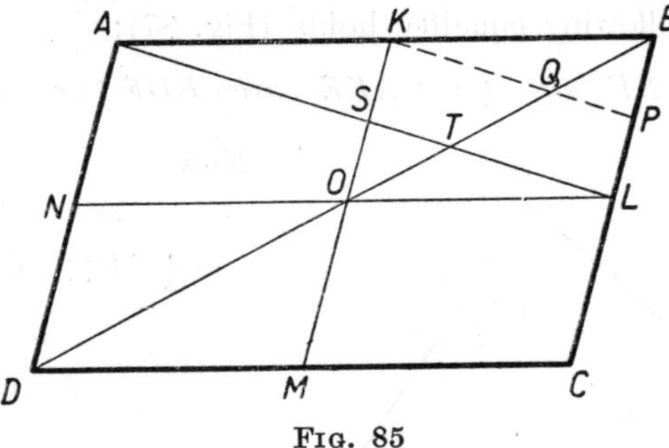

Fig. 85

the mid-point L of side BC intersects the segment joining the mid-points K and M of AB and CD at point S and the diagonal BD at a point T. Then

(a) $OS = \frac{1}{2}OK$ since S is the centre of the parallelogram $ABLN$;

(b) $OT = \frac{1}{3}OB$, which can be proved as follows. Let us draw a segment joining the mid-point K of segment AB with the mid-point P of segment BL. Then $KP \parallel SL$; thus if Q is the point of intersection of segments KP and OB, then in triangle KOQ we have $OT = TQ$ and in triangle BTL we have $TQ = QB$.

This implies that the area of triangle SOT is equal to $\frac{1}{6}$ the area of triangle KOB.

An analogous reasoning applies to any other segment joining a vertex of the parallelogram with the mid-point of one of the opposite sides. Parts of those segments, such as ST, are the bases of eight triangles forming together an octagon (Fig. 86).

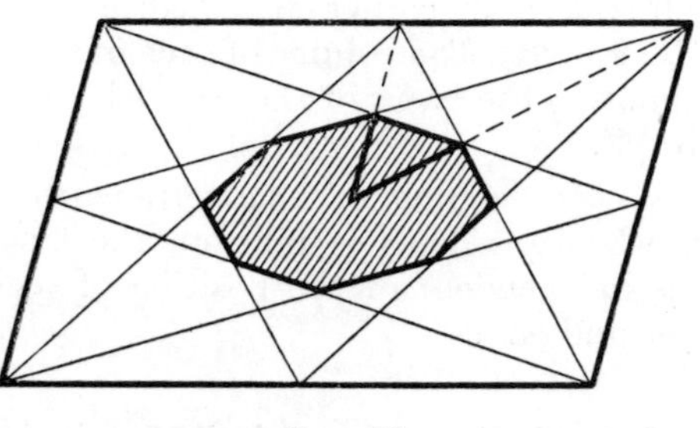

Fig. 86

The area of each of those triangles equals $\frac{1}{6}$ of the corresponding part of the parallelogram, whence the area of the octagon is equal to $\frac{1}{6}$ the area of the parallelogram.

96. The following equality holds (Fig. 87):

$$\text{area } DEF = S - (\text{area } AFE + \text{area } BDF + \text{area } CED). \quad (1)$$

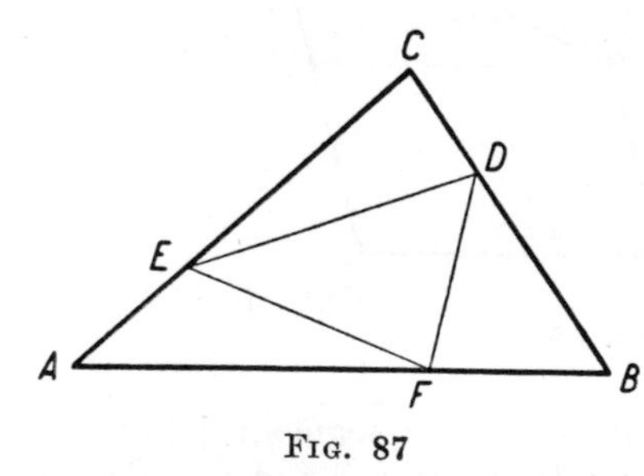

Fig. 87

Now

$$\text{area } AFE = \tfrac{1}{2} AF \times AE \times \sin A,$$

$$AF = \frac{k}{k+1} \times AB,$$

$$AE = \frac{1}{k+1} \times AC,$$

whence

$$\text{area } AFE = \frac{k}{(k+1)^2} \times \frac{1}{2} AB \times AC \times \sin A;$$

since $\frac{1}{2}AB\times AC\times \sin A = S$, we have

$$\text{area } AFE = \frac{k}{(k+1)^2}\times S. \tag{2}$$

Similarly

$$\text{area } BDF = \frac{k}{(k+1)^2}\times S, \quad \text{area } CED = \frac{k}{(k+1)^2}\times S. \tag{3}$$

Consequently, by formula (1) we have

$$\text{area } DEF = S - \frac{3k}{(k+1)^2}\times S = \frac{k^2-k+1}{(k+1)^2}\times S. \tag{4}$$

REMARK. The above argument can be shortened by using a theorem stating that the areas of triangles having an angle in common are in the same ratio as the products of the sides including that angle. We immediately obtain

$$\frac{\text{area } AFE}{S} = \frac{AE\times AF}{AC\times AB} = \frac{1}{k+1}\times\frac{k}{k+1} = \frac{k}{(k+1)^2},$$

i.e. formula (2).

97. The problem can be solved in a simple manner on the grounds of Thales' theorem. Let us draw $MK \parallel BN$ (Fig. 88); then in triangles BCN and MAK we have

$$\frac{NK}{KC} = \frac{BM}{MC} = m \quad \text{and} \quad \frac{AP}{PM} = \frac{AN}{NK}.$$

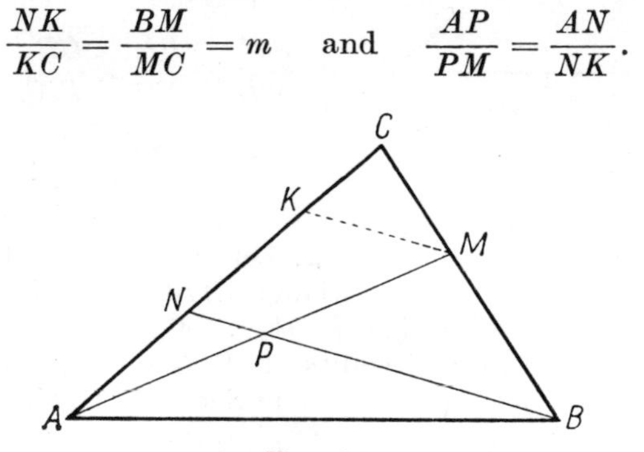

FIG. 88

Consequently,

$$\frac{AP}{PM} = \frac{AN}{NC}\times\frac{NC}{NK} = n\times\frac{NK+KC}{NK} = n\left(1+\frac{KC}{NK}\right) = n\left(1+\frac{1}{m}\right).$$

Analogously, we obtain

$$\frac{BP}{PN} = m\left(1+\frac{1}{n}\right).$$

Remark. The solution of problem 97 follows immediately from an important theorem of geometry, discovered by Menelaus of Alexandria (about the year 80 A.D.). It reads:

If the sides AB, BC, CA of a triangle ABC or those sides produced are intersected by a straight line k at points M, N and P respectively, then the product of the three ratios of division AM/MB, BN/NC, CP/PA, equals unity:

$$\frac{AM}{MB}\times\frac{BN}{NC}\times\frac{CP}{PA}=1. \qquad (1)$$

Various proofs of this theorem are known; we shall reproduce two of them.

1. Through the vertices A, B, C of triangle ABC we draw perpendiculars d_1, d_2, d_3 to the straight line k (Fig. 89). By Thales' theorem

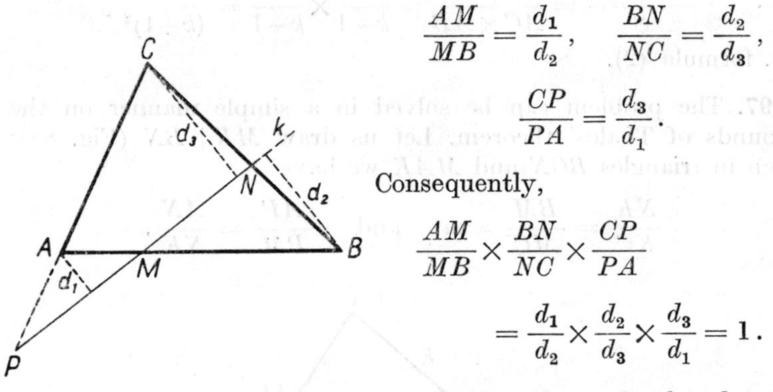

$$\frac{AM}{MB}=\frac{d_1}{d_2}, \qquad \frac{BN}{NC}=\frac{d_2}{d_3},$$

$$\frac{CP}{PA}=\frac{d_3}{d_1}.$$

Consequently,

$$\frac{AM}{MB}\times\frac{BN}{NC}\times\frac{CP}{PA}$$

$$=\frac{d_1}{d_2}\times\frac{d_2}{d_3}\times\frac{d_3}{d_1}=1.$$

Fig. 88

2. Let us choose in the plane of triangle ABC an arbitrary straight line l intersecting the line k (Fig. 90). Let A', B', C' be the parallel projections of points A, B, C on l in the direction of k, and let O' be the common projection of points M, N, P.

Since the ratio of segments of a straight line is equal to the ratio of their projections, we have

$$\frac{AM}{MB}=\frac{A'O'}{O'B'}, \qquad \frac{BN}{NC}=\frac{B'O'}{O'C'}, \qquad \frac{CP}{PA}=\frac{C'O'}{O'A'},$$

and therefore

$$\frac{AM}{MB}\times\frac{BN}{NC}\times\frac{CP}{PA}=\frac{A'O'}{O'B'}\times\frac{B'O'}{O'C'}\times\frac{C'O'}{O'A'}=1.$$

We invite the reader to verify that by either of methods 1 and 2 we can prove the following more general theorems, stated in 1801 by a well-known French mathematician, L. Carnot:

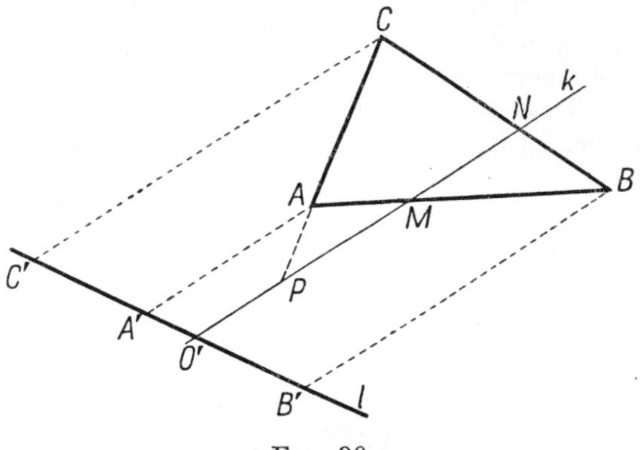

Fig. 90

I. *If a straight line intersects the sides $A_1A_2, A_2A_3, \ldots, A_nA_1$ of a plane polygon $A_1A_2 \ldots A_n$ or those sides produced at points $M_1, M_2, \ldots, M_n$ respectively, then*

$$\frac{A_1M_1}{M_1A_2} \times \frac{A_2M_2}{M_2A_3} \times \ldots \times \frac{A_nM_n}{M_nA_1} = 1. \quad (2)$$

II. *If a plane intersects the sides $A_1A_2, A_2A_3, \ldots, A_nA_1$ of a plane or skew polygon $A_1A_2 \ldots A_n$ or those sides produced at points $M_1, M_2, \ldots, M_n$ respectively, then we have equality* (2).

If we apply the theorem of Menelaus in problem 97, e.g. to triangle ACM intersected by the straight line NB (Fig. 88), we obtain

$$\frac{AN}{NC} \times \frac{CB}{BM} \times \frac{MP}{PA} = 1,$$

whence

$$\frac{AP}{PM} = \frac{AN}{NC} \times \frac{CB}{BM} = n \times \frac{CM+MB}{MB} = n\left(1+\frac{1}{m}\right);$$

we find the ratio BP/PN in an analogous way.

The theorem of Thales is a limiting case of the theorem of Menelaus. For example, if the point N (Fig. 89) is stationary, and point P recedes to infinity along the straight line CA, the straight line NP tends to a line parallel to CA and the ratio CP/PA in formula (1) tends to 1, whence formula (1) becomes

$$\frac{AM}{MB} \times \frac{BN}{NC} = 1,$$

expressing the theorem of Thales.

The theorem of Menelaus is often given a slightly different form, involving, instead of lengths of segments, relative measures of directed segments, i.e. vectors.

Let A, B, M be different points of a directed line, i.e. an axis, and let symbols AM, MB denote the relative measures of vectors on that axis. The ratio AM/MB is a positive number if the point M lies between points A and B, since then numbers AM and MB, as the relative measures of identically directed vectors, have the same sign; the ratio AM/MB is a negative number if point M lies outside segment AB.

Under this agreement Menelaus' theorem assumes the form

$$\frac{AM}{MB} \times \frac{BN}{NC} \times \frac{CP}{PA} = -1. \tag{3}$$

This is because only two cases are possible: either two of the points M, N, P lie on the sides of the triangle and the third on a side produced, or all three points M, N, P lie on the extensions of the sides of the triangle. In the first case two of the ratios AM/MB, BN/NC, CP/PA are positive and the third is negative; in the second case all three ratios are negative. Thus the product is always negative, its absolute value, as we have proved, being equal to 1.

It will be observed that the second of the proofs, given above, of formula (1) brings us immediately to formula (3) if we apply it to directed segments, since then

$$\frac{A'O'}{O'B'} \times \frac{B'O'}{O'C'} \times \frac{C'O'}{O'A'} = \frac{A'O'}{O'B'} \left(-\frac{O'B'}{O'C'} \right) \times \frac{O'C'}{A'O'} = -1.$$

We suggest to the reader the following easy exercises:

1. From the solution of problem 97 given at the beginning of this paragraph deduce a new proof of the theorem of Menelaus.

2. Prove the inverse of Menelaus' theorem:

If points M, N, P lie on straight lines AB, BC, CA and equation (3) *holds, then points M, N, P are collinear.*

3. Deduce from Menelaus' theorem the following theorem, stated in 1678 by an Italian mathematician Ceva:

If lines AS, BS, CS intersect, respectively, the sides BC, CA and AB of the triangle ABC or those sides produced at points N, P, M, point S lying on none of the lines AB, BC, CA, then

$$\frac{AM}{MB} \times \frac{BN}{NC} \times \frac{CP}{PA} = 1.$$

4. Formulate and prove the inverse of Ceva's theorem.

5. Investigate the limiting cases of Ceva's theorem if point S or one or two of the points M, N, P recede to infinity.

98. Suppose that in Fig. 91

$$\frac{BM}{MC} = \frac{CN}{NA} = \frac{AP}{PB} = k. \tag{1}$$

Denote the points of intersection of lines AM, BN and CP by X, Y, Z, as in Fig. 91.

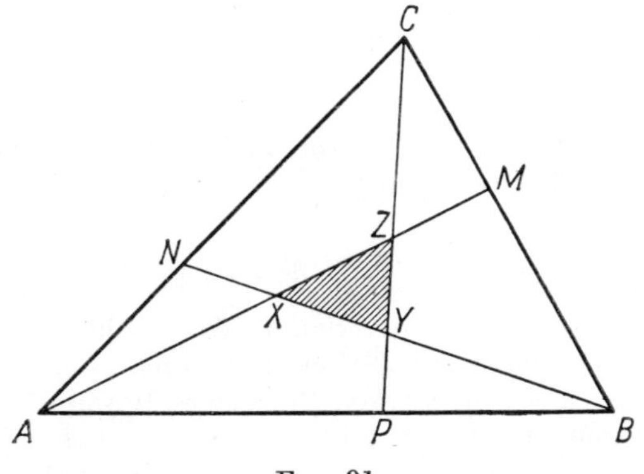

Fig. 91

Since the areas of triangles with equal altitudes are in the same ratio as their bases, we have

$$\text{area}\, ABM = \frac{BM}{BC} \times S, \qquad \text{area}\, ABX = \frac{AX}{AM} \times \text{area}\, ABM;$$

consequently

$$\text{area}\, ABX = \frac{BM}{BC} \times \frac{AX}{AM} \times S. \tag{2}$$

The ratios BM/BC and AX/AM are found by equation (1):

$$\frac{BC}{BM} = \frac{BM+MC}{BM} = 1 + \frac{1}{k} = \frac{k+1}{k},$$

whence

$$\frac{BM}{BC} = \frac{k}{k+1}. \tag{3}$$

The value of the ratio AX/AM can be obtained from the result of problem 97 provided we replace letter P in it by letter X (cf. Figs. 88 and 91); we obtain the equality

$$\frac{AX}{AM}=n\left(1+\frac{1}{m}\right),$$

where

$$n=\frac{AN}{NC}=\frac{1}{k}, \quad m=\frac{BM}{MC}=k,$$

whence

$$\frac{AX}{XM}=\frac{1}{k}\left(1+\frac{1}{k}\right)=\frac{k+1}{k^2}$$

and

$$\frac{AX}{AM}=\frac{AX}{AX+XM}=\frac{k+1}{k^2+k+1}. \tag{4}$$

Substituting the values of the ratios BM/BC and AX/AM from formulas (3) and (4) in formula (2), we obtain

$$\text{area } ABX=\frac{k}{k^2+k+1}S.$$

The value of the area of triangle BCY will be obtained by replacing in the above calculation of the area of ABX the letters A, B, C by B, C, A respectively, the letters M and N by N and P respectively, and the letter X by Y; the area of the triangle CAZ will be found in an analogous way. The results of the calculations will of course be the same: thus

$$\text{area } BCY=\text{area } CAZ=\text{area } ABX=\frac{k}{k^2+k+1}S.$$

Since

$$\text{area } XYZ=S-(\text{area } ABX+\text{area } BCY+\text{area } CAZ)$$

we have

$$\text{area } XYZ=S-\frac{3k}{k^2+k+1}\times S$$

and finally

$$\text{area } XYZ=\frac{(k-1)^2}{k^2+k+1}\times S. \tag{5}$$

REMARK 1. In the above reasoning formula (4) could also be obtained by applying Menelaus' theorem (see problem 97) to triangle ACM, intersected by the straight line BN, which gives

$$\frac{AN}{NC} \times \frac{CB}{BM} \times \frac{MX}{XA} = -1.$$

Writing this equation in the form

$$\frac{AN}{NC} \times \frac{CM+MB}{BM} \times \frac{MX}{XA} = -1,$$

i.e.

$$\frac{NA}{CN} \times \frac{MC+BM}{BM} \times \frac{MX}{XA} = 1,$$

and substituting in the last equality the values of the ratios

$$\frac{NA}{CN} = \frac{MC}{BM} = \frac{1}{k},$$

we obtain

$$\frac{1}{k}\left(\frac{1}{k}+1\right)\frac{MX}{XA} = 1,$$

and finally

$$\frac{AX}{XM} = \frac{k+1}{k^2},$$

which implies, as before, formula (4).

Remark 2. The condition that $k > 1$ can be replaced by a weaker one, namely that $k > 0$ and $k \neq 1$. The reasoning and result (5) remain the same. It will be observed that, if $k = 0$, points M, N, P coincide with points B, C, A respectively, and triangle XYZ coincides with triangle BCA. If $k = 1$, triangle XYZ is reduced to a single point—the centre of gravity of the triangle ABC. In both cases formula (5) gives correct values of the area: S and O.

99. Let us denote the required angles of triangle ABC by x and y and let the segment AD divide the triangle into two isosceles triangles, T_1 and T_2.

The segment AD cannot be the common base of the triangles T_1 and T_2, since then we should have $BD = BA$, $DC = AC$, whence $BC = BA+AC$, which is impossible.

Thus only the following two cases are possible.

I. The segment AD is, in both T_1 and T_2, one of the equal sides: consequently the vertex of each of these triangles can be either A or D.

Point A cannot be the common vertex of triangles T_1 and T_2, since the segments AB, AD, AC are not all equal (AD is shorter than at least one of the sides AB and AC). Thus two possibilities remain:

(a) Triangles T_1 and T_2 have D as their common vertex (Fig. 92). In this case point D is equidistant from the points A, B, C, i.e. it is the centre of the circumscribed circle of the triangle; angle A, inscribed in a semicircle, is a right angle. But we know that in every right-angled triangle the median drawn from the

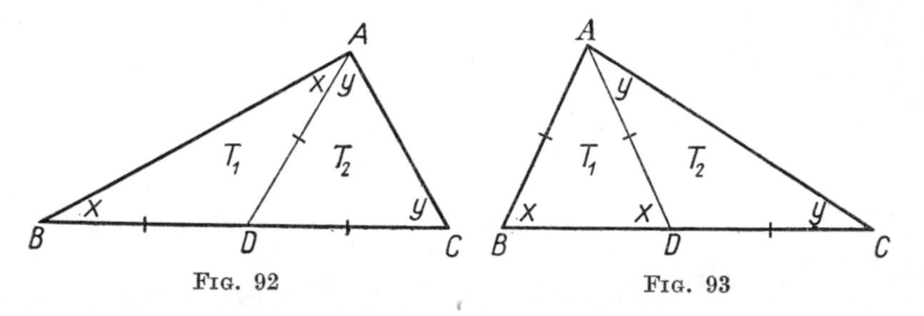

FIG. 92 FIG. 93

vertex of the right angle divides the triangle in two isosceles triangles. Consequently angles x and y can have arbitrary values, limited only by the condition: $x+y = 90°$.

(b) One of the isosceles triangles—say T_1—has the vertex A, and the other triangle, T_2, has the vertex D (Fig. 93).

In this case the angle at the base in T_1 is equal to the exterior angle of triangle T_2 at the vertex D, i.e. $x = 2y$.

From the above equation and from the equation $x+y = 180°-A$ we obtain

$$x = 120°-\tfrac{2}{3}A, \quad y = 60°-\tfrac{1}{3}A. \tag{1}$$

II. The segment AD is one of the equal sides in one triangle, say T_1, and the base in the other triangle, T_2.

In this case the vertex of triangle T_1 must be point D; for if it were point A, then point D would be an end-point of the base both in T_1 and in T_2, and each of the adjacent angles at vertex D would be an angle at the base of an isosceles triangle, i.e. both would be acute angles, which is impossible.

We thus have the figure represented in Fig. 94. The angle z at the base of triangle T_2, as an exterior angle of triangle T_1 at vertex D, equals $2x$; hence $\sphericalangle A = x+z = 3x$, and thus

$$x = \tfrac{1}{3}A, \quad y = 180°-\tfrac{4}{3}A. \tag{2}$$

It will be observed that the angle magnitudes defined by formulas (1) and (2) may be identical. This, as can easily be verified, takes place in two cases:

(a) if $\sphericalangle A = 90°$, and the remaining angles are $60°$ and $30°$, and

(b) if $\sphericalangle A = 120°$ and the remaining angles are 40° and 20°.

The answer to our question is thus as follows:

If $\sphericalangle A = 90°$, we can say nothing about the angles x and y except the fact that $x+y = 90°$.

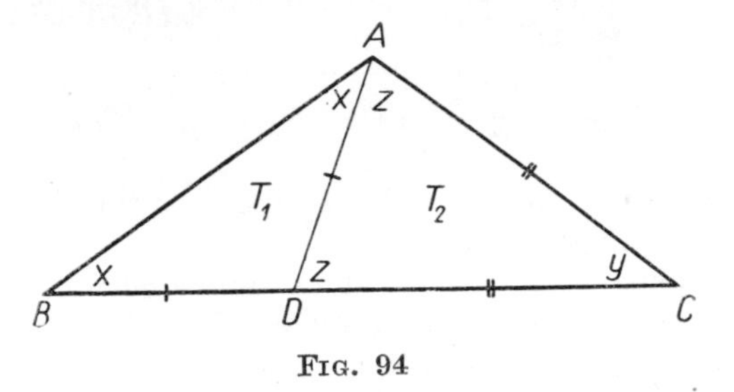

FIG. 94

If $\sphericalangle A \neq 90°$ and $\sphericalangle A \neq 120°$, then either one of the angles x, y equals twice the other angle—formulas (1), or one of them equals one third of angle A—formulas (2).

If $\sphericalangle A = 120°$, the remaining angles are 40° and 20°.

100. Let S, T, U (Fig. 95) denote the centres of the escribed circles of triangle ABC. These points are the vertices of the triangle formed by the bisectors of the exterior angles of triangle ABC.

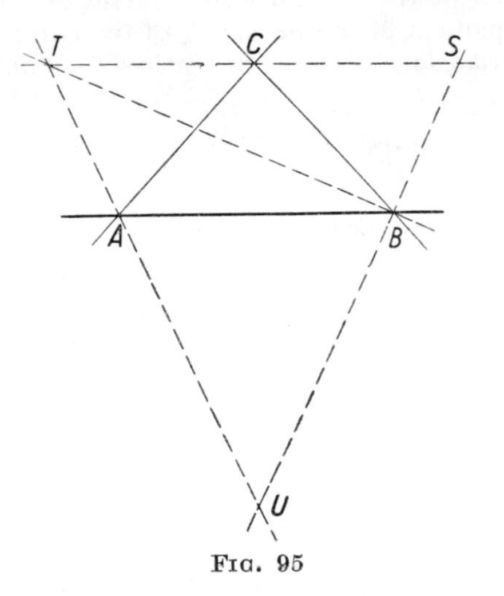

FIG. 95

In order to find the radius of the circumscribed circle of triangle STU we shall begin by determining the angles of this triangle.

Since

$$\measuredangle SBC = 90^\circ - \tfrac{1}{2}B, \qquad \measuredangle SCB = 90^\circ - \tfrac{1}{2}C,$$

we have

$$\measuredangle S = 180^\circ - (90^\circ - \tfrac{1}{2}B) - (90^\circ - \tfrac{1}{2}C) = \tfrac{1}{2}(B+C) = 90^\circ - \tfrac{1}{2}A.$$

Similarly

$$\measuredangle T = 90^\circ - \tfrac{1}{2}B, \qquad \measuredangle U = 90^\circ - \tfrac{1}{2}C.$$

We can see that triangle STU has angles of the same magnitude as those of triangles SBC, ABU and ATC; thus all these triangles are similar. The ratio of similitude equals the ratio of the corresponding sides, e.g. for triangles SBC and STU this ratio is SB/ST.

It will be observed that point T, being equally distant from the lines AB and BC, lies on the bisector BT of angle ABC, whence angle SBT, formed by the bisectors of adjacent angles, is a right angle.

The above-mentioned ratio of similitude is thus

$$\frac{SB}{ST} = \cos S = \cos(90^\circ - \tfrac{1}{2}A) = \sin\tfrac{1}{2}A.$$

The radius x of the circumscribed circle of triangle STU is equal to the quotient of the radius R_1 of the circle circumscribing the similar triangle SBC by the ratio of similitude of these triangles:

$$x = \frac{R_1}{\sin\frac{1}{2}A}.$$

Radius R_1 is determined by the side $BC = a$ and the opposite angle S of triangle SBC:

$$R_1 = \frac{a}{2\sin S} = \frac{a}{2\cos\frac{1}{2}A};$$

since $a = 2R\sin A$, where R is the radius of the circumcircle of triangle ABC, we have

$$R_1 = \frac{2R\sin A}{2\cos\frac{1}{2}A} = 2R\sin\tfrac{1}{2}A.$$

Substituting this value in the preceding expression for x, we obtain

$$x = 2R.$$

REMARK. This result can be obtained in an entirely different way by starting from the observation that points A, B, C are the feet of the altitudes of triangle STU (Fig. 95). The circle passing through these points is the so-called *nine-point circle*, or the *Feuerbach circle*†, for triangle STU. We now give the proof of this theorem—it contains also a solution of the above problem:

THEOREM ON THE FEUERBACH CIRCLE. *In every triangle the following nine points lie on the same circle*: (a) *the mid-points of the sides*, (b) *the feet of the altitudes*, (c) *the mid-points of those segments of the altitudes which join the orthocentre*‡ *of the triangle with its vertices. The centre of that circle is the mid-point of the line joining the orthocentre of the triangle with the centre of the circumcircle, and the radius is equal to half the radius of the circumcircle.*

Proof. We shall adopt the following notation (Fig. 96): K_1, K_2, K_3—feet of the altitudes, H—orthocentre, M_1, M_2, M_3—mid-points of the sides of triangle $A_1A_2A_3$; N_1, N_2, N_3—mid-points of segments A_1H, A_2H, A_3H.

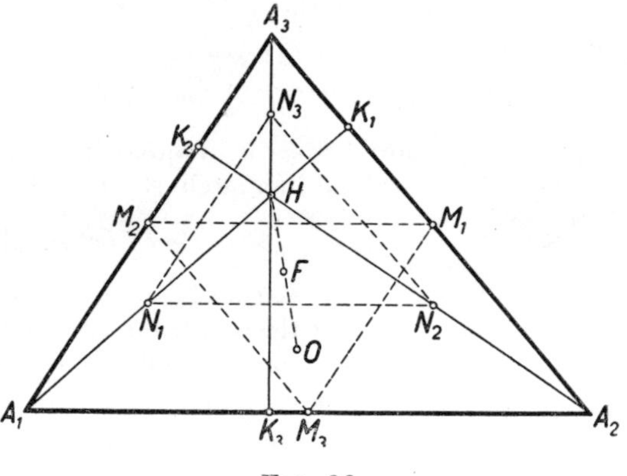

FIG. 96

The triangles $M_1M_2M_3$ and $N_1N_2N_3$ are symmetric with respect to a point F. Indeed, $M_2M_1 \parallel A_1A_2 \parallel N_1N_2$ and $M_2N_1 \parallel A_3H \parallel M_1N_2$; thus segments M_1N_1 and M_2N_2 bisect each other; the same holds for the pairs M_2N_2, M_3N_3 and M_3N_3, M_1N_1. Marking off $FO = HF$ on the line HF, we obtain point O symmetric with respect to point F to the orthocentre H of triangle $A_1A_2A_3$. But point H is

† Charles Feuerbach (1800–1834) a schoolmaster in Erlangen (Bavaria).

‡ The *orthocentre* of a triangle is the point of intersection of its altitudes.

also the orthocentre of triangle $N_1N_2N_3$, whence the symmetric point O is the orthocentre of the symetric triangle $M_1M_2M_3$.

Since the altitudes of triangle $M_1M_2M_3$ are the perpendicular bisectors of the sides of triangle $A_1A_2A_3$, point O is the centre of the circumcircle of triangle $A_1A_2A_3$.

The triangles $N_1N_2N_3$ and $A_1A_2A_3$ are homothetic in the ratio $1:2$ with respect to point H. In this homothety point F corresponds to point O because $HF = \frac{1}{2}HO$. And, since point O is the centre of the circumcircle of triangle $A_1A_2A_3$, point F is the centre of the circumcircle of triangle $N_1N_2N_3$, and thus, as the centre of symmetry of triangles $M_1M_2M_3$ and $N_1N_2N_3$, also the centre of the circumcircle of triangle $M_1M_2M_3$.

We have thus proved that points $M_1, M_2, M_3, N_1, N_2, N_3$ lie on a circle with centre F. The radius of this circle is equal to half the radius of the circumscribed circle of triangle $A_1A_2A_3$, since the ratio of homothety for the two circles is $\frac{1}{2}$.

Points K_1, K_2, K_3 lie on the same circle because from those points the diameters M_1N_1, M_2N_2, M_3N_3 appear at right angles.

101. Denoting the sides and the angles of the quadrilateral as in Fig. 97, we have

$$S = \text{area } ABD + \text{area } BCD = \tfrac{1}{2}(ad \sin A + bc \sin C), \tag{1}$$

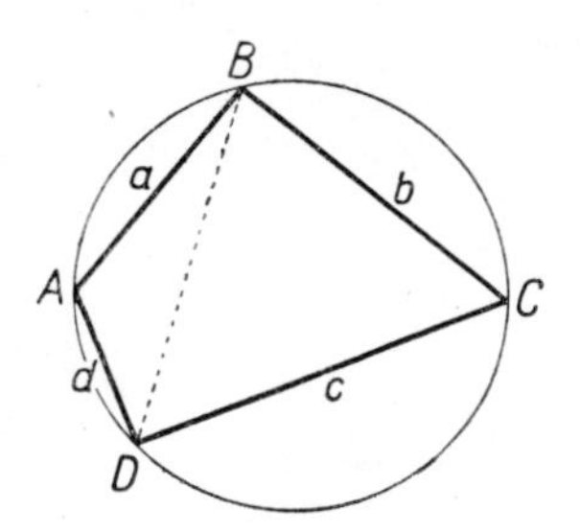

FIG. 97

and, since in a quadrilateral $ABCD$ inscribed in a circle $A+C = 180°$, we have

$$2S = (ad+bc) \sin A. \tag{2}$$

The relation between angle A and the sides of the quadrilateral will be obtained by finding the length BD from triangle ABD and from triangle BCD:

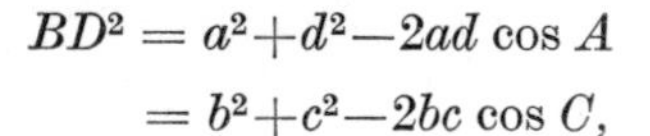

$$BD^2 = a^2+d^2-2ad \cos A$$
$$= b^2+c^2-2bc \cos C,$$

whence

$$a^2+d^2-b^2-c^2 = 2ad \cos A - 2bc \cos C \tag{3}$$

and, since $\cos C = -\cos A$, we have

$$a^2+d^2-b^2-c^2 = 2(ad+bc) \cos A. \tag{4}$$

We can eliminate angle A from relations (2) and (4). Accordingly, we multiply formula (2) by 2, square equalities (2) and (4) and add them, obtaining

$$16S^2+(a^2+d^2-b^2-c^2)^2 = 4(ad+bc)^2.$$

Hence

$$\begin{aligned}16S^2 &= 4(ad+bc)^2-(a^2+d^2-b^2-c^2)^2\\ &= [2(ad+bc)+(a^2+d^2-b^2-c^2)]\,[2(ad+bc)-(a^2+d^2-b^2-c^2)]\\ &= [(a+d)^2-(b-c)^2]\,[(b+c)^2-(a-d)^2]\end{aligned}$$

and finally

$$16S^2 = (a+d-b+c)(a+d+b-c)(b+c-a+d)(b+c+a-d). \quad (5)$$

We introduce the notation

$$a+b+c+d = 2p$$

and transform the factors appearing on the right-hand side of equality (5):

$$a+d-b+c = a+b+c+d-2b = 2(p-b).$$

Similarly

$$\begin{aligned}a+d+b-c &= 2(p-c),\\ b+c-a+d &= 2(p-a),\\ b+c+a-d &= 2(p-d).\end{aligned}$$

Substituting these expressions in equality (5) we obtain

$$16S^2 = 2^4(p-a)(p-b)(p-c)(p-d),$$

whence

$$S = \sqrt{[(p-a)(p-b)(p-c)(p-d)]}. \quad (6)$$

Remark 1. Formula (6) for the area of a quadrilateral inscribed in a circle is a particular case of a formula relating to the area of any (convex or concave) quadrilateral.

As can easily be seen, formulas (1) and (3) are applicable to every quadrilateral. They imply that

$$16S^2 = 4a^2d^2\sin^2 A+4b^2c^2\sin^2 C+8abcd\sin A\sin C,$$

$$(a^2+d^2-b^2-c^2)^2 = 4a^2d^2\cos^2 A+4b^2c^2\cos^2 C-8abcd\cos A\cos C.$$

Adding these equalities,

$$16S^2+(a^2+d^2-b^2-c^2)^2 = 4a^2d^2+4b^2c^2-8abcd\cos(A+C),$$

and substituting $\cos(A+C) = 2\cos^2\frac{1}{2}(A+C)-1$, we obtain

$$16S^2 = 4(ad+bc)^2-(a^2+d^2-b^2-c^2)^2-16abcd\cos^2\frac{A+C}{2}.$$

Accordingly, we shall first find the conditions which must be satisfied by numbers a, b, c, d if they are to be the lengths of the sides of a quadrilateral. Suppose that in a quadrilateral $ABCD$ sides AB, BC, CD and DA have the lengths a, b, c, d respectively. The diagonal BD is shorter than either of the sums $AB+DA$ and $BC+CD$ but greater than either of the differences $|AB-DA|$, $|BC-CD|$. Consequently

$$a+d > |b-c|, \quad b+c > |a-d|. \tag{8}$$

Conversely, if the numbers a, b, c, d satisfy inequalities (8), then those numbers are the lengths of the sides of a quadrilateral. Indeed, we can then choose a number e in such a way that

$$a+d > e > |a-d|, \quad b+c > e > |b-c|. \tag{9}$$

Constructing a triangle with sides a, d, e and a triangle with sides b, c, e, adjacent to the first one along the side e, we obtain the required quadrilateral with sides a, b, c, d.

The angles A and C of that quadrilateral, formed by the pairs of sides (a, d) and (b, c), are defined by the formulas

$$e^2 = a^2+d^2-2ad\cos A, \quad e^2 = b^2+c^2-2bc\cos C. \tag{10}$$

If the quadrilateral is inscribed in a circle, we have also

$$\cos C = -\cos A. \tag{11}$$

The proof of theorem (b) formulated above will be obtained by showing that under assumption (8) the system of equations (10) and (11) has a solution (e, A, C), with e satisfying conditions (9).

From (10) and (11) we obtain

$$\cos A = -\cos C = \frac{a^2+d^2-b^2-c^2}{2ad+2bc}. \tag{12}$$

Equalities (12) define the required angles A and C since the value of the right-hand side is contained in the interval $(-1, 1)$, which can be ascertained by (8):

$$1+\frac{a^2+d^2-b^2-c^2}{2ad+2bc} = \frac{(a+d)^2-(b-c)^2}{2ad+2bc} > 0$$

and

$$1-\frac{a^2+d^2-b^2-c^2}{2ad+2bc} = \frac{(b+c)^2-(a-d)^2}{2ad+2bc} > 0.$$

The value of e can be obtained from either of the formulas (10) by substituting for $\cos A$ or for $\cos C$ the value defined by formula (12).

The above theorem (b) is a particular case of a theorem of Cramer (Swiss mathematician, 1704–1752):

Among the polygons with given sides $a_1, a_2, \ldots, a_n$ (where $n \geqslant 3$) the polygon inscribed in a circle has the greatest area.

102. If a quadrilateral with sides a, b, c, d is inscribed in a circle, then the area S is expressed (see problem 101) by the formula

$$S = \sqrt{[(p-a)(p-b)(p-c)(p-d)]}, \tag{1}$$

in which

$$a+b+c+d = 2p.$$

If this quadrilateral is also circumscribed in a circle, then

$$a+c = b+d = p. \tag{2}$$

Equality (2) implies that in a quadrilateral of this kind

$$p-a = c, \quad p-b = d, \quad p-c = a, \quad p-d = b.$$

Substituting these values in formula (1) we obtain

$$S = \sqrt{(abcd)}. \tag{3}$$

REMARK. Let us ask whether the inverse theorem holds, i.e. whether the assumption that the area of a quadrilateral is expressed by formula (3) implies that the quadrilateral in question has a circumcircle and an inscribed circle. As can easily be ascertained, that is not so; e.g. if the quadrilateral is a rectangle with sides a, b, $c = a$, $d = b$, with $a \neq b$, then $S = ab = \sqrt{[(ab)^2]} = \sqrt{(abcd)}$, this quadrilateral has no inscribed circle although it has a circumscribed one.

This example can be modified so as to obtain a quadrilateral which would satisfy condition (3) without being either circumscribed on a circle or inscribed in one. Accordingly, instead of a rectangle let us take such a quadrilateral $ABCD$ inscribed in a circle that $AB = a$, $BC = b$, $CD = c$, $DA = d$, and $a = b$, $c \neq d$, B and D being right angles (Fig. 99).

The area of the quadrilateral $ABCD$ is expressed by the formula

$$S = \text{area } ABC + \text{area } ADC = \frac{ab+cd}{2}.$$

The right-angled isosceles triangle ABC has the same base as the right-angled triangle ADC but a greater altitude, whence area $ABC >$ area ADC, i.e. $ab > cd$; since the arithmetical mean of the unequal numbers ab and cd is greater than its geometrical mean, we have

$$S > \sqrt{(abcd)}.$$

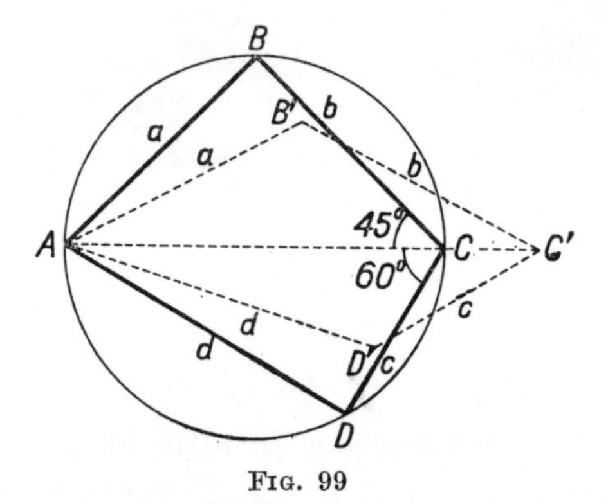

FIG. 99

If we increase the diagonal AC, leaving the lengths of the sides a, b, c, d unchanged, we obtain a quadrilateral $AB'C'D'$ (Fig. 99) with area S' smaller than S. It is essential that we should choose point C' in such a way that $S' = \sqrt{(abcd)}$. Accordingly we must determine the obtuse angles B' and D' so as to satisfy two conditions:

$$a^2+b^2-2ab\cos B' = c^2+a^2-2cd\cos D',$$

$$\tfrac{1}{2}ab\sin B'+\tfrac{1}{2}cd\sin D' = \sqrt{(abcd)}.$$

It is not necessary to carry out a general discussion of these equations; for our purpose it is sufficient to give the solution in some particular case. Suppose that the quadrilateral $ABCD$ is inscribed in a circle with radius 1 and that CD is a side of a regular hexagon inscribed in that circle, whence $a = b = \sqrt{2}$, $c = 1$, $d = \sqrt{3}$. Substituting these values in the above equations we obtain

$$2\cos B' = \sqrt{3}\cos D',$$

$$2\sin B'+\sqrt{3}\sin D' = 2\sqrt{(2\sqrt{3})}.$$

Squaring the first equation and then expressing the cosines of the angles in terms of their sines, we obtain the equation

$$4\sin^2 B'-3\sin^2 D' = 1.$$

Dividing this equation by the second equation of the preceding two, we have

$$2 \sin B' - \sqrt{3} \sin D' = \frac{1}{2\sqrt{(2\sqrt{3})}}.$$

Thus

$$\sin B' = \frac{1}{4}\left[2\sqrt{(2\sqrt{3})} + \frac{1}{2\sqrt{(2\sqrt{3})}}\right] = \frac{8\sqrt{3}+1}{8\sqrt{(2\sqrt{3})}},$$

$$\sin D' = \frac{1}{2\sqrt{3}}\left[2\sqrt{(2\sqrt{3})} - \frac{1}{2\sqrt{(2\sqrt{3})}}\right] = \frac{8\sqrt{3}-1}{4\sqrt{(6\sqrt{3})}}.$$

It can easily be verified that the values obtained are less than 1, i.e. they determine obtuse angles B' and D'.

The quadrilateral $AB'C'D'$ formed in this way has an area $S' = \sqrt{(abcd)} = \sqrt{(2\sqrt{3})}$, but is neither inscribed in a circle (since two opposite angles in it are obtuse) nor circumscribed about a circle (since the sums of its opposite sides are not equal).

Let us also observe the validity of a theorem *partially inverse* to the theorem from problem 102:

If the area S of a quadrilateral with sides a, b, c, d is expressed by the formula $S = \sqrt{(abcd)}$ and the quadrilateral has an inscribed circle, then it has also a circumscribed circle.

Indeed, the area S of the quadrilateral $ABCD$ with sides a, b, c, d is given by formula (7) of problem 101, namely

$$S^2 = (p-a)(p-b)(p-c)(p-d) - abcd \cos^2\frac{A+C}{2}. \quad (4)$$

From the assumption that the quadrilateral has a circumscribed circle follows, as we have found before, the relation

$$(p-a)(p-b)(p-c)(p-d) = abcd. \quad (5)$$

Formulas (3), (4) and (5) give

$$S^2 = S^2 - S^2 \cos^2\frac{A+C}{2},$$

whence

$$\cos\frac{A+C}{2} = 0,$$

and consequently

$$A+C = 180°,$$

which proves that the quadrilateral $ABCD$ has a circumcircle.

103. *Method I.* Let r denote the radius of a circle circumscribed about a regular polygon $ABCD\ldots$, and $2x$ the convex angle at the centre of that circle corresponding to the chord AB. Then (Fig. 100)

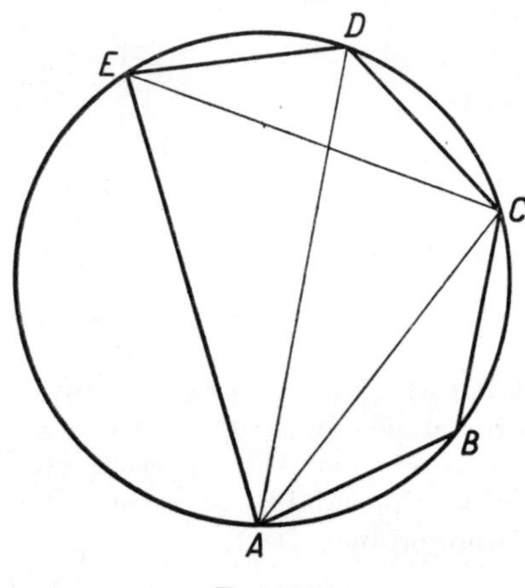

Fig. 100

$$AB = 2r \sin x, \qquad AC = 2r \sin 2x, \qquad AD = 2r \sin 3x.$$

Substituting these expressions in the equation

$$\frac{1}{AB} = \frac{1}{AC} + \frac{1}{AD} \tag{1}$$

we obtain the equation

$$\frac{1}{\sin x} = \frac{1}{\sin 2x} + \frac{1}{\sin 3x}. \tag{2}$$

In order to solve equation (2) we multiply it by

$$\sin x \times \sin 2x \times \sin 3x$$

and carry all the terms over to one side:

$$\sin 2x \times \sin 3x - \sin x \times \sin 3x - \sin x \times \sin 2x = 0.$$

The left-hand side of this equation can be factorized. We obtain successively

$$\sin 2x \times (\sin 3x - \sin x) - \sin x \sin 3x = 0,$$
$$\sin 2x \times 2 \sin x \cos 2x - \sin x \sin 3x = 0,$$
$$\sin x \times \sin 4x - \sin x \times \sin 3x = 0,$$
$$\sin x \times (\sin 4x - \sin 3x) = 0,$$

and finally

$$\sin x \times \sin \frac{x}{2} \times \cos \frac{7x}{2} = 0. \tag{3}$$

Equation (3) is not equivalent to equation (2). Namely, each solution x of equation (2) is a solution of equation (3) but not *vice versa* because only those solutions of equation (3) satisfy (2) for which $\sin x \times \sin 2x \times \sin 3x \neq 0$, i.e. which satisfy the condition

$$x \neq \frac{k\pi}{2} \quad \text{and} \quad x \neq \frac{k\pi}{3} \tag{4}$$

for any integer k.

Consequently, the roots of equations (2) are numbers satisfying the equation

$$\cos \frac{7x}{2} = 0$$

and condition (4), whence

$$x = \frac{\pi}{7} + \frac{2n\pi}{7}, \tag{5}$$

with n assuming integral values for which value (5) satisfies condition (4).†

The required polygon must have at least 4 vertices, whence $0 < x \leqslant \frac{1}{4}\pi$; in formula (5) we must then take $n = 0$ and the problem has a unique solution

$$x = \frac{\pi}{7},$$

which means that the regular polygon $ABCD\ldots$ is a heptagon.

Method II. To begin with, it will be observed that A, B, C, D cannot constitute all the vertices of the required polygon, since, if they did, the polygon would be a square, and for a square $ABCD\ldots$ equation (1) does not hold.

Let E denote the vertex of a polygon which follows D in the succession $ABCD\ldots$; according to the above remark, vertex E does not coincide with vertex A.

Equation (1) implies that

$$AC \times AD = AB \times AD + AB \times AC.$$

† This condition signifies that the integer n in formula (5) is not arbitrary but must satisfy the inequalities $n \neq (7k-2)/4$ and $n \neq (7k-3)/6$, where k is an arbitrary integer.

Since $AC = CE$, $AB = CD = DE$, this equality gives

$$CE \times AD = CD \times AD + DE \times AC. \qquad (\alpha)$$

On the other hand, by Ptolemy's theorem, with regard to the inscribed quadrilateral $ACDE$ we have:

$$CE \times AD = CD \times AE + DE \times AC. \qquad (\beta)$$

From equalities (α) and (β) we obtain

$$AE = AD.$$

This equality implies that points D and E lie symmetrically with respect to the diameter passing through point A; thus polygon $ABCD$ is a regular heptagon.

REMARK 1. In method 2 we assumed, in accordance with the wording of the problem, that for a certain regular polygon $ABCD$... equality (1) holds and we proved that such a polygon is a heptagon. In other words, the reasoning of method 2 is a proof of the following theorem: if there exists a regular polygon $ABCD$... with property (1), then that polygon is a heptagon. It does not follow from this theorem that a regular heptagon indeed has the property expressed by formula (1), just as from the sentence "if yesterday was Thursday, today is Friday" it does not follow that today is Friday. However, we can easily supplement our preceding argument and prove that for a regular heptagon $ABCDE$... equation (1) holds.

Namely, applying Ptolemy's theorem to the inscribed quadrilateral $ACDE$, formed by two adjacent sides and two diagonals of a regular heptagon, we have

$$CE \times AD = CD \times AE + DE \times AC.$$

Since $CE = AC$, $AE = AD$, $CD = DE = AB$, this equality gives

$$AC \times AD = AB \times AD + AB \times AC,$$

whence, dividing by $AC \times AD$, we obtain

$$\frac{1}{AB} = \frac{1}{AC} + \frac{1}{AD},$$

which is what was to be proved.

It will be observed that in method I this supplement is not necessary since according to the calculation performed the value $x = \frac{1}{7}\pi$ is the root of equation (2), and this means exactly that the regular heptagon $ABCDE$... has property (1).

REMARK 2. In method II we resorted to an important *theorem of Ptolemy*† on a quadrilateral inscribed in a circle; the theorem reads as follows:

The product of the diagonals of a quadrilateral inscribed in a circle is equal to the sum of the products of its opposite sides.

This theorem can be proved in different ways. We shall give here a short proof consisting in the calculation of the diagonals of an inscribed quadrilateral with given sides on the basis of the Cosine Rule.

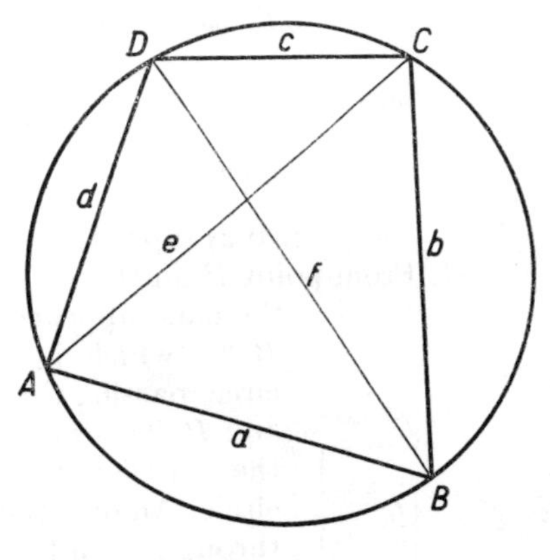

FIG. 101

Adopting the notation of Fig. 101 we obtain from triangles ABC and ADC

$$e^2 = a^2+b^2-2\,ab\cos\sphericalangle ABC,$$
$$e^2 = c^2+d^2-2\,cd\cos\sphericalangle ADC.$$

Since

$\sphericalangle ADC = 180°-\sphericalangle ABC$ we have $\cos\sphericalangle ADC = -\cos\sphericalangle ABC$;

multiplying the first of the preceding equalities by cd and the second by ab, and then adding them, we obtain

† Ptolemy of Alexandria (126–168 A. D.), mathematician and astronomer, author of a famous work, the *Almagest,* comprising an exposition of plane and spherical trigonometry. By means of his theorem on the quadrilateral Ptolemy devised the first tables of sines (the term "sine" appeared later). It is owing to the *Almagest* that the theorem of Menelaus has been handed down to us.

$$(ab+cd)\,e^2 = (a^2+b^2)\,cd+(c^2+d^2)\,ab$$
$$= (a^2cd+abc^2)+(b^2cd+abd^2)$$
$$= ac\,(ad+bc)+bd\,(bc+ad)$$
$$= (ad+bc)\,(ac+bd);$$
$$e^2 = \frac{(ac+bd)\,(ad+bc)}{ab+cd}.$$

Analogously

$$f^2 = \frac{(ab+cd)\,(ac+bd)}{ad+bc},$$

whence

$$e^2f^2 = (ac+bd)^2$$

and ultimately

$$ef = ab+bd.$$

104. Let the segment AB (Fig. 102) represent the tower and the segment BC—the mast. From point P on the surface of the earth the mast appears under the angle BPC, which is inscribed in the circle passing through points B, C and P. This angle is the greater the smaller the radius of the circle. Among the circles passing through points B and C and having common points with the earth, the circle $O(r)$ tangent to the surface of the earth has the least radius. The point of contact M of this circle with the earth is the point mentioned in the problem, i.e. $\sphericalangle BMC = \alpha$, $MA = a$. If $OD \perp BC$, then $\sphericalangle COD = \alpha$, being equal to half the angle at the centre subtended by the same arc as the inscribed angle BMC. Introducing the notation $AB = x$, $BC = y$, we obtain from the triangle COD and the rectangle $AMOD$ the relations

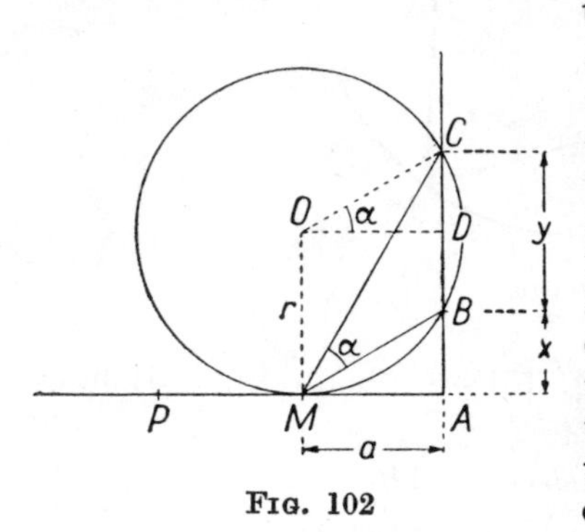

FIG. 102

$$y = 2a \tan\alpha, \tag{1}$$

$$r = \frac{a}{\cos\alpha}, \tag{2}$$

$$x = r-\frac{y}{2} = \frac{a}{\cos\alpha}-a\tan\alpha = a\times\frac{1-\sin\alpha}{\cos\alpha}, \tag{3}$$

containing the solution of the problem.

REMARK. In the above solution we have assumed that in the neighbourhood of the tower the surface of the earth is plane, which is justifiable in view of the fact that only a very small portion of that surface is involved. Taking into account the spherical shape of the earth we can find the unknowns x and y in a similar manner, using the figure in Fig. 103, in which S denotes the centre of the earth. From triangle COD, in which the segment OD will be denoted by b, we shall find

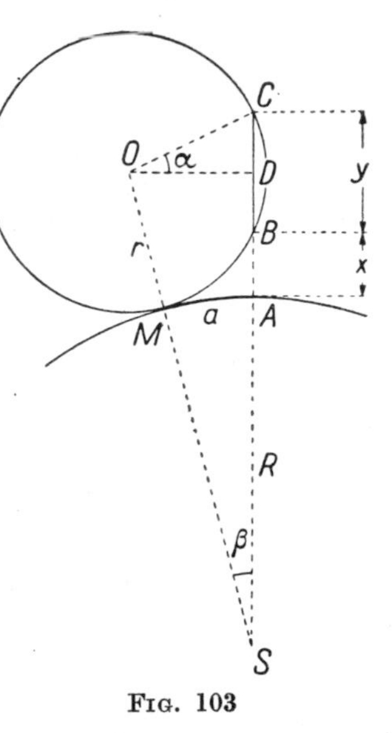

FIG. 103

$$y = 2r \sin \alpha, \quad (4)$$

$$b = r \cos \alpha, \quad (5)$$

and from the triangle DOS we have

$$b = (R+r) \sin \beta, \quad (6)$$

$$R+x+\frac{y}{2} = (R+r) \cos \beta, \quad (7)$$

where β is the angle under which the arc a appears from the centre of the earth.

Formulas (5) and (6) give

$$r = \frac{R \sin \beta}{\cos \alpha - \sin \beta}.$$

Substituting this value of r in formula (4), we obtain

$$y = \frac{2R \sin \alpha \sin \beta}{\cos \alpha - \sin \beta}, \quad (8)$$

and then find x from formula (7), obtaining after easy transformations the result

$$x = R \frac{\sin \beta (1-\sin \alpha) - \cos \alpha (1-\cos \beta)}{\cos \alpha - \sin \beta}. \quad (9)$$

The angle β is, in the circle with radius R, an angle at the centre subtended by an arc of length a; the circular measure of angle β is thus a/R and formulas (8) and (9) can be written as

$$y = \frac{2R \sin \alpha \sin \dfrac{a}{R}}{\cos \alpha - \sin \dfrac{a}{R}}, \quad (10)$$

$$x = R\frac{\sin\frac{a}{R}(1-\sin\alpha)-\cos\alpha\left(1-\cos\frac{a}{R}\right)}{\cos\alpha-\sin\frac{a}{R}}. \tag{11}$$

From these rather complicated formulas we can find x and y for given a, α and R. But instead of these formulas we could use approximate ones, which are obtained in the following way:

It is proved in trigonometry that for an angle with circular measure β we have the approximate equalities†

$$\sin\beta \approx \beta, \qquad \cos\beta \approx 1-\frac{\beta^2}{2};$$

for $\sin\beta$ the value β is too large, the error being less than $\beta^3/6$; for $\cos\beta$ the value $1-\beta^2/2$ is too small, the error being less than $\beta^4/24$.

For example if the distance a in our problem does not exceed 100 km, then $\measuredangle\beta$ (under the assumption that $R = 6\,370$ km) is less than $1°$, whence, taking the ratio a/R instead of $\sin\beta$, we introduce an error of less than

$$\frac{\pi^3}{6\times 180^3},$$

i.e. less than 0·000001‡ and, taking number $1-\frac{1}{2}(a/R)^2$ instead of $\cos\beta$ we introduce an error of less than

$$\frac{\pi^4}{24\times 180^4},$$

i.e. less than 0·00000001. Substituting these approximate values of $\sin(a/R)$ and $\cos(a/R)$ in formulas (10) and (11) we obtain the formulas

$$y = \frac{2a\sin\alpha}{\cos\alpha-a/R}, \tag{10a}$$

† The sign $\approx$ means "equals approximately".

‡ In order make sure of this fact it is not at all necessary to calculate the fraction $\pi^3/(6\times 180^3)$: it suffices to reason as follows:

$$\frac{1}{6}\times\frac{\pi^3}{180^3} < \frac{1}{6}\left(\frac{3\times 15}{180}\right)^3 = \frac{1}{6\times 10^6}\left(\frac{315}{180}\right)^3 = \frac{1}{6\times 10^6}\left(\frac{7}{4}\right)^3 = \frac{1}{10^6}\times\frac{343}{6\times 64} < \frac{1}{10^6};$$

we find in a similar way that

$$\frac{\pi^4}{24\times 180^4} < \frac{1}{10^8}.$$

$$x = \frac{a(1-\sin\alpha)-\dfrac{a^2\cos\alpha}{2R}}{\cos\alpha-a/R}. \tag{11a}$$

We observe that, if R increases indefinitely, or "tends to infinity", whereas a and α are constant, then the variables x and y, defined by formulas (10a) and (11a), tend to the values defined by formulas (1) and (3).

The question arises whether formulas (10a) and (11a) have any practical significance, i.e. whether, for certain values of a and α which might occur in reality, the accuracy of formulas (1) and (3) might prove insufficient, making it necessary to use formulas (10a) and (11a) or even the still more accurate formulas (10) and (11). In order to answer this question we should have to estimate the differences between the values obtained from all these formulas, taking also into account the errors of measurement which necessarily affect the given magnitudes. Calculations of this kind are often performed in various branches of applied mathematics, e.g. in geodesy.

105. In Fig. 104, representing a portion of the cross-section of a ball-bearing, S denotes the centre of one of the small circles,

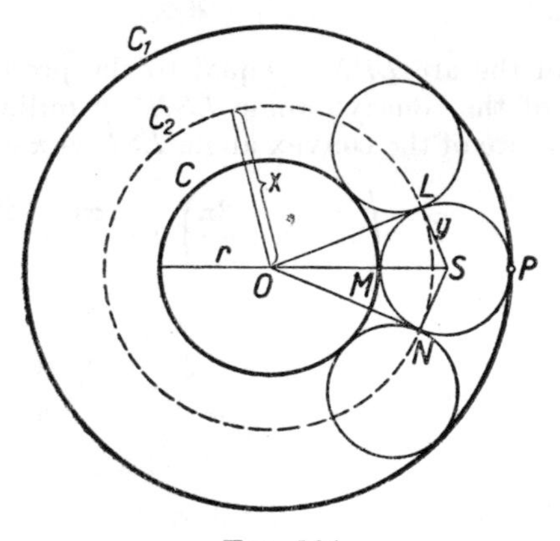

FIG. 104

L and N denote the points of contact of this circle with the adjacent two small circles, and M and P denote its points of contact with the circles C and C_1 with the common centre O. According to the conditions of the problem, $\sphericalangle LON = 2\pi/n$. In triangle LOS the following relations hold between the sides $OL = x$, $LS = y$, the hypotenuse $OS = r+y$ and $\sphericalangle LOS = \pi/n$

$$y = x\tan\frac{\pi}{n}, \tag{1}$$

$$x^2+y^2 = (r+y)^2. \tag{2}$$

Simplifying equation (2) and substituting in it value (1) for y, we obtain the equation

$$x^2-2r\tan\frac{\pi}{n}\times x-r^2 = 0.$$

This equation has two roots of different signs. The required length of the radius x of circle C_2 is the positive root, i.e.

$$x = r\tan\frac{\pi}{n}+\sqrt{\left(r^2\tan^2\frac{\pi}{n}+r^2\right)} = r\left(\tan\frac{\pi}{n}+\frac{1}{\cos\frac{\pi}{n}}\right)$$

or

$$x = r\times\frac{1+\sin\frac{\pi}{n}}{\cos\frac{\pi}{n}}. \tag{3}$$

The length of the arc LPN is equal to the product of the circular measure of the concave angle LSN by radius y and, since the circular measure of the convex angle LSN is $\pi-2\pi/n$, we have

$$\text{length of arc } LPN = \left(2\pi-\pi+\frac{2\pi}{n}\right)y = \frac{\pi(n+2)}{n}y$$

$$= \frac{n+2}{n}\pi x\tan\frac{\pi}{n} = \frac{n+2}{n}\pi r\tan\frac{\pi}{n}\times\frac{1+\sin\frac{\pi}{n}}{\cos\frac{\pi}{n}}.$$

The required sum s of the lengths of those arcs of the small circles which lie outside circle C_2 is n times as great as the length of the arc LPN; consequently

$$s = \pi r(n+2)\tan\frac{\pi}{n}\times\frac{1+\sin\frac{\pi}{n}}{\cos\frac{\pi}{n}} \tag{4}$$

or, more simply,

$$s = \pi r \frac{(n+2)\sin\dfrac{\pi}{n}}{1-\sin\dfrac{\pi}{n}}. \tag{5}$$

Formulas (3) and (5) give the solution of the problem.

REMARK. Number s given by formula (4) is the length of the curve formed by all such arcs of the small circles as the arc LPN. It is a closed curve surrounding point O and lying in the ring between circles C and C_1. The distance of each point of that curve from the corresponding point of circle C (i.e. from the point lying on the same half-line starting from point O) is at most equal to the difference between the radii of circles C_1 and C. Suppose that the difference is very small, i.e. that the ring between C and C_1 is very narrow. Then the points of that closed curve lie very close to the corresponding points of circle C and it might be supposed that the length s differes very little from the length of circle C, i.e. from $2\pi r$. However, we find that that is not so and that the length s is always more than $\frac{3}{2}$ of the length of circle C, even for a ring as narrow as we wish. In order to ascertain this fact let us observe that in formula (4) we have

$$\frac{1+\sin(\pi/n)}{\cos(\pi/n)} > 1$$

and that the tangent of an acute angle is greater than the circular measure of that angle, whence $\tan(\pi/n) > \pi/n$; thus formula (4) implies the inequality

$$s > \pi r(n+2)\times\frac{\pi}{n}.$$

Considering that $(n+2)/n > 1$, we obtain the inequality

$$s > \pi^2 r \quad \text{or} \quad \frac{s}{2\pi r} > \frac{\pi}{2} \approx 1{\cdot}57\ldots$$

106. In solving geometrical problems a correctly executed drawing is an important aid. Figures in space are shown by means of mappings or projections upon the plane of the drawing. There are various methods of such a mapping. In elementary geometry we usually draw *oblique projections* of figures; in many cases it is convenient to use the *method of orthogonal projections* on two perpendicular planes, i.e. the so called *Monge method.*

We shall present the solution of our problem in two variants, using first one and then the other of the above-mentioned methods of representation.

Method I. Using the method of oblique projection we shall adopt as the plane of projection the plane passing through the beam AB and the suspension points M and N. We draw the quadrilateral $ABNM$ "life size" (Figs. 105 and 106). Let S denote the centre of the beam. When twisted, the beam will assume the position CD. The mid-point of the segment CD lies on the plane of projection; suppose that it is point T.

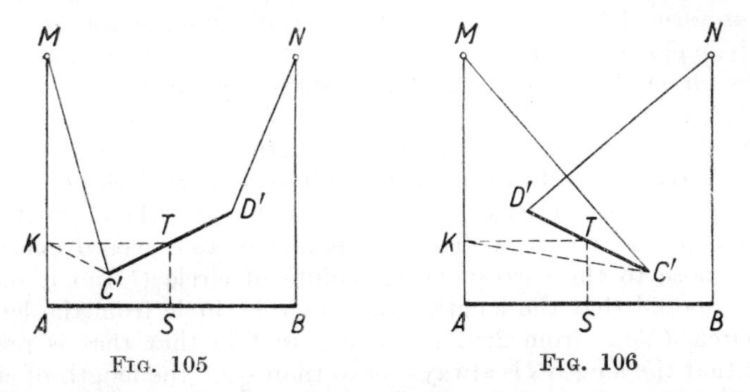

FIG. 105 FIG. 106

The position of the projection of point C depends on the direction of projecting; we can regard any point C' as an oblique projection of point C, for example as in Figs. 105 or 106. The projection D' of point D will be a point symmetric to point C' with respect to point T.

The finding of the required length $ST = x$ is simple. We draw a segment TK, parallel and equal to SA; then

$$x = AK = AM - KM.$$

Now $AM = b$, while the segment KM is a side of the right-angled triangle KMC with the hypotenuse $MC = b$ and the other side KC. The segment KC is the base of the isosceles triangle KTC, in which $TK = TC = \frac{1}{2}AB = \frac{1}{2}a$, $\sphericalangle KTC = \varphi$.

Consequently

$$KC = a \sin\frac{\varphi}{2}, \qquad KM = \sqrt{\left(b^2 - a^2 \sin^2\frac{\varphi}{2}\right)}.$$

Finally

$$x = b - \sqrt{\left(b^2 - a^2 \sin^2\frac{\varphi}{2}\right)}.$$

If $b < a$, the torsion angle φ cannot be greater than the angle φ_0 defined by the formula

$$\sin\frac{\varphi_0}{2} = \frac{b}{a} \qquad \text{where} \qquad \varphi_0 < 180°.$$

For the value $\varphi = \varphi_0$ we have $x = b$. A further enlargement of the angle is not possible without stretching out the ropes. If $b \geqslant a$, the greatest value of φ is 180°. For $\varphi = 180°$ the ropes cross each other if $b > a$ and coincide if $b = a$.

In the above solution we were concerned with finding the elevation of the beam when twisted. The drawing of the figure in the parallel projection was only a relatively simple illustration necessary for the calculation. If we want the drawing to constitute the graphic solution of the problem, i.e. to give the correct length of the segment ST for given lengths a, b and a given angle φ, we must execute it in a different way. Namely, point T, which in Figs. 105 and 106 was fixed arbitrarily, must be determined by construction from the given magnitudes a, b, φ.

Accordingly, it will be observed that in the right-angled triangle KMC we know the hypotenuse $MC = MA = b$ and the side KC, equal to the base of the isosceles triangle KCT, in which $TK = TC = \frac{1}{2}a$ and $\sphericalangle KTC = \varphi$. From these data we can construct a triangle in order to find the length KM and the length $ST = AM - KM$.

The construction is represented in Fig. 107.

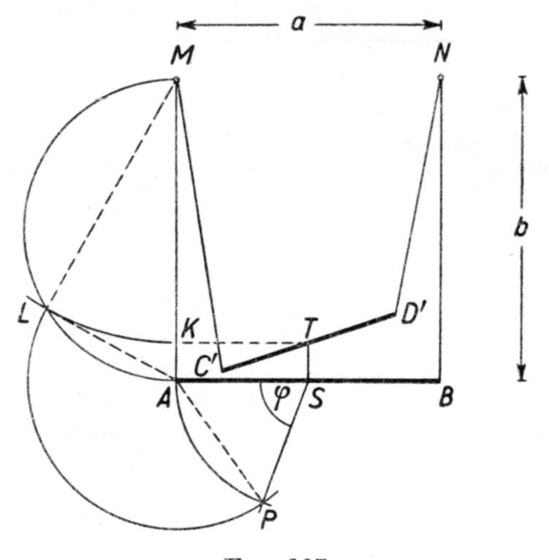

Fig. 107

We construct a triangle ASP in which $AS = SP = \frac{1}{2}a$, $\sphericalangle ASP = \varphi$. We draw a semicircle with diameter AM and a chord $AL = AP$ in it. We mark off on MA a segment MK equal to ML. Point K determines the level of point T, which will be found by drawing a segment $TS = KA$ parallel to AK.

The projection $C'D'$ of the twisted beam will be drawn as before by choosing point C' in an arbitrary manner.

Method II. Figures 108–111 represent the figure in question in the Monge projections for different values of angle φ. The vertical plane of projection is the plane $ABNM$, and the horizontal

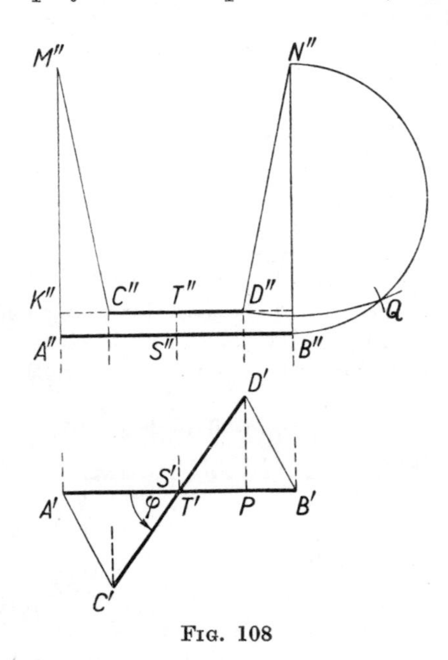

Fig. 108

plane of projection is an arbitrary plane perpendicular to $ABNM$. We shall describe the execution of Fig. 108.

The projection $C'D'$ is obtained by rotating the segment $A'B'$ about its mid-point S' through the angle φ. The vertical projection of point D lies at such a point D'' of the perpendicular drawn from point D' to $A'B'$ that the length of the segment ND is equal to b, i.e. to the length of $N''B''$.

In order to determine point D'' we consider the right-angled triangle $N''DD''$ formed by the segment $ND = N''D = b$, its vertical projection $N''D''$ and the segment DD'', equal to the distance of point D from the vertical plane of projection, i.e. equal to the distance $D'P$ of the projection D' from $A'B'$. We construct such a triangle taking the segment $N''B'' = b$ as the hypotenuse, describing a circle with diameter $N''B''$ and drawing in it a chord $B''Q = D'P$. Point D'' will lie at the intersection of the circle described from point N'' with radius $N''Q$ and the

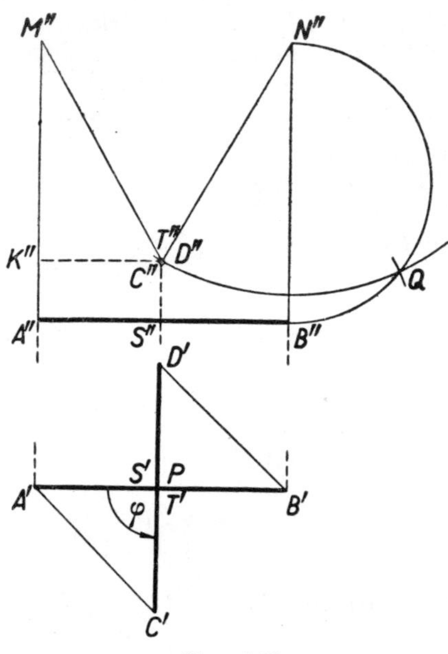

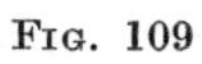

Fig. 109

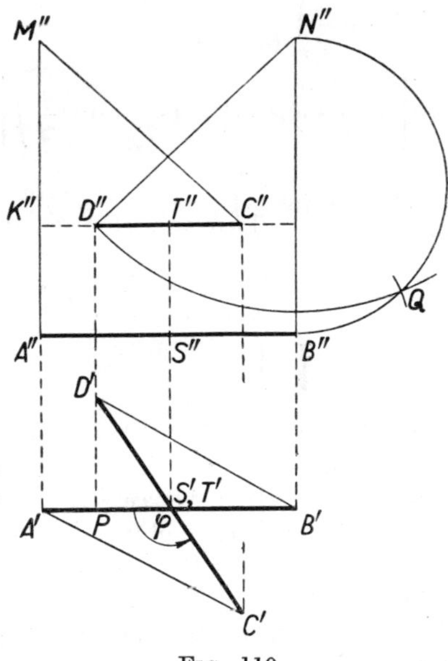

Fig. 110

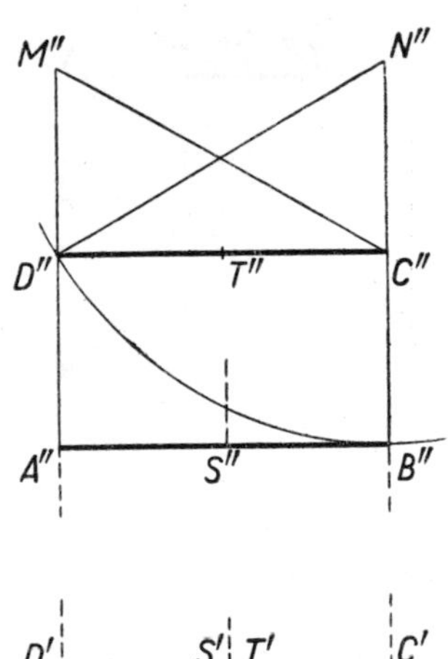

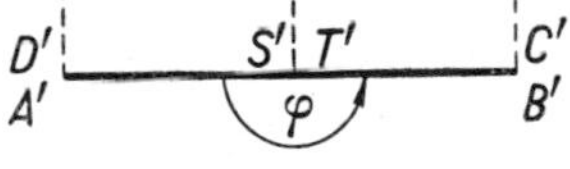

Fig. 111

straight line $D'P$. Point C'' lies symmetrically to point D'' on the other side of the drawing.

Figure 109 for $\varphi = 90°$, Fig. 110 for an obtuse φ, and Fig. 111 for $\varphi = 180°$ are executed in a similar way.

A drawing in the Monge projection executed (on a suitable scale) as shown above gives the graphic solution of our problem, since we obtain in it the required length $x = ST = S''T''$ according to given lengths a, b and a given angle φ.

In a drawing of this kind we can also calculate x. Using Fig. 108 we have

$$x = S''T'' = A''K'' = A''M''-K''M'' = b-K''M'',$$

$$K''M'' = \sqrt{[(M''C'')^2-(K''C'')^2]},$$

$$(M''C'')^2 = (N''D'')^2 = (N''Q)^2$$

$$= b^2-(B''Q)^2 = b^2-(D'P)^2 = b^2-\frac{a^2}{4}\sin^2\varphi,$$

$$(K''C'')^2 = (PB')^2 = \left(\frac{a}{2}-\frac{a}{2}\cos\varphi\right)^2 = a^2\sin^4\frac{\varphi}{2}.$$

Consequently

$$K''M'' = \sqrt{\left(b^2-\frac{a^2}{4}\sin^2\varphi-a^2\sin^4\frac{\varphi}{2}\right)}$$

$$= \sqrt{\left[b^2-a^2\sin^2\frac{\varphi}{2}\left(\cos^2\frac{\varphi}{2}+\sin^2\frac{\varphi}{2}\right)\right]}$$

$$= \sqrt{\left(b^2-a^2\sin^2\frac{\varphi}{2}\right)}.$$

Finally

$$x = b-\sqrt{\left(b^2-a^2\sin^2\frac{\varphi}{2}\right)}.$$

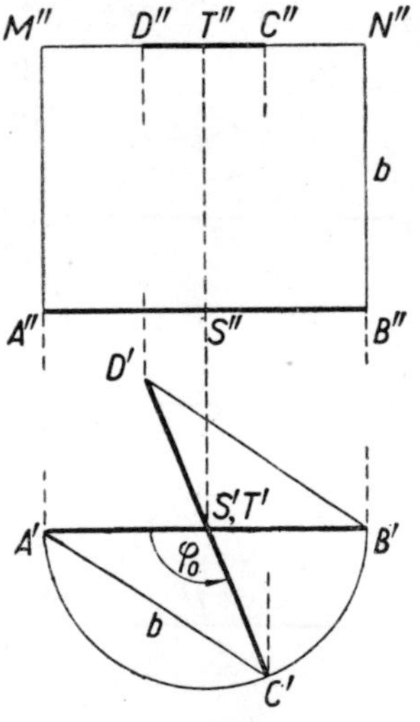

Figure 112 corresponds to the case where $b < a$ and the angle $\varphi = \varphi_0$ satisfies the condition $\sin(\varphi_0/2) = b/a$.

In a circle with diameter $A'B'$ we draw chords $A'C' = B'D' = b$. Then $\sphericalangle A'S'C' = \varphi_0$.

The projection $C''D''$ lies on the straight line $M''N''$.

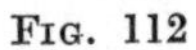

FIG. 112

107. The resultant of forces p_1 and p_2 pointing in the same direction passes through a point M of the segment AB (Fig. 113). Since the disc, when loaded with the weights, retains its equilibrium, the resultant of the three forces p_1, p_2, p_3 passes through point O. Consequently, point M is different from point O; the point of application of force p_3, i.e. point C, lies at the point of intersection of the edge of the disc with the half-line MO, and the sum of the convex angles AOB, BOC, COA is equal to 360°.

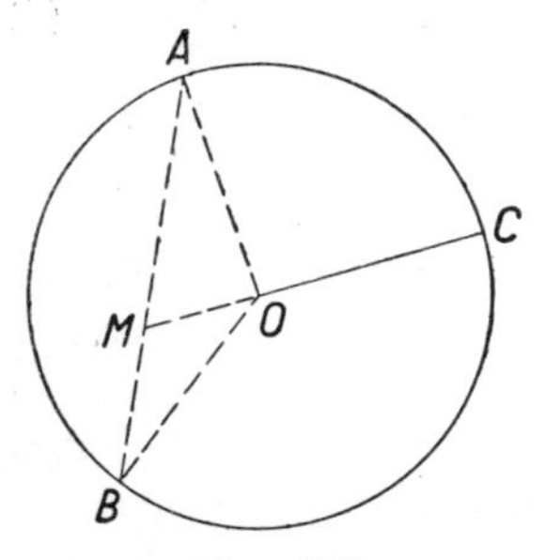

FIG. 113

We know from statics that the position of point M is defined by

$$\frac{AM}{MB} = \frac{p_2}{p_1}. \tag{1}$$

In triangles AOM and BOM

$$\frac{AM}{\sin \measuredangle AOM} = \frac{OM}{\sin \measuredangle A},$$

$$\frac{MB}{\sin \measuredangle MOB} = \frac{OM}{\sin \measuredangle B},$$

and since $\measuredangle A = \measuredangle B$, the above equalities imply that

$$\frac{AM}{\sin \measuredangle AOM} = \frac{MB}{\sin \measuredangle MOB}. \tag{2}$$

Equations (1) and (2) give

$$\frac{p_2}{\sin \measuredangle AOM} = \frac{p_1}{\sin \measuredangle MOB},$$

and, since $\measuredangle AOM = 180° - \measuredangle COA$, $\measuredangle MOB = 180° - \measuredangle BOC$, we have

$$\frac{p_1}{\sin \measuredangle BOC} = \frac{p_2}{\sin \measuredangle COA}.$$

Analogous equalities hold for the remaining pairs of the required angles, i.e.

$$\frac{p_1}{\sin \measuredangle BOC} = \frac{p_2}{\sin \measuredangle COA} = \frac{p_3}{\sin \measuredangle AOB}. \tag{3}$$

Equation (3) permits us to determine the angles BOC, COA and AOB. Indeed, let $180° - \measuredangle BOC = \alpha$, $180° - \measuredangle COA = \beta$, $180° - \measuredangle AOB = \gamma$; since $\measuredangle BOC + \measuredangle COA + \measuredangle AOB = 360°$, we have $\alpha + \beta + \gamma = 180°$. Equations (3) assume the form

$$\frac{p_1}{\sin \alpha} = \frac{p_2}{\sin \beta} = \frac{p_3}{\sin \gamma},$$

whence angles α, β, γ are the angles of a triangle with sides p_1, p_2, p_3. We can find angles α, β, γ by applying the Cosine Rule to that triangle, whereas the angles BOC, COA, AOB are calculated as supplementary angles to α, β, γ. We thus obtain

$$\cos \sphericalangle BOC = \frac{p_1^2-p_2^2-p_3^2}{2p_2p_3},$$

$$\cos \sphericalangle COA = \frac{p_2^2-p_3^2-p_1^2}{2p_3p_1}, \tag{4}$$

$$\cos \sphericalangle AOB = \frac{p_3^2-p_1^2-p_2^2}{2p_1p_2}.$$

Each of the required angles being contained between 0° and 180°, formulas (4) determine them in a unique manner.

REMARK. According to the conditions of the problem, we have assumed in the above solution that, when loaded with weights p_1, p_2, p_3, the disc remains in equilibrium. The question arises whether the weights p_1, p_2, p_3 can have arbitrary values, i.e. whether arbitrary weights p_1, p_2, p_3 can be placed at three different points of the disc in such a way that the disc retains its equilibrium. Obviously that is not so. In the above reasoning it has been found that numbers p_1, p_2, p_3 express the lengths of the sides of a triangle, and as such they must satisfy the inequalities:

$$p_1+p_2 > p_3, \quad p_2+p_3 > p_1, \quad p_3+p_1 > p_2. \tag{5}$$

We shall show that the necessary conditions (5) are also sufficient. Suppose that numbers p_1, p_2, p_3 satisfy conditions (5). Then there exists a triangle with sides p_1, p_2, p_3. If the angles of that triangle are α, β, γ, then we can determine on the edge of the disc three different points A, B, C in such a way that $\sphericalangle BOC = 180°-\alpha$, $\sphericalangle COA = 180°-\beta$, $\sphericalangle AOB = 180°-\gamma$, because $(180°-\alpha)+(180°-\beta)+(180°-\gamma) = 360°$. Angles BOC, COA, AOB then satisfy equations (3). Let us place at points A, B, C the weights p_1, p_2, p_3; we shall prove that the disc will then be in equilibrium. Since each of the angles BOC, COA, AOB is less than 180°, the radius CO produced intersects the side AB of the triangle AOB at a point M. In that case equality (2) holds, and, since $\sphericalangle AOM = 180°-\sphericalangle COA$, $\sphericalangle MOB = 180°-\sphericalangle BOC$, we have

$$\frac{AM}{\sin \sphericalangle COA} = \frac{MB}{\sin \sphericalangle BOC}. \tag{6}$$

It follows from equalities (6) and (3) that

$$\frac{AM}{MB} = \frac{p_2}{p_1},$$

whence point M is the point of application of the resultant of the parallel forces p_1 and p_2 applied at points A and B. Since this resultant is equal to p_1+p_2, it remains to prove that point O is the point of application of the resultant of the parallel forces p_1+p_2 and p_3 applied at points M and C respectively. Now equality (3) implies that

$$\begin{aligned}\frac{p_1+p_2}{p_3} &= \frac{\sin \sphericalangle BOC+\sin \sphericalangle COA}{\sin \sphericalangle AOB} \\ &= \frac{\sin \sphericalangle MOB+\sin \sphericalangle AOM}{\sin \sphericalangle AOB}.\end{aligned} \tag{7}$$

Applying the Sine Rule to triangles AOM, MOB and AOB, we have

$$\sin \sphericalangle MOB = \frac{MB}{MO} \sin \sphericalangle B,$$

$$\sin \sphericalangle AOM = \frac{AM}{MO} \sin \sphericalangle A,$$

$$\sin \sphericalangle AOB = \frac{AB}{OA} \sin \sphericalangle B.$$

Substituting these values in equation (7) and considering that $\sphericalangle A = \sphericalangle B$, $AM+MB = AB$, $OA = OC$, we obtain

$$\frac{p_1+p_2}{p_3} = \frac{OC}{MO}.$$

This equality shows that the resultant of the parallel forces p_1+p_2 and p_3 applied at points M and C indeed passes through O. Consequently the resultant of the parallel forces p_1, p_2, p_3 applied at points A, B, C respectively passes through point O, which is what we were to prove.

108. Suppose that points A, B, C, D do not lie on the same plane. The mid-points M, N, P, Q, R, S of the segments AB, BC, AC, AD, BD, CD (Fig. 114) are then different from one another. By the theorem on the line joining the mid-points of two sides of a triangle we have $MR \| AD \| PS$ and $MR = PS = \frac{1}{2}AD$, and similarly $MP \| BC \| RS$ and $MP = RS = \frac{1}{2}BC$. The quadrilateral $MPSR$ is a parallelogram; by the theorem on the sum of the squares of the diagonals of a parallelogram we have

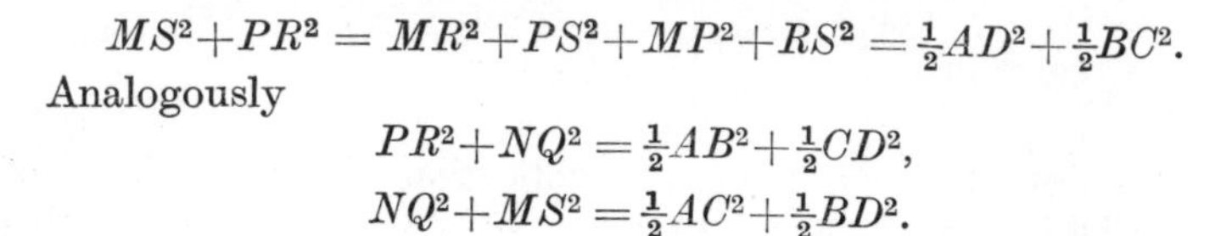

$$MS^2+PR^2 = MR^2+PS^2+MP^2+RS^2 = \tfrac{1}{2}AD^2+\tfrac{1}{2}BC^2.$$

Analogously

$$PR^2+NQ^2 = \tfrac{1}{2}AB^2+\tfrac{1}{2}CD^2,$$

$$NQ^2+MS^2 = \tfrac{1}{2}AC^2+\tfrac{1}{2}BD^2.$$

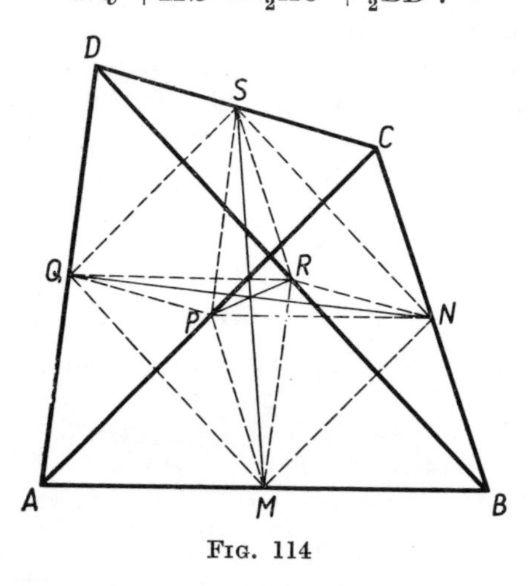

FIG. 114

From the above three equations it is easy to find the lengths MS, PR, NQ. For example we obtain

$$MS^2 = \tfrac{1}{4}(AC^2+AD^2+BC^2+BD^2-AB^2-CD^2). \qquad (1)$$

This is the required formula for the distance between the mid-points of AB and CD, the mutual distances of points A, B, C, D being given.

The above reasoning remains unchanged if points A, B, C, D are co-planar but no three of them are collinear, the points M, N, P, Q, R, S being all different from one another.

It is easy to ascertain that formula (1) is general, i.e. that it is true irrespective of the position of points A, B, C, D. Some of those points, or even all of them, could coincide. We leave it to the reader to prove the validity of formula (1), by modifying the above reasoning in a suitable manner, for the following cases, which, together with those already considered, exhaust all the possibilities.

(a) Three of the points A, B, C, D are collinear, and the fourth lies apart;

(b) points A, B, C, D are collinear;

(c) no three of the points A, B, C, D are collinear but points M and S, or N and Q, or P and R, coincide.

It would be possible, however, to prove the generality of formula (1) in a different way, namely by regarding cases (a), (b), (c) as *limiting cases* of the essential case, in which points A, B, C, D are not co-planar. We shall explain this for case (a). Suppose that A, B, C are three different points of a straight line p and point D is not of p. Let M be the mid-point of segment AB and S the mid-point of segment CD.

Let us choose point A' in the neighbourhood of point A but not belonging to the plane of points A, B, C, D. Let M' denote the mid-point of segment $A'B$. Points A', B, C, D are not co-planar, whence, as has been proved before,

$$M'S^2 = \tfrac{1}{4}(A'C^2+A'D^2+BC^2+BD^2-A'B^2-CD^2). \qquad (1a)$$

Let point A' tend to point A along, say, a straight line. Then point M' tends to point M, the lengths $M'S$, $A'C$, $A'D$, $A'B$ tend to the lengths MS, AC, AD, AB and from formula (1a) we obtain formula (1) *passing to the limit* (as we say in mathematics). This reasoning is based on the elementary facts of the so-called theory of limits, which can be found in the initial chapters of any textbook of mathematical analysis.

109. *Method I.* Let us draw through vertex A of the tetrahedron $ABCD$ a plane α perpendicular to the edge AB (Fig. 115) and

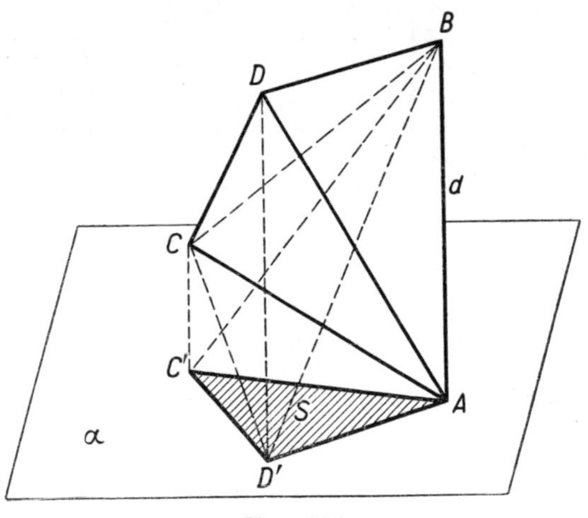

FIG. 115

through points C and D—lines parallel to the line AB which intersect the plane α at points C' and D'. The orthogonal projection of the tetrahedron $ABCD$ upon α is the triangle $AC'D'$.

The volume V of the tetrahedron $ABCD$ is equal to the volume of the tetrahedron $ABCD'$ since both have the same base ABC and the line DD' on which their vertices lie is parallel to AB, i.e. it is also parallel to the plane ABC. Similarly, the volume of the tetrahedron $ABCD'$ is equal to that of the tetrahedron $ABC'D'$ because the vertices C and C' of those tetrahedrons lie on a line parallel to their common base ABD'. Consequently the volume of the tetrahedron $ABCD$ is equal to the volume of the tetrahedron $ABC'D'$. Since $AB \perp \alpha$, we have

$$\text{volume } ABC'D' = \tfrac{1}{3}(\text{area } AC'D')\,AB = \tfrac{1}{2}Sd$$

and finally

$$V = \tfrac{1}{3}Sd.$$

Method II. The volume of the tetrahedron $ABCD$ is expressed by the formula

$$V = \tfrac{1}{3}(\text{area } BCD) \times h, \tag{1}$$

in which h denotes the length of the altitude AH of the tetrahedron drawn from vertex A (Fig. 116).

The projection $A'B'C'D'$ of the tetrahedron on a plane perpendicular to the line AB coincides with the projection $B'C'D'$ of the triangle BCD (since the projection A' of point A coincides with the projection B' of point B); therefore

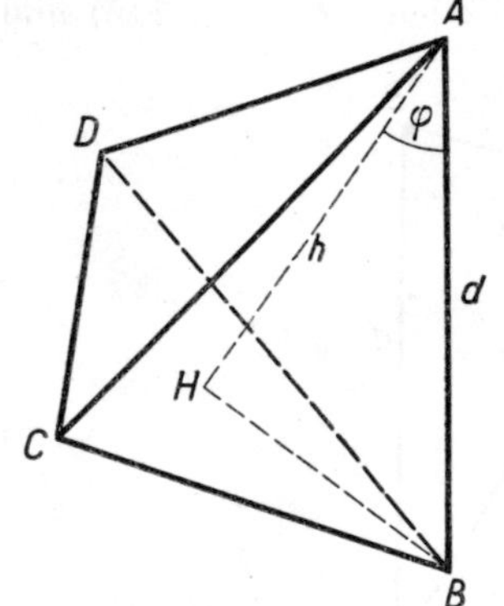

FIG. 116

$$S = \text{area } B'C'D'.$$

The area of a projection of a plane figure is equal to the product of the area of that figure and the cosine of the acute angle φ between the plane of the figure and the plane of projection; hence

$$\text{area } B'C'D' = \text{area } BCD \times \cos\varphi,$$

and consequently

$$\text{area } BCD = \frac{S}{\cos\varphi}. \tag{2}$$

Since the straight lines AH and AB form the same acute angle as planes perpendicular to them, in the right-angled triangle ABH the angle BAH equals φ and

$$h = d\cos\varphi. \tag{3}$$

If follows from equalities (1), (2), (3) that

$$V = \tfrac{1}{3}Sd.$$

Method III. Let the segment $DH = h$ be the altitude of the tetrahedron $ABCD$ drawn from vertex D upon the face ABC, and the segment $CK = k$ the altitude of the triangle ABC drawn from vertex C upon side AB. Then

$$V = \tfrac{1}{3}(\text{area } ABC)h = \tfrac{1}{3}\times\tfrac{1}{2}AB\times k\times h = \tfrac{1}{3}d(\tfrac{1}{2}kh).$$

Let us denote by A', B', C', D' the projections of the vertices A, B, C, D of the tetrahedron upon a plane α perpendicular to the edge AB. Points A' and B' coincide and the projection of the tetrahedron is the triangle $A'C'D'$. The projection of the edge AB is point A', and the projection of the face ABC is the segment $A'C'$, whence the projection K' of point K coincides with point A', and the projection H' of point H lies on the segment $A'C'$ (Fig. 117).

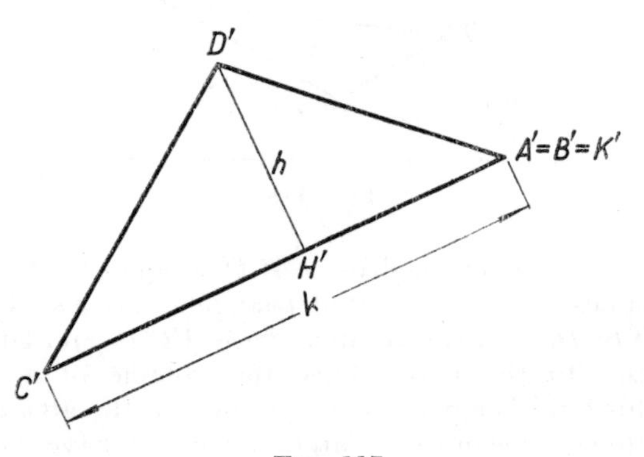

FIG. 117

The projection of the segment CK is the segment $C'A'$; since $CK \perp AB$, we have $CK \| \alpha$, and thus $C'A' = CK = k$, because the projection of a segment parallel to the plane of projection is a segment of the same length.

The segment DH is perpendicular to the plane ABC, and the plane ABC is perpendicular to the plane of projection α, whence $DH \| \alpha$, and consequently $D'H' = DH = h$. Since $\sphericalangle AHD$ is a right angle whose arm DH is parallel to the plane of projection, the projection of that angle, i.e. $\sphericalangle A'H'D'$ is also a right angle, and the segment $D'H'$ is an altitude of triangle $A'C'D'$.

This implies that the product $\frac{1}{2}kh$ is equal to the area of the triangle $A'C'D'$, i.e. is equal to S, and the preceding formula gives

$$V = \tfrac{1}{3}Sd.$$

Method IV. Let us place the projection plane α as shown in Fig. 118, i.e. in such a way that the tetrahedron $ABCD$ should lie on one side of plane α and point A should be at a greater distance from plane α than point B.

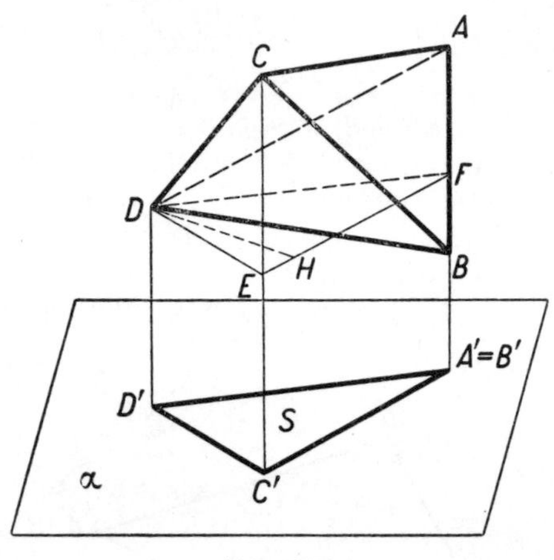

FIG. 118

The volume V of tetrahedron $ABCD$ is equal to the difference of the volumes of two truncated triangular prisms, $A'C'D'ACD$ and $A'C'D'BCD$, with the common base $A'C'D'$ and lateral edges perpendicular to the base. Since the volume of a truncated prism of this kind is equal to the product of the area of its base by the arithmetic mean of its lateral edges, we have

$$\text{vol. } A'C'D'ACD = S \times \frac{AA' + CC' + DD'}{3},$$

$$\text{vol. } A'C'D'BCD = S \times \frac{BA' + CC' + DD'}{3}.$$

Consequently

$$V = S \times \frac{AA' - BA'}{3} = \frac{1}{3} Sd.$$

REMARK. The formula for the volume of a truncated triangular prism, which we have used in method IV, can be proved as follows:

Let $ACDA'C'D'$ (Fig. 118) be a prism of this kind and let DD' be its shortest lateral edge. Drawing through point D a plane parallel to the plane $A'C'D'$ we cut the truncated prism into the straight triangular prism $DEFD'C'A'$ and the pyramid $DACEF$.

The base $ACEF$ of this pyramid is a trapezium with bases AF and CE and altitude EF, and the altitude DH of the pyramid is at the same time an altitude of the triangle DEF. Consequently

$$V = \text{vol. } DEFD'C'A' + \text{vol. } DACEF$$

$$= (\text{area } D'C'A') \times A'F + \frac{1}{3} (\text{area } ACEF) \times DH$$

$$= \frac{1}{3} (\text{area } D'C'A') \times 3A'F + \frac{1}{3} \times \frac{AF+CE}{2} \times EF \times DH$$

$$= \frac{1}{3} (\text{area} D'C'A')(D'D+C'E+A'F) + \frac{1}{3} (\text{area } DEF)(AF+CE)$$

$$= \frac{1}{3} (\text{area } D'C'A')(D'D+C'C+A'A).$$

110. To make easier the task of finding the solution, let us consider first an analogous problem on the plane.

Let us draw through each vertex of triangle ABC a straight line parallel to the opposite side of the triangle. These lines form a triangle $A_1B_1C_1$ homothetic to triangle ABC (Fig. 119). The centre of homothety is point S at which the lines AA_1, BB_1, CC_1 intersect; point S is the centre of gravity of triangle ABC and also of triangle $A_1B_1C_1$. The homothety is inverse, since the corresponding points, e.g. A and A_1, lie on half-lines SA and SA_1 of opposite directions. The homothety ratio is

$$\frac{SA_1}{SA} = \frac{A_1B_1}{AB} = 2.$$

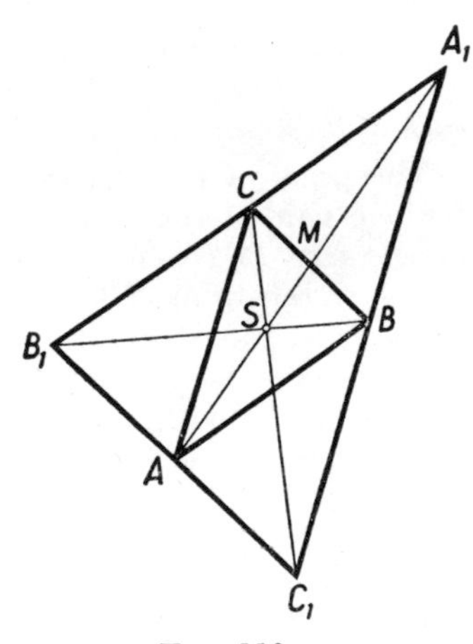

FIG. 119

Since the ratio of the areas of homothetic figures is equal to the square of the homothety ratio, the area of triangle $A_1B_1C_1$ is four times as great as the area of triangle ABC, which, in fact, can be seen at once.

Returning to our proper subject, let us briefly recall the properties of homothetic figures in space. Suppose that F is a given figure in space with points A, B, C, ... Let us choose any point S and a positive number k. If we mark off on the half-lines SA, SB,

SC, ... segments $SA' = k \times SA$, $SB' = k \times SB$, $SC' = k \times SC$, ..., then points A', B', C', ... will form a figure F' directly homothetic to figure F with respect to point S in the ratio k. If we mark off segments $SA_1 = k \times SA$, $SB_1 = k \times SB$, $SC_1 = k \times SC$, ... on half-lines SA, SB, SC, ... produced, we obtain a figure F_1 inversely homothetic to figure F with respect to point S in the ratio k (Fig. 120). It is easy to infer from this definition that segment A_1B_1 is a figure homothetic to segment AB, the straight lines AB and A_1B_1 being parallel (or coinciding, which takes place if AB passes through point S). Triangle $A_1B_1C_1$ is thus a figure homothetic to triangle ABC, the planes of the triangles being parallel (or coinciding if the plane ABC contains point S). Tetrahedron $A_1B_1C_1D_1$ is a figure homothetic to tetrahedron $ABCD$. The ratio of homothetic segments is equal to the homothety ratio, the ratio of the areas of homothetic triangles (or generally: of homothetic plane figures) is equal to the square of the homothety ratio, and the ratio of the volumes of homothetic tetrahedrons (or generally: of homothetic solids) is equal to the cube of the homothety ratio.

In order to solve our problem we need also the following theorem:

The segments joining the vertices of a tetrahedron with the centres of gravity of the opposite faces intersect at one point, which divides each of those segments in the ratio $3:1$.

The proof of this theorem is simple; it suffices to carry it out for one pair of those segments. Let M be the centre of gravity of the face BCD and N the centre of gravity of the face ACD of the tetrahedron $ABCD$ (Fig. 120). The straight lines BM and AN intersect at the mid-point P of the edge CD.

The segments AM and BN, joining the vertices A and B of triangle ABP with points M and N of the sides BP and AP, intersect at point S inside this triangle. Since $AP = 3 \times NP$ and $BP = 3 \times MP$, we have $MN \| AB$ and $AB = 3 \times MN$; hence $AS = 3 \times SM$ and $BS = 3 \times SN$, which is what we were to prove.

Point S is the centre of gravity of the tetrahedron.

Using the above data we can formulate the solution of the problem very briefly: We construct a figure inversely homothetic in the ratio 3 to the tetrahedron $ABCD$ with respect to its centre of gravity S. That figure is the tetrahedron $A_1B_1C_1D_1$ whose vertices lie on the extensions of the half-lines SA, SB, SC, SD, and $SA_1 = 3 \times SA$, $SB_1 = 3 \times SB$, $SC_1 = 3 \times SC$, $SD_1 = 3 \times SD$.

The plane $B_1C_1D_1$ is parallel to the corresponding plane BCD and passes through vertex A, since point A is homothetic to the centre of gravity M of face BCD.

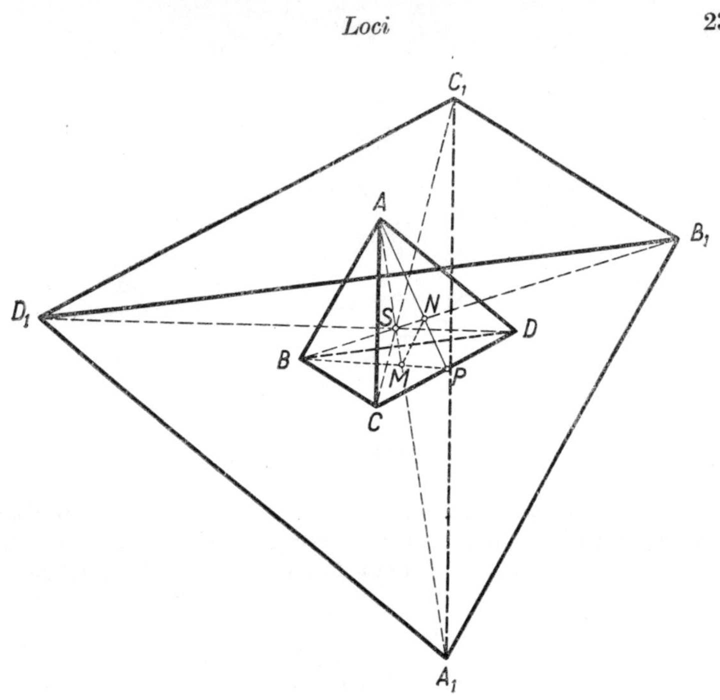

FIG. 120

Consequently the tetrahedron $A_1B_1C_1D_1$ is the one required in the problem. Since it is homothetic to tetrahedron $ABCD$ in the ratio 3, its volume is $3^3 \times V$, i.e. $27V$.

§ 7. Loci

111. Suppose that given straight lines m and n intersect at point O (Fig. 121).

Suppose that point K belongs to the required locus. We draw $KM \perp m$, $KN \perp n$; the distances KM and KN satisfy the condition

$$KM + KN = a.$$

Let us extend the segments MK and NK and mark off $KP = KN$ and $KR = KM$; then $MK + KP = NK + KR = a$, whence points P and R lie, respectively, on lines m_1 and n_1 drawn parallel to m and n at a distance a. Point K, being equidistant from m and n_1, and likewise from n and m_1, lies on the diagonal AB of the rhombus $OAQB$ formed by the two pairs of parallels.

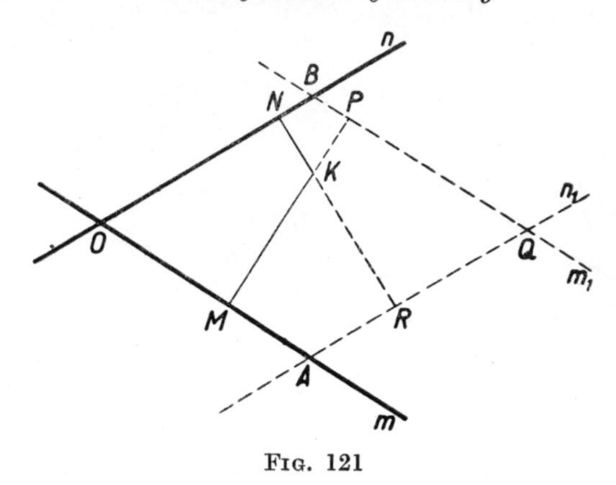

FIG. 121

Conversely, if point L lies on the diagonal AB (Fig. 122), then, in view of the symmetry of the rhombus with respect to AB, the sum of the distances of point L from m and n is equal to the

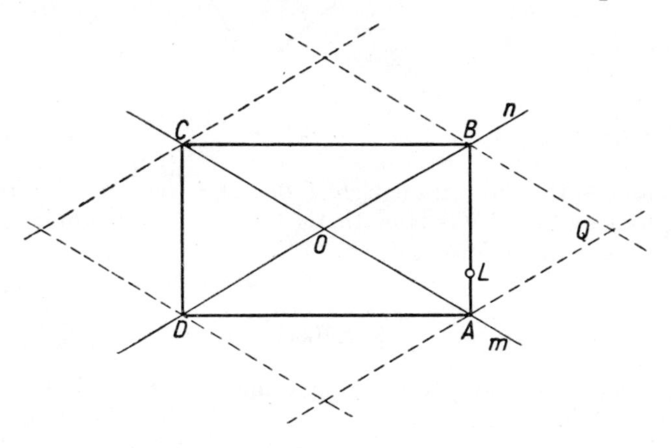

FIG. 122

sum of its distances from m and m_1, i.e. is equal to a. It can thus be seen that the locus of points lying in the angle AOB such that the sum of their distances from m and n is equal to a is the segment AB. In the angles BOC, COD, DOA those loci are the segments BC, CD, DA. Consequently the required locus is the perimeter of the rectangle $ABCD$.

112. To begin with, we shall determine the locus of the centre L of a rectangle $EFGH$ with two vertices, E and F, lying on the side AB of the triangle ABC, vertex G lying on the side BC and vertex H on the side AC. Such rectangles exist only if neither of the angles A and B of the triangle is obtuse. In the sequel we shall assume that the angles A and B are both acute; in the case of one of them being a right angle the reasoning below requires a slight modification, which we leave to the reader.

Point L, whose locus we are seeking, is the mid-point of the segment PQ joining the mid-points of sides EH and FG of the rectangle (Fig. 123). Let us draw the altitude CD of triangle ABC.

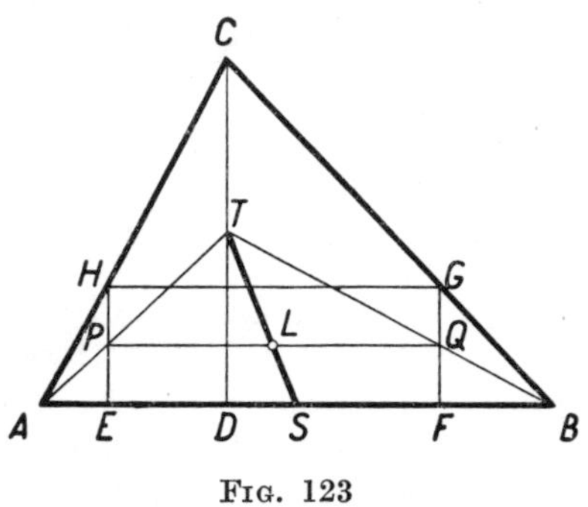

FIG. 123

Since triangles AEH and ADC are homothetic with respect to point A and triangles BFG and BDC are homothetic with respect to point B, the straight lines AP and BQ pass through the mid-point T of the segment CD. The triangles PQT and ABT are homothetic with respect to point T, whence the straight line TL passes through the mid-point S of the segment AB. Consequently the centre L of the rectangle $EFGH$ lies inside the segment ST joining the mid-point of the side AB with the mid-point of the altitude CD of triangle ABC.

Conversely, every point L lying inside the segment ST is the centre of a certain rectangle $EFGH$ placed in the above position. Indeed, drawing through point L a segment PQ homothetic to segment AB with respect to point T, and then drawing through points P and Q segments EH and FG homothetic to segment CD with respect to points A and B respectively, we obtain a quadrilateral $EFGH$; and it follows from the properties of homothetic segments that points P and Q are the mid-points of segments EH and FG and point L is the mid-point of segment PQ; hence we conclude that $EFGH$ is a rectangle and point L is its centre. The required locus is the interior of the segment ST.

A certain variant of the above reasoning is based on the fact that point L is the mid-point of segment MN joining the mid-points of sides EF and GH of the rectangle (Fig. 124). Triangles CHG and CAB are homothetic with respect to point C, whence CN intersects segment AB at its mid-point S. From the homothety of triangles SMN and SDC with respect to point S it follows that SL passes through the mid-point T of the altitude CD. Point L thus lies inside the segment ST.

Conversely, if L is any point lying inside the segment ST then drawing first a segment MN homothetic to segment DC with respect to point S, then a segment HG homothetic to segment AB with respect to point C, and finally segments HE and GF perpendicular to HG, we obtain a rectangle $EFGH$ with centre L.

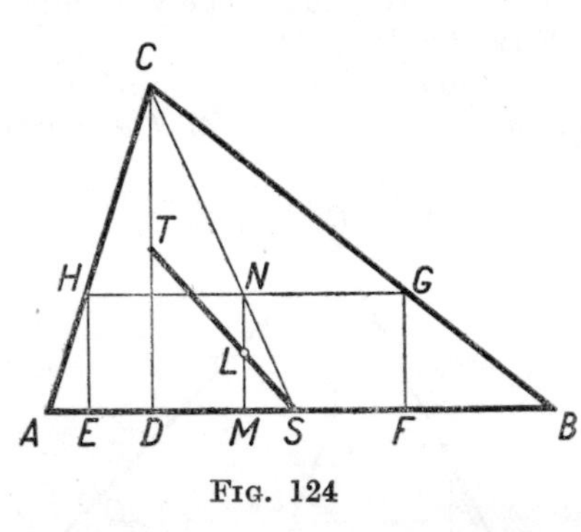

Fig. 124

So far we have established the following result:

The locus of the centre L of a rectangle with two vertices lying on the side AB of a triangle ABC and the remaining two vertices lying on the sides AC and BC is the interior of the segment joining the mid-point of side AB with the mid-point of the corresponding altitude of the triangle ABC.

If the given triangle is acute-angled, this theorem holds for each of its three sides. The locus mentioned in the problem then consists of the interior points of three segments joining the mid-points of the sides of the triangle with the sides of the corresponding altitudes. We shall prove that those three segments intersect at one point.

We shall adopt the notation of Fig. 125: D, E, F are the feet of the altitudes, M, N, P—the mid-points of the sides, and Q, R,

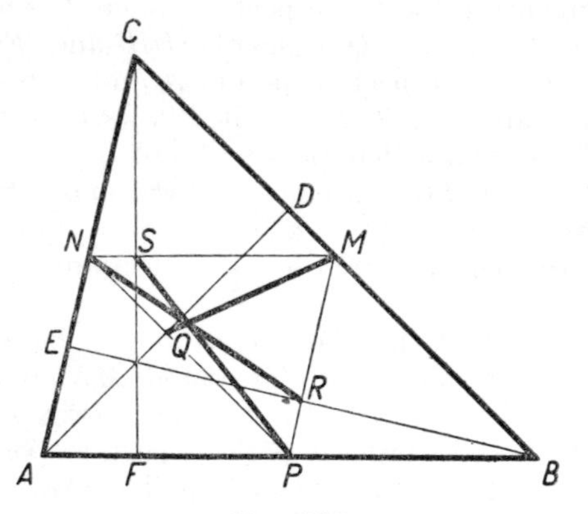

Fig. 125

S—the mid-points of the altitudes of triangle ABC. Applying Ceva's theorem† to this triangle and its three altitudes, we obtain:

$$\frac{AF \times BD \times CE}{FB \times DC \times EA} = 1.$$

Since the sides of triangle MNP are parallel to the corresponding sides of triangle ABC, we have:

$$\frac{NS}{SM} = \frac{AF}{FB}, \quad \frac{PQ}{QN} = \frac{BD}{DC}, \quad \frac{MR}{RP} = \frac{CE}{EA},$$

whence

$$\frac{NS \times PQ \times MR}{SM \times QN \times RP} = 1.$$

By the inverse of Ceva's theorem, we infer from the above equation that the lines MQ, NR and PS intersect at one point.

If the given triangle is right-angled, two of the three segments constituting the locus in question coincide and we have Fig. 126.

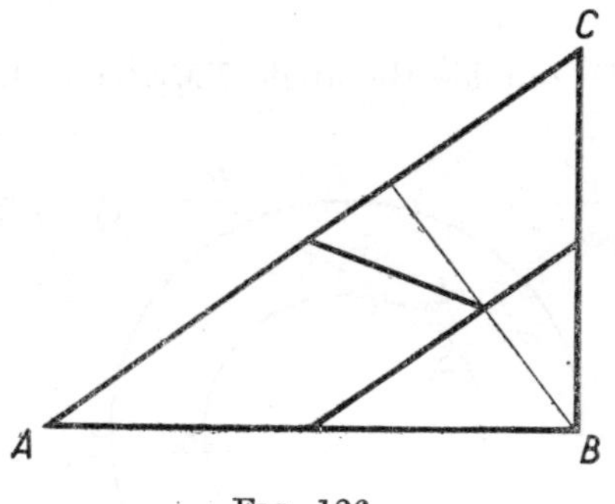

Fig. 126

In an obtuse-angled triangle the required locus is the interior of one segment.

113. *Hint*: Find the angle AMB. The required locus is formed by two circular arcs joining points A and B. The points A and B themselves do not belong to the locus.

114. Suppose that from point M a tangent MT to the circle K_1 and a tangent MS to circle K_2 have been drawn. Two cases are possible:

---

† See remark to problem 97.

(a) the centre O of the given circles lies inside the angle TMS (Fig. 127),

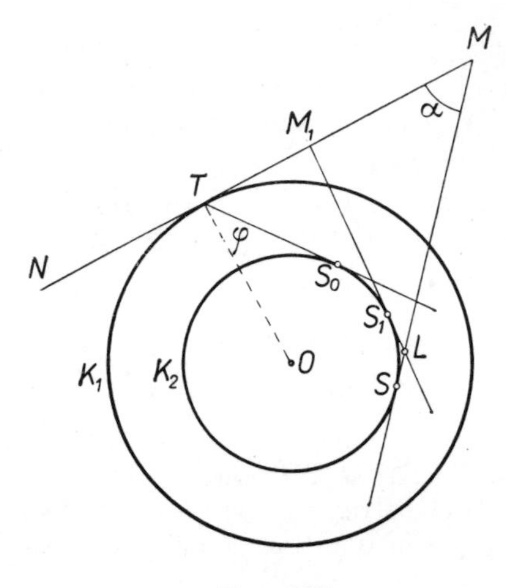

Fig. 127

(b) point O lies outside the angle TMS (Fig. 128).

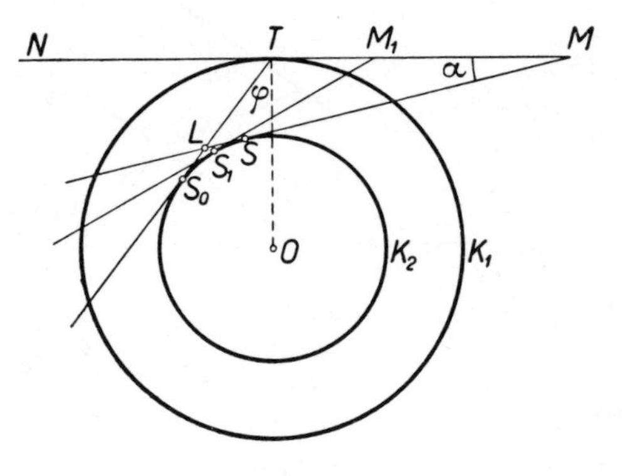

Fig. 128

We shall begin by determining the required locus for case (a). If point M belongs to that locus, i.e. if in Fig. 127 we have $\sphericalangle TMS = \alpha$, then the locus includes also the whole circle described by point M when the figure is rotated about point O, since in this rotation $\sphericalangle TMS$ remains unchanged. We shall prove that the circumference of that circle is the required locus, i.e. that apart from it there are no points satisfying the required condition.

Let us consider any point lying nearer to point O than point M, e.g. point M_1 of the segment TM. Let the tangent M_1S_1 to circle K_2 intersect the tangent MS to the same circle at point L. In triangle MM_1L the exterior angle TM_1L is greater than the interior angle TML, i.e. $\sphericalangle TM_1S_1 > \alpha$, whence neither point M_1 nor any other point lying on the circle with centre O and radius OM_1 belongs to the locus under consideration. No point M_2 lying farther away from point O than point M belongs to this locus either, since then $\sphericalangle TM_2S < \alpha$. The theorem is thus proved. It can be seen at the same time that the angle α has the greatest value when the point M lies on circle K_1, e.g. at point T; then $\alpha = \sphericalangle NTS_0 = 90°+\varphi$, where φ is half the angle between the tangents drawn from point T of circle K_1 to circle K_2.

In case (b) the reasoning is analogous; it is illustrated in Fig. 128. The required locus is another circle with the same centre O. The greatest possible value of angle α is then $\sphericalangle NTS_0 = 90°-\varphi$; where φ has the same value as before.

Let us list the results obtained:

(1) If $\alpha \leqslant 90°-\varphi$, the required locus consists of two circles with centre O.

(2) If $90°-\varphi < \alpha \leqslant 90°+\varphi$, the locus is a circle with centre O.

(3) If $\alpha > 90°+\varphi$, the locus does not exist.

REMARK 1. To construct the above-mentioned circles it suffices to find one point of each of them. Accordingly, it is best to begin by drawing the angle TOS, which is equal to $180°-\alpha$ in case (a) and to α in case (b).

REMARK 2. The above reasoning requires a certain supplement; namely it should be proved that the half-lines MS and M_1S_1 always intersect. Now, if point M_1 lies on MN between points T and M, then in case (a) circle K_2 is in the angle TM_1S_1, and point M lies outside that angle; consequently the segment MS must intersect the half-line M_1S_1; in case (b) point M_1 and circle K_2 lie on opposite sides of MS, whence the segment M_1S_1 must intersect the half-line MS.

115. *Answer*: The required locus is the circle with the centre at the point of intersection of the given straight lines and with a radius equal to $d/\sin\alpha$, where α denotes one of the angles between the given lines.

116. Consider a certain position of the moving lines m and n; let the straight lines A_1B_2 and A_2B_1 intersect at point Q. Suppose that point P is outside the strip of the plane determined by the parallel lines a and b; in that case point Q lies in this strip

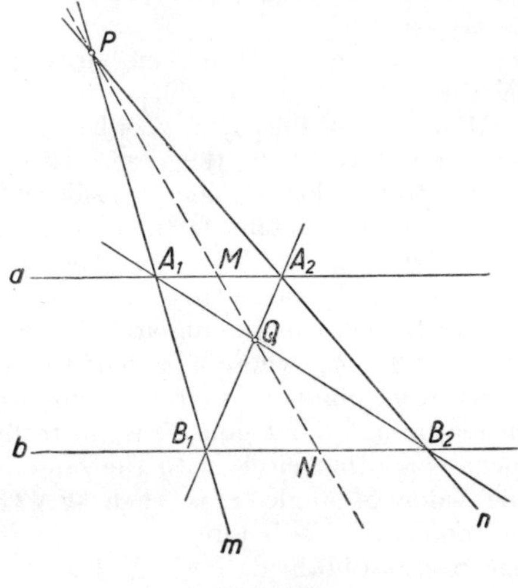

FIG. 129

(Fig. 129). Let h_1 and h_2 denote the distances of point P from a and from b, and k_1 and k_2—the distances of the moving point Q from a and from b.

From the fact that a and b are parallel it follows that triangles $Q_1A_1A_2$ and QB_1B_2 are homothetic with respect to centre Q; the ratio of the corresponding altitudes k_1 and k_2 of these triangles is thus equal to the ratio of the corresponding sides A_1A_2 and B_1B_2, i.e.

$$\frac{k_1}{k_2} = \frac{A_1A_2}{B_1B_2}.$$

Similarly, the homothety of triangles PA_1A_2 and PB_1B_2 with respect to centre P implies that

$$\frac{h_1}{h_2} = \frac{A_1A_2}{B_1B_2}.$$

Consequently

$$\frac{k_1}{k_2} = \frac{h_1}{h_2}. \tag{1}$$

Conversely, let Q be a point of the strip between the lines a and b whose distances k_1 and k_2 from a and b satisfy equation (1). Let us draw through point Q an arbitrary straight line intersecting

a and b at points A_1 and B_2; let the straight line m passing through points P and A_1 intersect b at point B_1 and let the straight line n passing through points P and B_2 intersect a at point A_2. Thus point Q', at which A_2B_1 intersects A_1B_2, must coincide with point Q, since we know from the above reasoning that the distances of point Q' from a and b must be in the ratio $h_1:h_2$, and on the segment A_1B_2 point Q is the only one to have this property. It follows that point Q belongs to the required locus.

We have proved that the required locus coincides with the locus of those points lying in the strip between the lines a and b whose distances from a and b are in the given ratio $h_1:h_2$. Thus the locus in question is a straight line parallel to a and b, lying in the strip between a and b and dividing the width of that strip in the ratio $h_1:h_2$.

If point P is between the lines a and b but is not equidistant from them, the same reasoning as before leads to the conclusion that the required locus is a straight line parallel to a and b, lying outside the strip between a and b and such that its distances from a and b are in the ratio $h_1:h_2$.

If point P is equidistant from a and b, the required locus does not exist because then the straight lines A_1B_2 and A_2B_1 are parallel.

REMARK 1. Using the above theorem we can easily solve the following problem: We are given on a plane two parallel lines a and b and a point P. Draw through P a line parallel to a and b using only a straight edge.

REMARK 2. Let M and N be the points of intersection of the straight line PQ with a and b (Fig. 129). Since triangles PA_1A_2 and PB_1B_2 are homothetic with respect to the centre P and triangles QA_1A_2 and QB_1B_2 are homothetic with respect to the centre Q, the homothety ratio $A_1A_2:B_1B_2$ being the same in both cases, we have

$$\frac{PM}{PN}=\frac{QM}{QN}{}^{\dagger}.$$

This equality signifies that the pairs of points P, Q and M, N divide one another harmonically, i.e. that point Q is harmonically conjugate to point P with respect to points M and N.

† We consider here non-directed segments. For directed segments (vectors) we should have to write $PM:PN=-QM:QN$ since, if the segments PM and PN have the same direction, then the segments QM and QN have opposite directions.

It will be observed that points M and N are the centres of the segments A_1A_2 and B_1B_2. Indeed, the triangles PA_1M and PB_1N are homothetic with respect to centre P, and the triangles QMA_2 and QB_1N are homothetic with respect to centre Q, the homothety ratio being the same in both cases; consequently

$$\frac{A_1M}{B_1N} = \frac{MA_2}{B_1N},$$

whence $A_1M = MA_2$ and analogously $B_1N = NB_2$.

Conversely, on an arbitrary straight line passing through point P and intersecting a and b at points M and N let us choose a point Q harmonically conjugate to point P with respect to points M and N. Let us then draw through point P a straight line m intersecting a and b at points A_1 and B_1; mark off $MA_2 = A_1M$ and draw a straight line n through points P and A_2: it will intersect b at a point B_2 such that $B_1N = NB_2$. We shall prove that lines A_1B_2 and A_2B_1 intersect at point Q. Indeed, if the point of intersection of A_1B_2 and A_2B_1 is Q_1, then, as we have proved before, the straight line PQ_1 passes through the mid-points M and N of the segments A_1A_2 and B_1B_2, whence point Q_1 is harmonically conjugate to point P with respect to points M and N; consequently point Q_1 coincides with point Q.

It follows that the locus determined previously is also the locus of points Q which are harmonically conjugate to point P with respect to the points of intersection of PQ with a and b.

REMARK 3. The above consideration can be extended to the case where a and b intersect. We then obtain Fig. 130. We are to find the locus of point Q when the straight lines m and n are variable.

The figure formed by the lines m, n, A_1B_2 and A_2B_1 is a complete quadrilateral with vertices A_1, A_2, B_1, B_2, P, Q and diagonals PQ, a, b. As we know, the pair of vertices lying on one diagonal of a complete quadrilateral separates harmonically the pair of points at which that diagonal intersects the remaining two diagonals. (See problem 79.)

Consequently point Q is harmonically conjugate to point P with respect to points M and N. Conversely, if a point Q is harmonically conjugate to point P with respect to the points of intersection M and N of the straight line PQ with a and b, then point Q belongs to the required locus. To prove this let us draw through P the line m, intersecting a and b at points A_1 and B_1, let us determine point A_2 harmonically conjugate to point A_1 with respect to points M and R and let us draw through P and A_2 the line n, intersecting b at point B_2. Then the lines

m, n, A_1B_2 and A_2B_1 will form a complete quadrilateral, in which one diagonal is the line A_1A_2 and the second diagonal is the line B_1B_2, whence the third diagonal must pass through a point harmonically conjugate to R with respect to A_1 and A_2, i.e. through point M. Consequently lines A_1B_2 and A_2B_1 intersect on PM; since their point of intersection is harmonically conjugate to P with respect to M and N and point Q is the only one to have this property, the straight lines A_1B_2 and A_2B_1 intersect at point Q. Point Q is thus a point of the required locus.

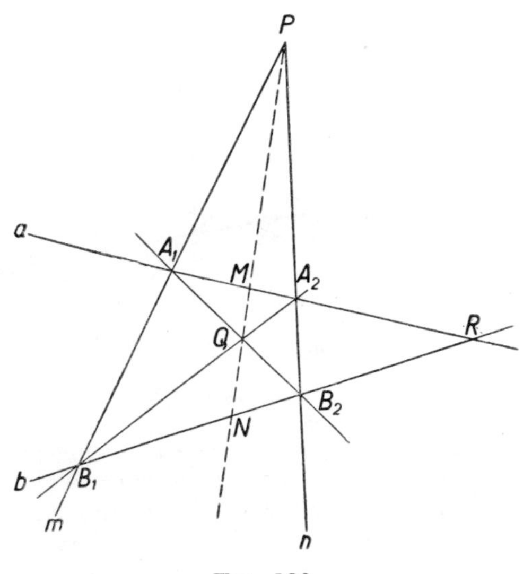

Fig. 130

If follows that the locus in question coincides with the locus of points Q harmonically conjugate to P with respect to the points of intersection of PQ with a and b. We shall prove that the required locus is a straight line passing through the intersection R of lines a and b.

Let Q (Fig. 131) be the point of our locus determined by the lines m and n, and T any other point of that locus; point T is harmonically conjugate to P with respect to points A_3, B_3 at which PT intersects a and b. We are to prove that point T lies on RQ. Let S be the intersection of RQ with A_1B_1. Investigating the complete quadrilateral formed by the straight lines a, b, A_1B_2 and A_2B_1 with vertices A_1, B_1, A_2, B_2, Q, R and diagonals A_1B_1, A_2B_2 and QR, we find that S is harmonically conjugate to P with respect to A_1 and B_1. Consequently, in the complete quadrilateral formed by the lines a, b, A_1B_3, A_3B_1 the diagonal passing through vertex R must intersect the diagonal A_1B_1 at point S; that diagonal is thus the line RQ, whence it follows further that the point of intersection of RQ with the third diagonal, A_3B_3, of the complete quadrilateral in question is point T, as the one harmonically conjugate to P with respect to A_3 and B_3. The theorem is thus proved.

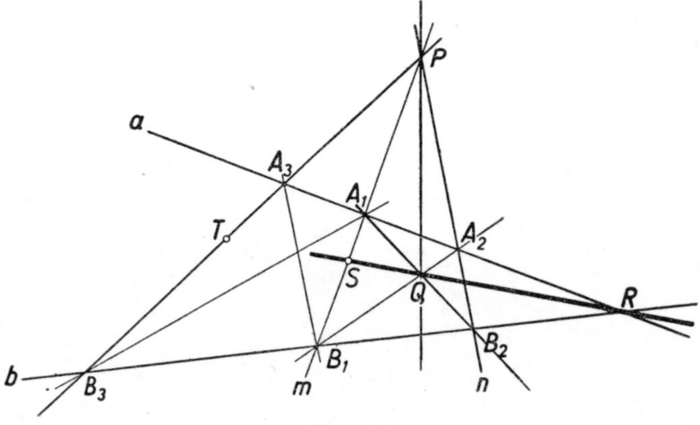

FIG. 131

Using this theorem we can solve the following problem. Given straight lines a and b intersecting at an inaccessible point R (e.g. a point lying outside our drawing space) and a point P, draw the straight line PR using only a straight edge.

117. When the disc rolls along the rim, the points of the edge of the disc become one after another the points of contact of the disc with the rim. The condition of rolling without sliding means that the length of the arc PQ between two points, P and Q, of the edge of the disc is equal to the length of that arc of the rim upon which the points of the arc PQ successively fall when the disc is rolling. Since the radius of the disc is equal to half the radius of the rim, the edge of the disc always passes through the centre O of the rim.

Let us choose a point P on the edge of the disc and let P coincide with point A of the rim in the initial position of the disc.

When the disc rolls so far that its point of contact with the rim runs over the quarter AB of the rim, the corresponding points of the edge of the disc, i.e. those points which successively touch the rim at the points of the arc AB, will form a semi-circle; accordingly, when the point of contact reaches point B, point

P will coincide with point O. We shall prove that during that motion P runs over the radius AO. Suppose that at a certain moment the disc touches the rim at point Q of the arc AB. The centre S of the disc is then the mid-point of the segment OQ. Point P is then so placed that the lengths of the arcs AQ and QP (less than semi-circles) are equal. It follows that

$$\sphericalangle AOQ = \tfrac{1}{2} \sphericalangle PSQ. \tag{1}$$

On the other hand, by the theorem on the exterior angle of a triangle, $\sphericalangle PSQ = \sphericalangle POS + \sphericalangle OPS$, and since $\sphericalangle OPS = \sphericalangle POS$ we have

$$\sphericalangle POS = \tfrac{1}{2} \sphericalangle PSQ. \tag{2}$$

Equations (1) and (2) give

$$\sphericalangle POS = \sphericalangle AOQ,$$

whence point P lies on the radius AO (Fig. 132).

Conversely, if P' is an arbitrary interior point of the segment AO, then there exists a position of the disc in which P coincides with P': the centre of the disc then lies at the intersection of the perpendicular bisector of segment OP' with that arc of the circle with centre O and radius r which lies in the angle AOB.

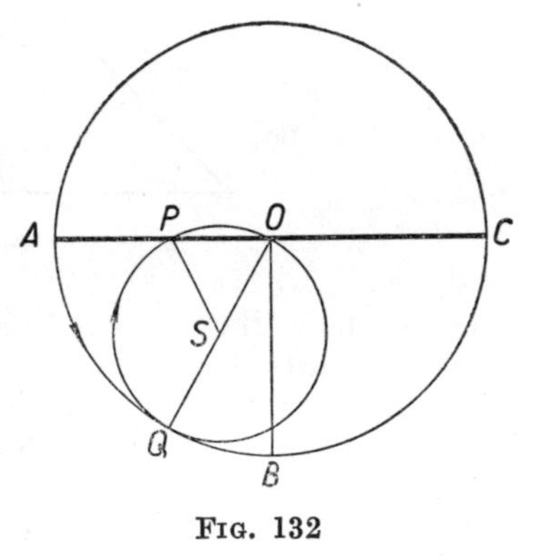

Fig. 132

We have proved that, when the disc rolls over the quarter AB of the rim, point P runs over the radius AO. When rolling over the quarter BC, the disc assumes positions symmetrical to those it assumed before with respect to the straight line OB; thus point P runs over the radius OC, symmetrical to OA, i.e. forming together with OA the diameter AC of the rim. Rolling over the remaining quarters of the rim, point P will run over the diameter CA, since it will then assume positions symmetrical to those it assumed before with respect to the straight line AC.

We have obtained the following result: when a disc is rolling inside a rim with a radius twice as great, each point of the disc runs over a diameter of that rim.

118. Denote by M and N the points of intersection of the straight line m with the planes α and β, by S—the mid-point of the segment MN and by s—the intersection line of the planes

α and β. We are to find the figure formed by the mid-points of all segments XY parallel to MN and having their end-points X and Y on the planes α and β. Accordingly let us consider the pencil of planes with edge m, i.e. the set of all planes passing through the line m. Every segment XY, being parallel to MN, lies on one of the planes of that pencil. The required locus will thus be determined if we find its points on every plane of the pencil. We shall distinguish two cases.

(1) Suppose that a plane of the pencil in question intersects the straight line s at point P, i.e. that it intersects the planes

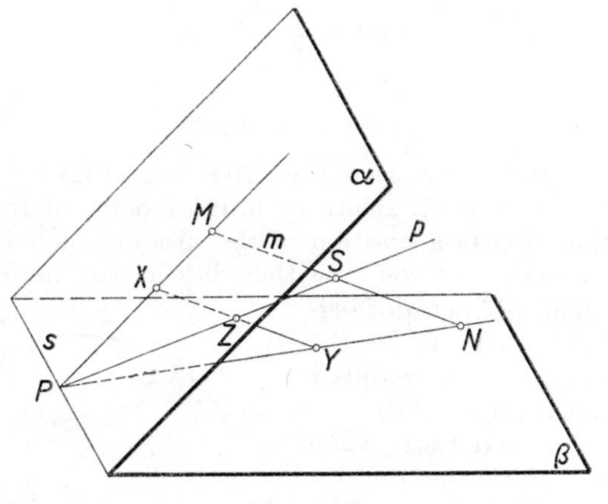

Fig. 133

α and β along the straight lines PM and PN (Fig. 133). Those of the segments XY under consideration which lie on the plane MPN have their end-points on the lines PM and PN; the mid-points Z of those segments thus form a straight line p passing through points P and S.

(2) Suppose that a plane of the pencil in question is parallel to the line s, i.e. that it intersects the planes α and β along the lines MK and NL, parallel to s (Fig. 134).

Those of the segments XY under consideration which lie on the plane MKS have their end-points on the lines MK and NL, and the mid-points Z of those segments form a straight line q passing through point S and parallel to s.

Thus the required locus consists of the points of all straight lines drawn through point S to all points P of line s and of the points of line q drawn through point S parallel to s.

Those points form a plane, determined by point S and the straight line s.

REMARK. We might be in doubt whether the straight line s belongs to our locus or not, since a straight line drawn parallel to m from a point of s intersects the two planes α and β at the same point and there is no segment XY on it. However, in geometry it is convenient to assume that, if points X and Y coincide, they form a *zero segment*, i.e. a segment consisting of one point only, which is then in itself the initial point, the mid-point and the end-point of the segment.

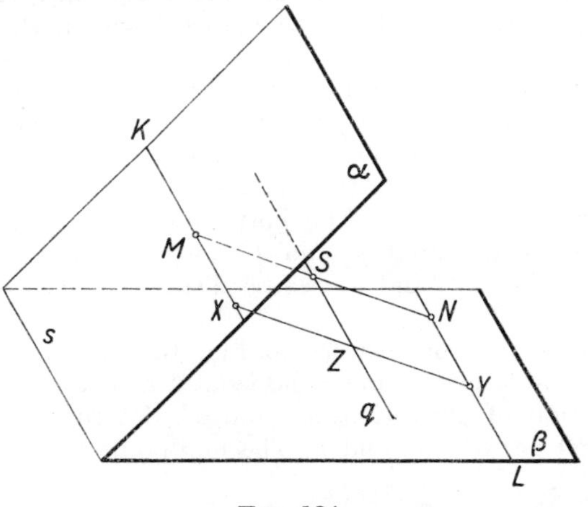

FIG. 134

According to this agreement the straight line s belongs to the locus which we have determined.

119. (1) Let m and n be perpendicular lines intersecting at point 0 (Fig. 135).

Let point S be the mid-point of a segment MN of length a with end-points lying on m and n. In the rectangle $OMPN$ (which is reduced to the segment MN if M or N coincides with O), we have

$$OS = SM = \tfrac{1}{2}a.$$

Thus point S lies on a circle lying in the plane of m and n with centre O and radius $\frac{1}{2}a$.

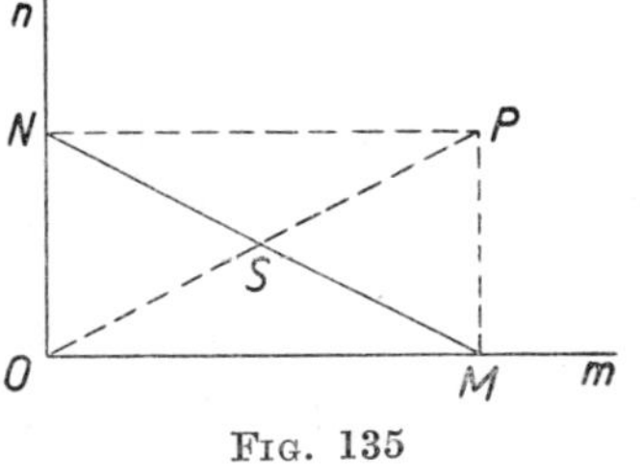

FIG. 135

Conversely, let S be a point of that circle, i.e. $OS = \frac{1}{2}a$. Point S is then the mid-point of a segment of length a with end-points lying on m and n. This segment will be obtained by marking off on segment OS produced a segment $SP = OS$ and constructing the rectangle $OMPN$; then

$$MN = OP = 2 \times OS = a.$$

If point S lies on one of the given straight lines, the rectangle $OMPN$ is reduced to the segment OP.

We have shown that, *if m and n intersect, then the required locus is the circle with centre O and radius $\frac{1}{2}a$ lying in the plane of the given straight lines.*

(2) Let us now consider the case where the lines m and n are skew. Suppose that the shortest distance between them is equal to d.

In a drawing it is easiest to represent m and n by means of orthogonal projections on two perpendicular planes. As the horizontal plane of projection we shall take a plane α parallel to both m and n, and as the vertical plane of projection—any plane perpendicular to α. The horizontal projections of m and n are perpendicular lines m' and n', and the vertical projections are parallel lines m'' and n'' whose distance from each other is equal to d.

These projections are shown in Fig. 136. Points P and Q of lines m and n, whose common horizontal projection is the point of intersection of projections m' and n', determine the shortest distance PQ between m and n. The mid-point O of segment PQ

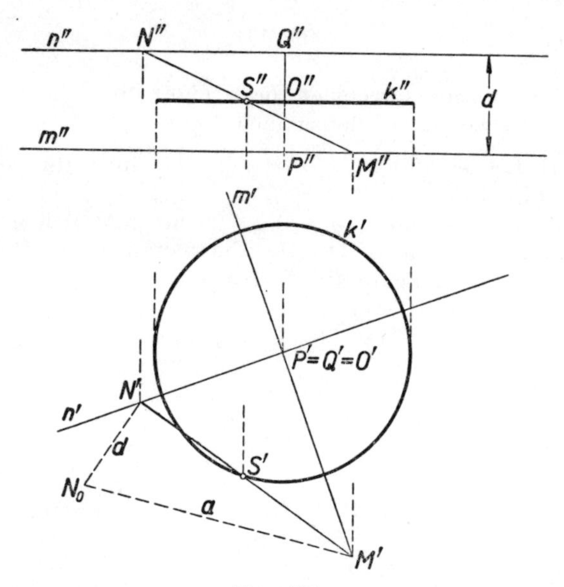

FIG. 136

has the projections: $O' = P' = Q'$; O'' is the mid-point of the segment $P''Q''$.

Let us join a point M of m with a point N of n. The projections of the mid-point S of segment MN lie at the mid-points S' and S'' of the projections $M'N'$ and $M''N''$ of that segment. The actual length of the segment MN will be determined from its projections by constructing a right-angled triangle with sides $M'N'$ and $N'N_0 = d$; the hypotenuse $M'N_0$ is equal to the segment MN.

If $MN = a$, then

$$M'N' = \sqrt{(M'N_0^2 - N'N_0^2)} = \sqrt{(a^2 - d^2)}.$$

When the segment MN runs its end-points along the lines m and n retaining its constant length a, the segment $M'N'$, as can be seen from the above formula, also retains the constant length $\sqrt{(a^2-d^2)}$.

Therefore the locus of the mid-point S' of the segment $M'N'$ is the circle k' with centre O' and radius $\frac{1}{2}\sqrt{(a^2-d^2)}$.

When point S' describes circle k', point S'' runs over a segment k'' of the straight line equidistant from m'' and n'', and point S describes a circle k whose projections are k' and k''. This circle is the required locus; it lies on a plane equidistant from m and n; its centre is at point O and its radius is equal to $\frac{1}{2}\sqrt{(a^2-d^2)}$.

Remark. The investigation of case (2) can also be carried out on a drawing made in the oblique projection. As the plane of projection we shall take the plane determined by one of the given skew lines m and n, say m, and by the segment PQ—the shortest distance between m and n. For simplicity, we shall denote the projections of points and lines on this plane by the letters which denote those points and lines themselves.

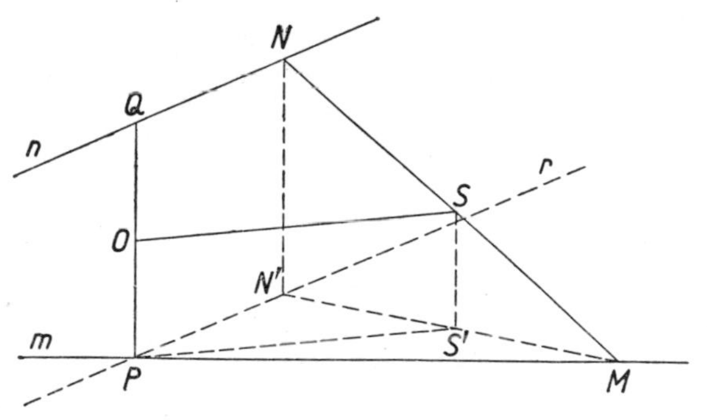

Fig. 137

We draw the straight line m and the segment PQ of length d perpendicular to m (Fig. 137). As the projection of n we can take any straight line passing through point Q. Let MN be an arbitrary segment of length a whose end-points lie on m and n. Let S be the mid-point of this segment and O the mid-point of segment PQ.

Let us draw through point P a straight line r parallel to n and consider the orthogonal projection of segment MN on the plane (m, r) of lines m and r. Since $PQ \perp m$ and $PQ \perp r$, the segment PQ is perpendicular to the plane (m, r).

Consequently the orthogonal projection N' of point N on this plane will be obtained by drawing a parallel to PQ from point N to the point of intersection with r. The orthogonal projection S' of points S is found in a similar way by drawing a parallel to PQ from point S to the point of intersection with MN'; point S' is the mid-point of segment MN'.

In the right-angled triangle MNN' we have $MN = a$, $NN' = d$, whence

$$MN' = \sqrt{(MN^2 - NN'^2)} = \sqrt{(a^2 - d^2)}.$$

When segment MN of constant length a runs its end-points along the lines m and n, segment MN' keeps a constant length

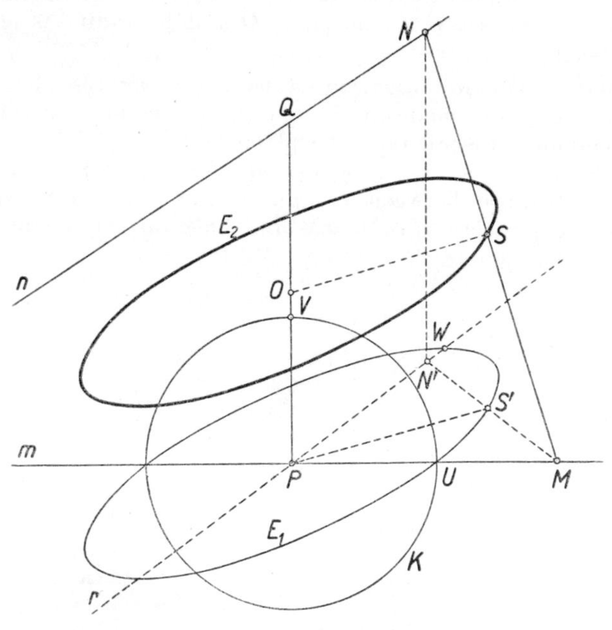

Fig. 138

$\sqrt{(a^2-d^2)}$. Consequently, the locus of point S' is a circle with centre P and radius $PS' = \frac{1}{2}\sqrt{(a^2-d^2)}$ lying in the plane (m, r).

Point S is obtained through translating point S' perpendicularly to the plane (m, r) by the length $SS' = PO = \frac{1}{2}d$.

Thus the locus of point S is a circle with centre O and radius $\frac{1}{2}\sqrt{(a^2-d^2)}$ lying in a plane parallel to both m and n.

In the oblique projection upon the plane MPQ both loci are ellipses. Those ellipses have been drawn in Fig. 138 in the following manner: Ellipse E_1 has been drawn as the oblique projection of circle K with radius $PU = \frac{1}{2}\sqrt{(a^2-d^2)}$, segments PU and PW being taken as the projections of two perpendicular radii PU and PV of that circle. Ellipse E_2 has been obtained through a translation of ellipse E_1 determined by vector PO.

§ 8. Constructions

120. Let MN be the required segment with end-point M on a given circle k, end-point N on the given straight line n and the mid-point at a given point A (Fig. 139).

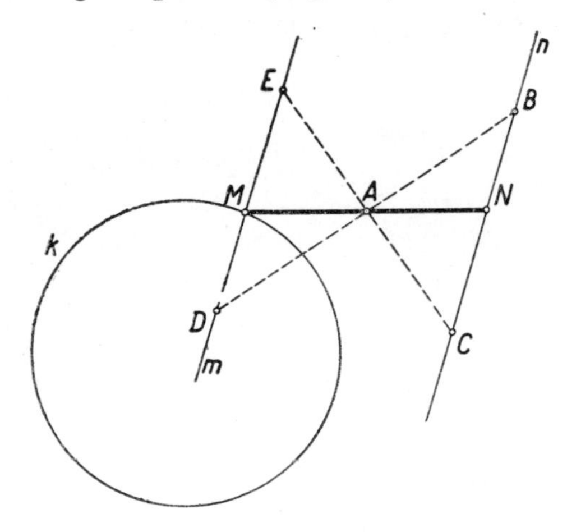

FIG. 139

Since point M is placed symmetrically to point N with respect to centre A, it lies on the line symmetrical to n with respect to A. This observation leads to the following construction.

We draw line m symmetrical to n with respect to A. It is most convenient to choose on n two arbitrary points B and C, draw the straight lines BA and CA and mark off on them the segments

$AD = BA$ and $AE = CA$, as shown in Fig. 139; DE is the required line m. Suppose that m intersects the circle k at point M. Drawing the straight line MA until it intersects n at point N, we obtain the required segment MN, points M and N being the corresponding points of lines m and n symmetrical with respect to A; consequently $MA = AN$.

The problem has two solutions, one solution or no solutions according to whether m intersects the circle k, is tangent to it or lies outside it.

REMARK. Instead of drawing a straight line m symmetrical to n with respect to point A we could describe a circle l symmetrical to the circle k with respect to A and determine the required point N at the intersection of l with the given straight line n.

In the same way we can solve any problem of the following kind: Given any two lines l_1 and l_2 (straight lines, circles or any other curves) and a point A, draw a segment with mid-point A, one end-point on l_1 and the other on l_2.

121. (1) On the given straight line p let us mark off a segment $MN = d$ (Fig. 140). The path $AM+MN+NB$ will be shortest when the sum of the segments $AM+NB$ is the least. Let us move

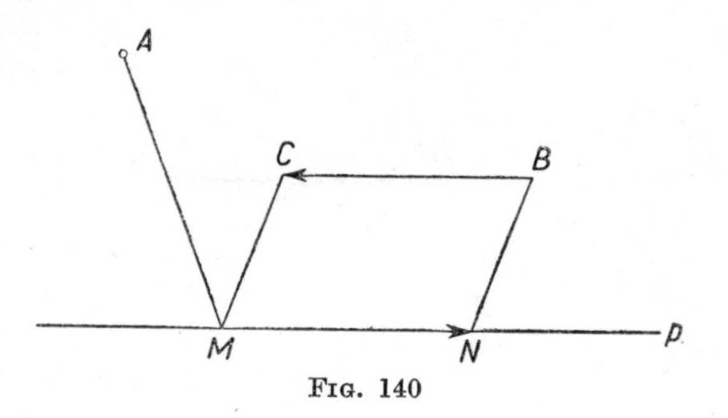

FIG. 140

point B parallel to p the distance $BC = d$ in the opposite direction to the direction MN; then $AM+NB = AM+MC$. The problem is thus reduced to determining on p a point M for which the sum of its distances from the given points A and C, lying on the same side of p, is the least. We solve this problem in a well-known way: point M is obtained as the point of intersection of p with the segment joining point A with a point C', symmetrical to C with respect to p.

From the above analysis of the problem we derive the construction represented in Fig. 141. We move point B parallel to p through the distance d. This translation can be made in two opposite directions; thus we obtain two points C_1 and C_2. We find points C_1' and C_2' symmetrical to points C_1 and C_2 with respect to p and the points of intersection M_1 and M_2 of the segments AC_1' and AC_2' with the straight line p. If points A and B do not

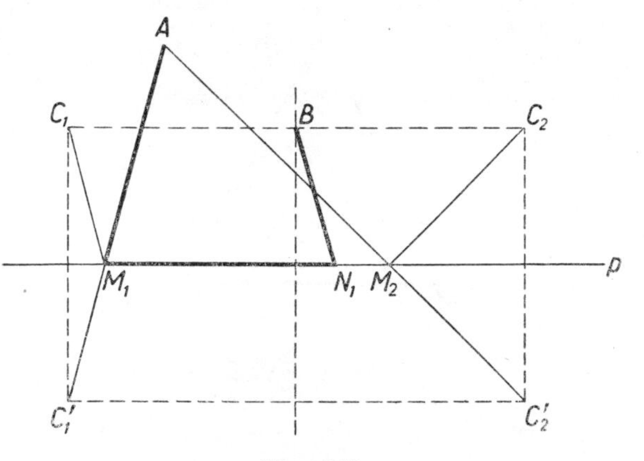

FIG. 141

lie on the same perpendicular to p, then only one of the points C_1 and C_2 is suitable, namely that one which lies together with point A on the same side of a perpendicular drawn through B to the straight line p; suppose it is point C_1 (as in Fig. 141). Indeed

$$AM_1+M_1C_1 = AM_1+M_1C_1' = AC_1' < AC_2'$$
$$= AM_2+M_2C_2' = AM_2+M_2C_2,$$

whence the sum $AM_2+M_2C_2$, being greater than the sum AM_1+ $+M_1C_1$, does not give the required minimum.

Marking off on the straight line p a segment $M_1N_1 = d$ in the direction of C_1B, we obtain the required path $AM_1+M_1N_1+$ $+N_1B$. In this case the problem has one solution.

If points A and B lie on a perpendicular to the straight line p, the problem has two solutions symmetrical with respect to the line AB:

$$AM_1+M_1N_1+N_1B \quad \text{and} \quad AM_2+M_2N_2+N_2B \text{ (Fig. 142).}$$

EXERCISE. We suggest that the reader should solve the same problem in the case where points A and B lie on opposite sides of the straight line p.

(2) Suppose that MN is a segment of given length d lying on p for which $AM = BN$ (Fig. 143). Let us move point B parallel

to p through the distance $BC = d$ in the direction opposite to that of MN; then the quadrilateral $MNBC$ is a parallelogram, whence $CM = BN$, and consequently also $CM = AM$. Thus point M lies on the perpendicular bisector of the segment AC.

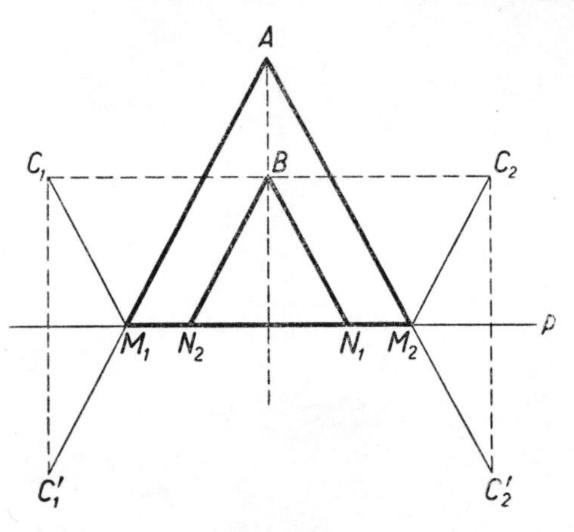

FIG. 142

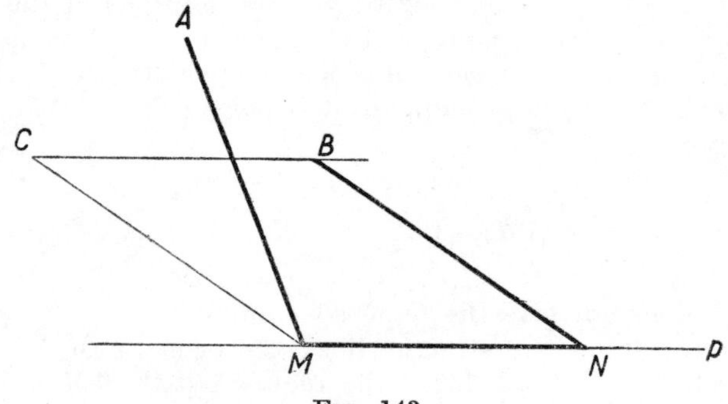

FIG. 143

We derive hence the solution of the problem represented in Fig. 144. We move point B parallel to p the given distance d. Since this translation can be made in two opposite directions, we obtain two points C_1 and C_2. We draw the perpendicular bisectors of the segments AC_1 and AC_2. If those perpendicular bisectors intersect p at points M_1, M_2, we mark off on p a segment $M_1N_1 = d$ in the direction of C_1B and a segment $M_2N_2 = d$ in the direction of C_2B; we obtain the required polygonal lines AM_1N_1B and AM_2N_2B. The problem has two solutions with the exception of the case where one of the points C_1, C_2, say point C_1, lies on a perpendicular drawn from point A to p. In that case, if point C_1 is different from point A, the problem has one solution AM_2N_2B, since the perpendicular bisector of the segment AC_1 does not intersect p; if point C_1 coincides with point A, the problem has infinitely many solutions. Namely, a solution is provided by any polygonal line AM_1N_1B for which the segment M_1N_1 is of length d and direction AB and by one more polygonal line AM_2N_2B, in which M_2 and N_2 are the projections of points B and A, respectively, upon the straight line p.

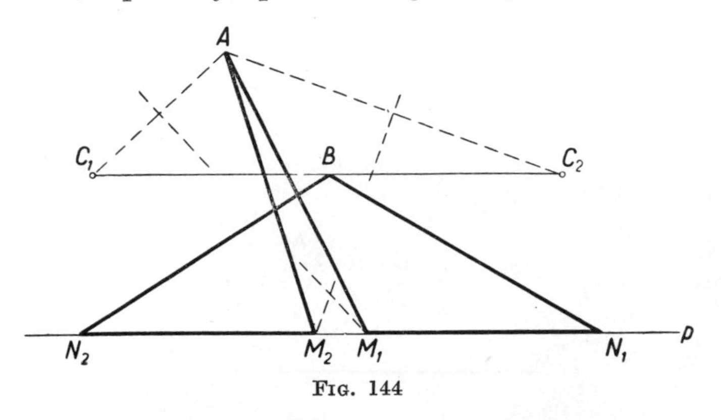

FIG. 144

REMARK. In problem (2) the condition that points A and B should lie on the same side of p is inessential. If those points lie on opposite sides of p or if p passes through one of them (or even through both), the solution of the problem is the same.

122. Let us denote the lengths of the given sides of the quadrilateral as follows: $AB = a$, $CD = c$. Consider the sums of the successive angles of the quadrilateral:

$$A+B, \quad B+C, \quad C+D, \quad A+D$$

and put these sums in pairs:

$$(A+B, C+D)$$

and

$$(A+D, B+C).$$

In each of these pairs one of the sums is not greater than 180°, since the sum of all the four angles $A+B+C+D$ is 360°.

It suffices to consider the case where $A+B \leqslant 180°$ and $A+D \leqslant 180°$; the argument which we

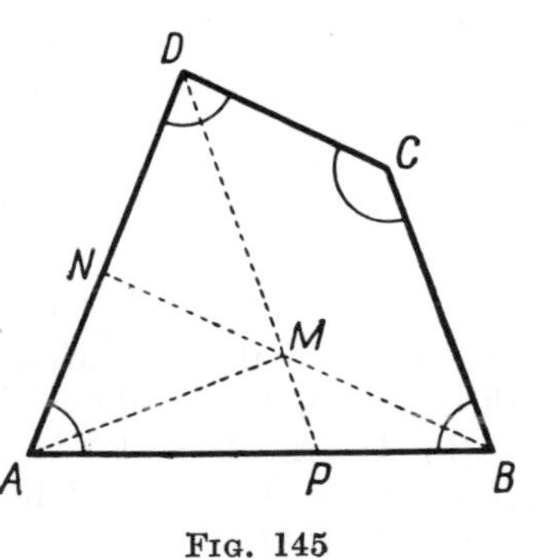

FIG. 145

shall develop can be applied to any other case by changing the letter notation in a suitable way.

It is convenient to consider two possibilities:

I. $A+B < 180°$, whence $C+D > 180°$. Let $ABCD$ be the required quadrilateral. This quadrilateral is convex (Fig. 145) or concave (Fig. 146) according to whether $C < 180°$ or $C > 180°$.

Let us draw through the vertex B a line parallel to CD, and through the vertex D a line parallel to BC; a parallelogram $BCDM$ will be formed.

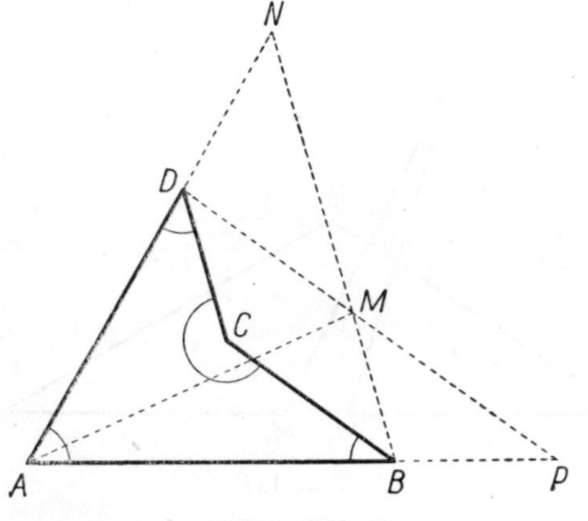

FIG. 146

(a) If $A+D = 180°$, i.e. if the quadrilateral $ABCD$ is a trapezium (Fig. 147), point M lies on the side AB, $MB = c$.

In this particular case the construction of the quadrilateral is easy.

We draw a segment $AB = a$. We construct $\sphericalangle BAD = A$ and $\sphericalangle ABC = B$ on the same side of the line AB. We then mark off $BM = c$. We draw through point M a parallel to BC as far as the intersection with the half-line AD at point D. Through point D we draw the line DC as far as the intersection with BC at point C. It is easy to verify that the quadrilateral $ABCD$ satisfies the conditions of the problem. This solution exists if point M lies inside the segment AB, i.e. if $c < a$.

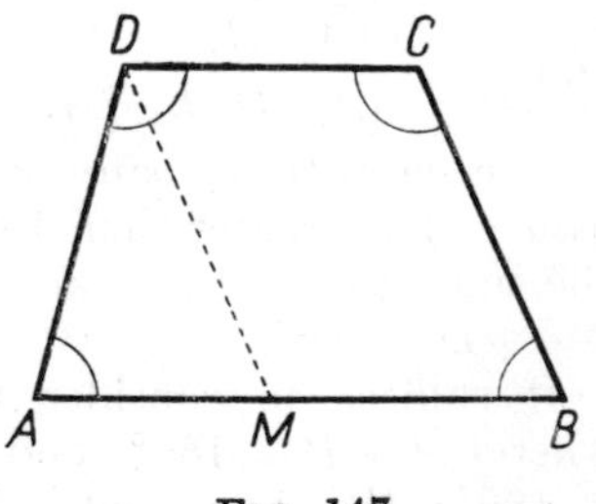

FIG. 147

(b) If $A+D < 180°$ (whence $B+C > 180°$), point M lies inside the angle BAD.

Indeed, if the quadrilateral $ABCD$ is convex (Fig. 145), the half-line DM lies inside angle ADC since $\sphericalangle MDC = 180° - C < D$; consequently the half-line DM intersects the straight line AB at point P lying inside the segment AB. The half-line BM lies inside angle ABC because $\sphericalangle CBM = 180° - C < B$, whence the half-line BM intersects the straight line AD at a point N lying inside the segment AD. The segments DP and BN intersect at point M lying inside the angle BAD.

Now if the quadrilateral $ABCD$ is concave (Fig. 146), then the half-line DM intersects the side AB produced at point P lying beyond point B; similarly, the half-line BM intersects the side AD produced at point N lying beyond point D. The segments DP and BN intersect at point M lying inside angle BAD.

Let us calculate the angle ABM. In the case of a convex polygon

$$\sphericalangle ABM = B - \sphericalangle MBC = B - (180° - C) = B + C - 180°.$$

In the case of a concave polygon

$$\sphericalangle ABM = B + \sphericalangle CBM = B + 180° - (360° - C) = B + C - 180°.$$

In both cases we know in the triangle ABM the sides $AB = a$ and $BM = c$ and the angle included between them; it is therefore possible to construct triangle ABM, and thus find point M, as follows.

We draw a segment $AB = a$ and angles A and B at the half-lines AB and BA, respectively; we then construct triangle ABM and draw through point M a line parallel to the arm of angle B which we have constructed. If that parallel line intersects the arm of angle A at point D, then the vertex C of the quadrilateral will be found at the intersection of the arm of angle B with the straight line drawn through point D parallel to BM. The construction makes it clear that the quadrilateral $ABCD$ satisfies the conditions of the problem. This solution exists if point M happens to be inside angle BAD since it is only in that case that the construction is feasible. This condition can easily be expressed as a relation between the data of the problem. Namely, the half-line BM intersects in this case the arm of angle A which has been constructed and forms triangle ABN because

$$A + \sphericalangle ABM = A + (B + C - 180°) = 180° - D < 180°.$$

Point M lies inside angle BAD if $BM < BN$. Now

$$BM = c, \quad BN = AB \times \frac{\sin \sphericalangle NAB}{\sin \sphericalangle ANB} = a \times \frac{\sin A}{\sin D};$$

thus the condition of solvability of the problem is

$$c < \frac{a \sin A}{\sin D},$$

i.e.

$$a \sin A > c \sin D. \tag{1}$$

It will be observed that, if $A+D = 180°$, then $\sin A = \sin D$ and inequality (1) gives the condition $a > c$, which has been found for this case before. We can thus formulate the result of the whole reasoning as follows:

If $A+B < 180°$, *the problem has a solution on condition that inequality* (1) *is satisfied; there is only one solution.*

II. $A+B = 180°$. In this case the required quadrilateral is a trapezium with bases AD and BC ($AD \geqslant BC$, since we assumed at the beginning that $A+D \leqslant 180°$).

If $A+D = 180°$, the quadrilateral is a parallelogram, whence $AB = CD$ (Fig. 148). The problem can then be solved only if the given sides satisfy the condition $a = c$; there are infinitely many solutions.

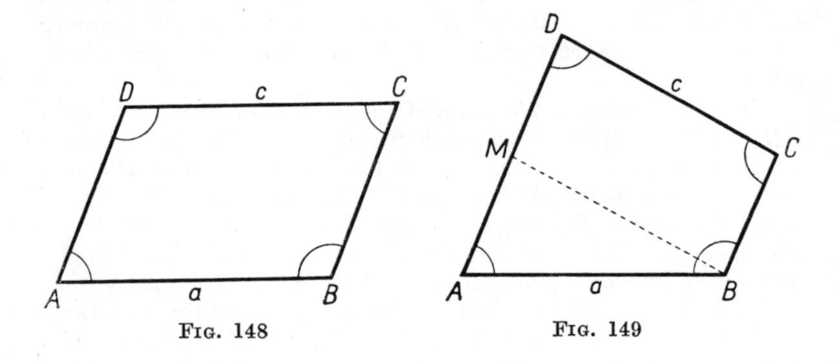

FIG. 148 FIG. 149

If $A+D < 180°$, then the parallel drawn from point B to the side CD cuts off from the quadrilateral a triangle ABM (Fig. 149) in which $AB = a$, $BM = c$, $\sphericalangle AMB = D$, $\sphericalangle BAM = A$. The given sides and angles must therefore satisfy the following relation (the Sine Rule):

$$\frac{a}{\sin D} = \frac{c}{\sin A},$$

i.e.

$$a \sin A = c \sin D. \tag{2}$$

If this condition is satisfied, the problem has infinitely many solutions, which are obtained by constructing the triangle ABM and translating the segment BM in the direction of AM.

It will be observed that, if $A+D = 180°$, equation (2) gives the condition $a = c$, which has been found for this case before. We can thus say that:

If $A+B = 180°$, *the problem has solutions provided equation* (2) *is satisfied; there are infinitely many solutions.*

123. *Method I. Analysis.* Let B_1, B_2, B_3, B_4, B_5 be vertices of a pentagon the mid-points of whose sides are the vertices A_1, A_2, A_3, A_4, A_5 of a given pentagon (Fig. 150).

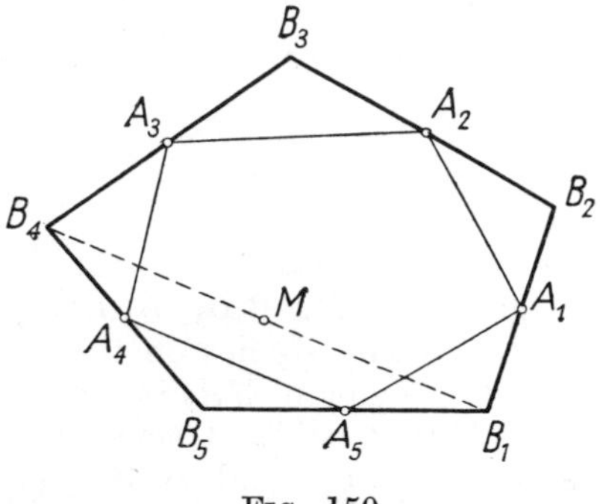

FIG. 150

Let us draw the diagonal B_1B_4 and denote its mid-point by M. The mid-points A_1, A_2, A_3, M of the sides of the quadrilateral $B_1B_2B_3B_4$ are, as we know, the vertices of a parallelogram. Point M can thus be determined as the fourth vertex of a parallelogram whose three successive vertices are A_1, A_2, A_3. Next it will be observed that, since A_4 and A_5 are the mid-points of the segments B_4B_5 and B_5B_1, the segment A_4A_5 is parallel to the diagonal B_4B_1 and equal to half that diagonal. Thus, having point M, we shall find point B_1 by drawing through point M a line parallel to the side A_4A_5 and marking a segment $MB_1 = A_4A_5$ in the direction of A_4A_5. We shall then determine successively the points B_2, B_3, B_4, B_5.

Construction (Fig. 151). We construct the parallelogram A_1A_2 A_3M. Through point M we draw the parallel to the side A_4A_5 and mark off $MB_1 = A_4A_5$ in the direction of A_4A_5. We then join point B_1 with point A_1, and on B_1A_1 produced beyond point A_1 we mark off $A_1B_2 = B_1A_1$. In the same way we determine points B_3, B_4, B_5. Finally we join point B_5 with point B_1 and obtain the polygon $B_1B_2B_3B_4B_5$. We shall prove that it is the required one.

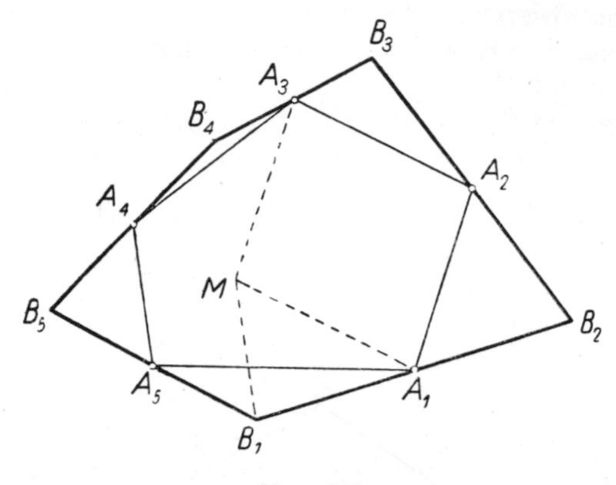

FIG. 151

Since—according to the construction—points A_1, A_2, A_3, A_4 are the mid-points of segments B_1B_2, B_2B_3, B_3B_4, B_4B_5, respectively, it suffices to prove that point A_5 lies on the segment B_5B_1 and is its mid-point. Now the segment MA_3 is parallel to the segment B_1B_3 and equal to half that segment since the same properties characterize the segment A_1A_2 equal to MA_3 and parallel to it. Therefore point M is the mid-point of the segment B_1B_4. The segment A_4A_5 is thus parallel to the segment B_1B_4 and equal to half that segment, whence point A_5 is the mid-point of the segment B_1B_5.

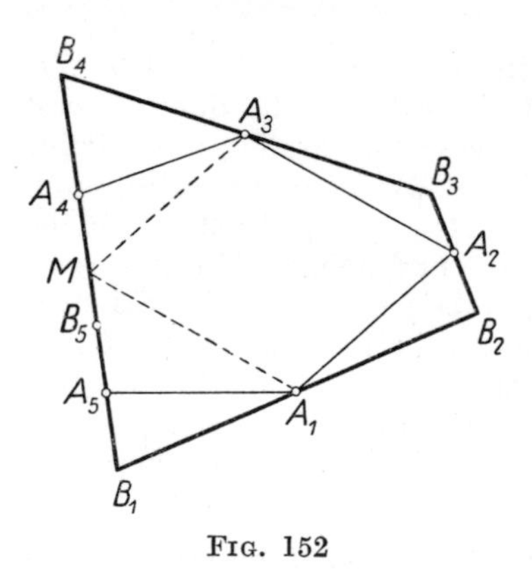

FIG. 152

In performing the above construction we can encounter certain particular cases. Namely, three successive vertices of the required polygon may occur on one straight line (containing a side of the given polygon) as shown in Fig. 152.

It may also happen that one of the points B_i will coincide with one of those vertices A_i with which, in our construction, it is to be joined by a segment. This case is shown in Fig. 153, where point B_3 happens to coincide with A_3. We must then regard point B_3 as point B_4 as well (obtaining equal "zero segments" B_3A_3 and A_3B_4) and perform the rest of the construction as before.

In the constructions shown in Figs. 152 and 153 the figure $B_1B_2B_3B_4B_5$ obtained is not a pentagon in the usual sense of the word; thus the problem has no solution. However, we can extend the meaning of the term "pentagon" to cover the above cases, e.g. by calling those figures "deformed pentagons".

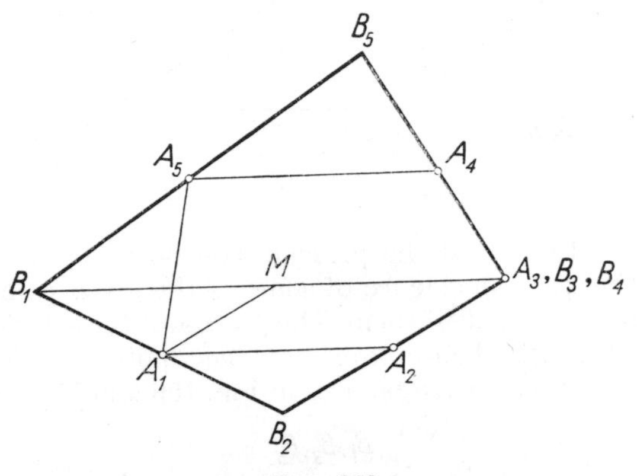

Fig. 153

If we agree to do this, the problem always has a solution (an ordinary pentagon or a deformed one) and the solution is unique because the construction determines the vertices B_1, B_2, B_3, B_4, B_5 in a unique manner.

Remark. (1) The condition that points A_1, A_2, A_3, A_4, A_5 are the vertices of a convex pentagon can be disregarded and a more general problem posed:

In a plane, points A_1, A_2, A_3, A_4, A_5 are given. Find points B_1, B_2, B_3, B_4, B_5 such that point A_1 is the mid-point of the segment B_1B_2, point A_2 the mid-point of the segment B_2B_3, etc., and finally, point A_5 the mid-point of the segment B_5B_1.

The solution proceeds as before, except that points A_1, A_2, A_3 can now lie on a straight line. We shall now determine point M by marking off on that line a segment A_3M, equal to the segment A_2A_1 and identically directed, which replaces our former construction of the parallelogram $A_1A_2A_3M$ (Fig. 154).

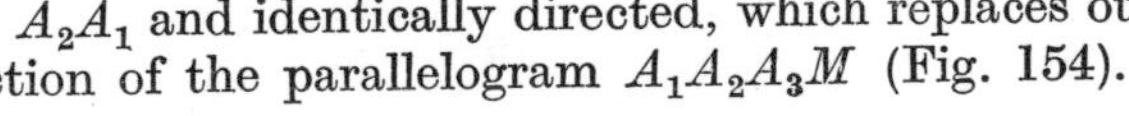

M A3 A1 A2

Fig. 154

This problem always has one and only one solution if we admit deformed solutions in which some of the points B_1, B_2, B_3, B_4, B_5 can coincide, e.g. as in Fig. 155.

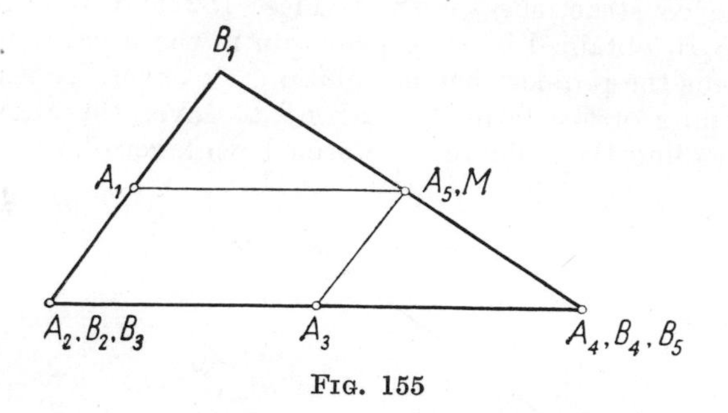

Fig. 155

The generalization of the problem can be pushed still further, by admitting the coincidence of some of the given points A_1, A_2, A_3, A_4, A_5, or even all of them. The process of the solution remains the same but the deformations of the "polygon" $B_1B_2B_3B_4B_5$ will go still further, as can be seen in Fig. 156 and 157 for instance.

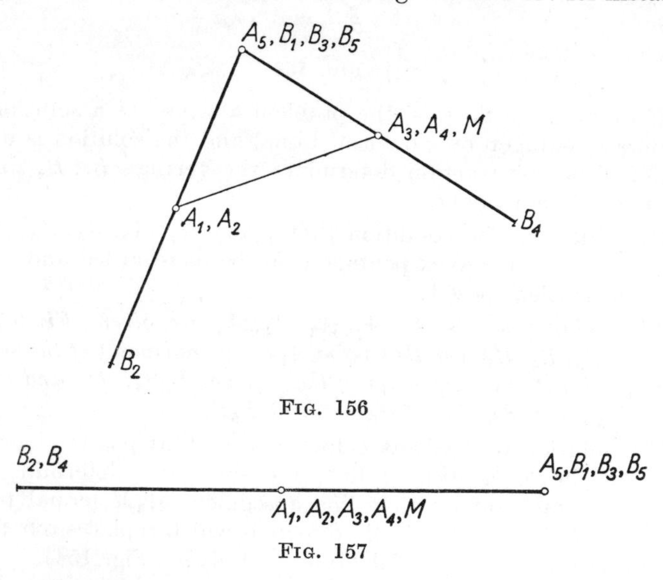

Fig. 156

Fig. 157

(2) A further generalization of the problem will be obtained by assuming that we are given in a plane an arbitrary odd number of points A_1, A_2, ..., A_{2n+1} (n—a natural number), not necessarily

different, and we seek points $B_1, B_2, \ldots, B_{2n+1}$ such that point A_i is for $i = 1, 2, \ldots, 2n$ the mid-point of segment B_iB_{i+1} and point A_{2n+1} is the mid-point of segment $B_{2n+1}B_1$.

We know how to solve this problem for $n = 2$ (and also for $n = 1$). The general solution will be obtained by induction. Namely, we shall show how to solve the problem for any arbitrarily chosen value of n greater than 2 under the assumption that the solution for the value of n less by unity is already known.

We determine point M, as before, by drawing a segment A_3M equal to the segment A_2A_1 and identically directed. Then, for the system of points $M, A_4, A_5, \ldots, A_{2n+1}$, whose number is $2(n-1)+1$, we find the required system of points, in which point M is the mid-point of segment B_1B_4, point A_4 is the mid-point of the segment B_4B_5, etc. Finally, we determine points B_2 and B_3.

Under the agreement concerning the coincidence of points the problem always has one solution.

If the number of given points is even, the problem generally has no solution. For example for a given convex quadrilateral $A_1A_2A_3A_4$ the quadrilateral $B_1B_2B_3B_4$ exists only if the quadrilateral $A_1A_2A_3A_4$ is a parallelogram.

Method II. Analysis. Let $B_1B_2B_3B_4B_5$ be the required pentagon (Fig. 158). Let us draw the diagonals B_1B_3 and B_1B_4. In the

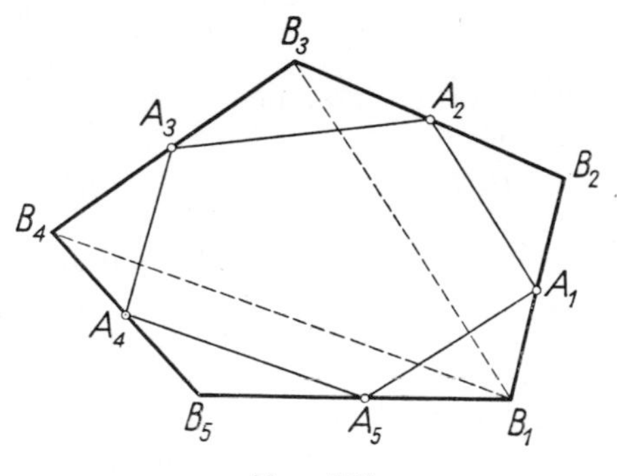

FIG. 158

triangle $B_1B_2B_3$ the segment A_1A_2 joins the mid-points of the sides B_3B_2 and B_1B_2, whence $B_1B_3 \| A_1A_2$ and $B_1B_3 = 2A_1A_2$; also $B_1B_4 \| A_5A_4$ and $B_1B_4 = 2A_5A_4$.

Thus we know the lengths and the directions of two sides of the triangle $B_1B_3B_4$, whence we can find the length and the direction of the third side, i.e. determine points B_3 and B_4.

Construction (Fig. 159). From an arbitrary point M, say from point A_1, we draw parallels to the sides A_1A_2 and A_5A_4 of the given pentagon and mark off $MN = 2A_1A_2$ and $MP = 2A_5A_4$. We then draw through point A_3 the parallel to the line NP and mark off $A_3B_3 = A_3B_4 = \frac{1}{2}NP$. Next we find the vertex B_1 as the point of intersection of the straight line drawn from point B_3 parallel to the side A_2A_1 with the straight line drawn from point B_4 parallel to the side A_4A_5, the vertex B_2 at the intersection of B_3A_2 and B_1A_1, and finally the vertex B_5 at the intersection of B_4A_4 and B_1A_5.

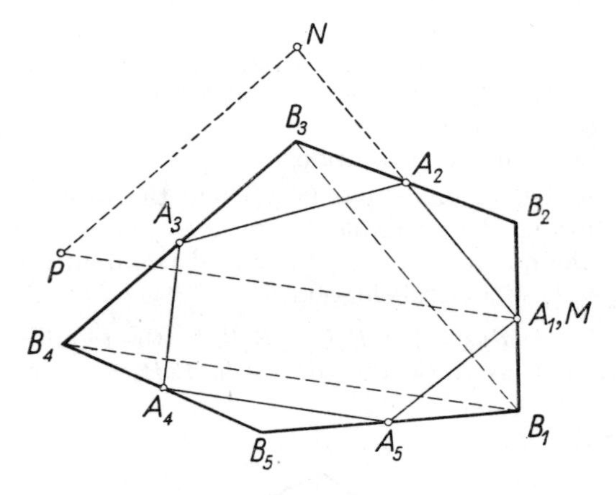

FIG. 159

Since triangle $B_1B_3B_4$ is congruent to triangle MNP, we have $B_1B_3 = MN = 2A_1A_2$, and, since $B_1B_3 || A_1A_2$, points A_1 and A_2 are the mid-points of the sides B_1B_2 and B_2B_3 and similarly points A_4 and A_5 are the mid-points of the sides B_4B_5 and B_5B_1.

In the construction performed above we can encounter similar particular cases to those occurring in the construction of method I. If $A_1A_2 || A_5A_4$, then instead of a triangle MNP we obtain three segments of the same straight line. If, moreover, $A_1A_2 = A_5A_4$, then points N and P coincide; we must then assume that points B_3 and B_4 coincide with point A_3, as in Fig. 153. Keeping to our previous agreement as regards deformed pentagons, we conclude that the problem always has a solution and that solution is unique.

REMARK. The above method can also be applied to the more general problems discussed in method I. Given $2n+1$ points, where $n > 2$, we replace the construction of the triangle $B_1B_3B_4$ by the construction of the "polygon" $B_1B_3 \ldots B_{2n-1}B_{2n+1}$, in which the segments $B_1B_3, B_3B_5, \ldots, B_{2n-1}B_{2n+1}$ are parallel to the segments $A_1A_2, A_3A_4, \ldots, A_{2n-1}A_{2n}$ respectively and are twice the length of those segments. This is illustrated in Fig. 160 for $n = 3$.

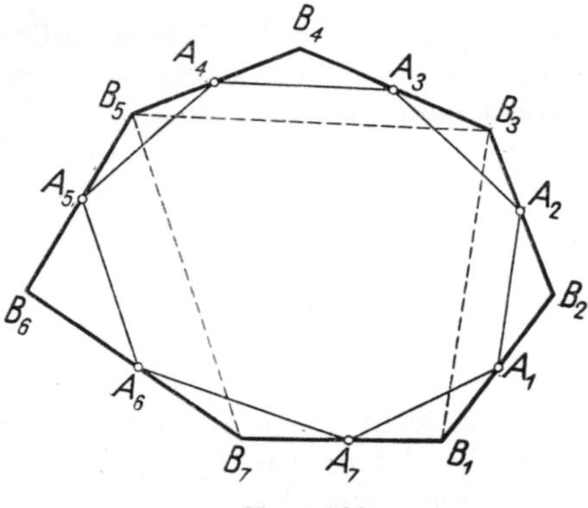

FIG. 160

Method III (calculatory). Let us choose an arbitrary system of orthogonal coordinates: given the points

$$A_1(x_1, y_1),\ A_2(x_2, y_2),\ A_3(x_3, y_3),\ A_4(x_4, y_4),\ A_5(x_5, y_5);$$

we seek points

$$B_1(X_1, Y_1),\ B_2(X_2, Y_2),\ B_3(X_3, Y_3),\ B_4(X_4, Y_4),\ B_5(X_5, Y_5).$$

Since the coordinates of the mid-point of a segment are equal to the arithmetical means of the corresponding coordinates of the end-points of the segment, for the determination of the required coordinates we have two systems of equations:

$$\begin{aligned}
X_1+X_2 &= 2x_1, & Y_1+Y_2 &= 2y_1,\\
X_2+X_3 &= 2x_2, & Y_2+Y_3 &= 2y_2,\\
X_3+X_4 &= 2x_3, & Y_3+Y_4 &= 2y_3,\\
X_4+X_5 &= 2x_4, & Y_4+Y_5 &= 2y_4,\\
X_5+X_1 &= 2x_5, & Y_5+Y_1 &= 2y_5.
\end{aligned}$$

Let us add the equations of each system:

$$X_1+X_2+X_3+X_4+X_5 = x_1+x_2+x_3+x_4+x_5,$$
$$Y_1+Y_2+Y_3+Y_4+Y_5 = y_1+y_2+y_3+y_4+y_5.$$

Hence, taking into account the preceding equations, we shall obtain the coordinates of point B_1,

$$X_1 = x_1 - x_2 + x_3 - x_4 + x_5,$$

$$Y_1 = y_1 - y_2 + y_3 - y_4 + y_5,$$

and analogous formulas for the coordinates of points B_2, B_3, B_4, B_5.

The construction of the required points according to these formulas presents of course no difficulties but is not very interesting from the geometrical point of view.

Keeping to our former agreement as regards deformed pentagons, we always obtain one solution. The condition that the given points are to be the vertices of a convex pentagon is of no significance here and the points can be chosen arbitrarily. The same applies to the general problem if an arbitrary odd number of points is given.

If the number of given points is even, the problem generally has no solution, since it leads to an inconsistent system of equa tions. For example, in the case of four points we would have to solve the following system of equations:

$$X_1 + X_2 = 2x_1, \quad X_2 + X_3 = 2x_2, \quad X_3 + X_4 = 2x_5, \quad X_4 + X_1 = 2x_4.$$

Adding the first equation to the third one, and the second to the fourth, we should obtain

$$X_1 + X_2 + X_3 + X_4 = 2(x_1 + x_3),$$

$$X_1 + X_2 + X_3 + X_4 = 2(x_2 + x_5).$$

These equations are consistent only if

$$x_1 + x_3 = x_2 + x_4.$$

124. We can arrange the given squares to form a rectangle in one way only. We shall prove this by the following inference.

To begin with, it is clear that squares filling up a rectangle must have sides parallel to the sides of the rectangle.

The area of the rectangle must be equal to the sum of the areas of all the squares, i.e. the area of the required rectangle, is

$$1^2 + 4^2 + 7^2 + 8^2 + 9^2 + 10^2 + 14^2 + 15^2 + 18^2 = 1056.$$

Number $1056 = 2^5 \times 3 \times 11$ is the product of the lengths of two adjacent sides of the rectangle. In order to find those lengths we must decompose 1056 into two factors, each of them not less than 18, i.e. than the length of the side of the largest square. There exist three such decompositions:

$$1056 = 22 \times 48 = 24 \times 44 = 32 \times 33.$$

The decompositions 22×48 and 24×44 must be rejected; for, if we place a square with side 18 in a rectangle of width 22 or 24, the part of the rectangle remaining above or below the square will be a strip of length 18 and width not greater than 6. That strip cannot be filled up with the given squares: we can place in it squares with sides 1 and 4 at most, and the sum $1+4 = 5$ is less than 18. The required rectangle, if it exists, has sides 32 and 33. We shall show that a rectangle $ABCD$ with sides $AB = 32$

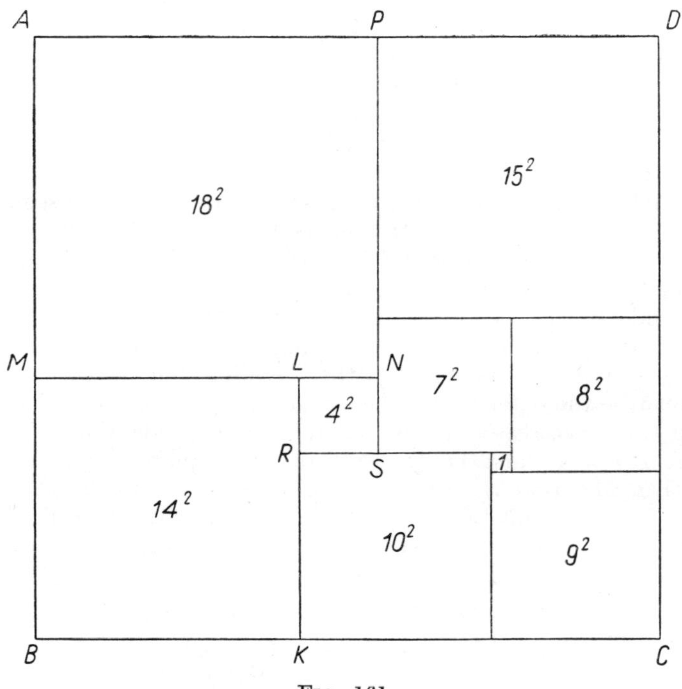

Fig. 161

and $AD = 33$ (Fig. 161) can indeed be made up from the given nine squares in one way only, disregarding of course rotations about the centre of the rectangle and reflections on its axes of symmetry.

The square with side 18 must be placed in a corner of the rectangle. Otherwise there would remain above and below that square strips of length 18 and joint width 14 or 15. One of those strips would thus be of width not greater than 7, and a strip like that could not be filled up by the given squares: we could place in it squares with sides 1, 4, 7 at most, the sum $1+5+7$ being less than 18.

Thus we draw a square $AMNP$ with side $AM = 18$.

We shall now decide which squares should be placed alongside the segment $MB = 14$. Number 14 must be either the length of the side of one square or the sum of the lengths of the sides of several squares; in the second case there are only two possibilities:

$$14 = 9+4+1 \quad \text{or} \quad 14 = 10+4.$$

These possibilities must both be rejected; for, if we place alongside the segment MB a square with side 9 or 10, there will remain a strip of length 9 or 10 and width not greater than 5, which again cannot be filled up with the remaining squares. Thus along the side MB we must place a square $MBKL$ with side 14.

Let us now consider point N. From this point there should run into the interior of the figure $PNLKCD$ a line dividing the rectangle into squares—either as an extension of the segment MN or as an extension of the segment PN. The first alternative cannot take place: the side $PN = 18$ would then be the sum of the sides of several squares, including side 15, since for the square with side 15 there would be no room elsewhere. But $18 = 15+3$, and 3 is neither the side nor the sum of the sides of any of the given squares. Consequently, the dividing line starting from N must be the extension of the side PN.

In this case, reasoning as before, we conclude that the segment $LN = 4$ must be the side of one square only, $LRSN$; we then find that the segment $KR = 10$ must also be the side of one square only, after which we shall easily determine the position of the remaining squares and obtain the solution of the problem represented in Fig. 161.

REMARK. It follows from the above that a rectangle with sides 32 and 33 can be divided into 9 unequal squares. The question arises whether every rectangle can be divided into squares that are all different. Now it can be proved that if the sides of the rectangle are incommensurable, the rectangle cannot be divided into squares at all. A rectangle with commensurable sides, however, can be divided into squares that are all different, and this can be done in infinitely many ways. Such divisions have been discovered quite recently; until a short time ago they were believed to be impossible. Figure 162 represents the division of a square into 24 unequal squares given by the Canadian mathematician Tutte in 1950; the numbers placed in the squares denote the lengths of their sides. So far we do not know whether it is possible to divide a square into fewer than 24 squares.

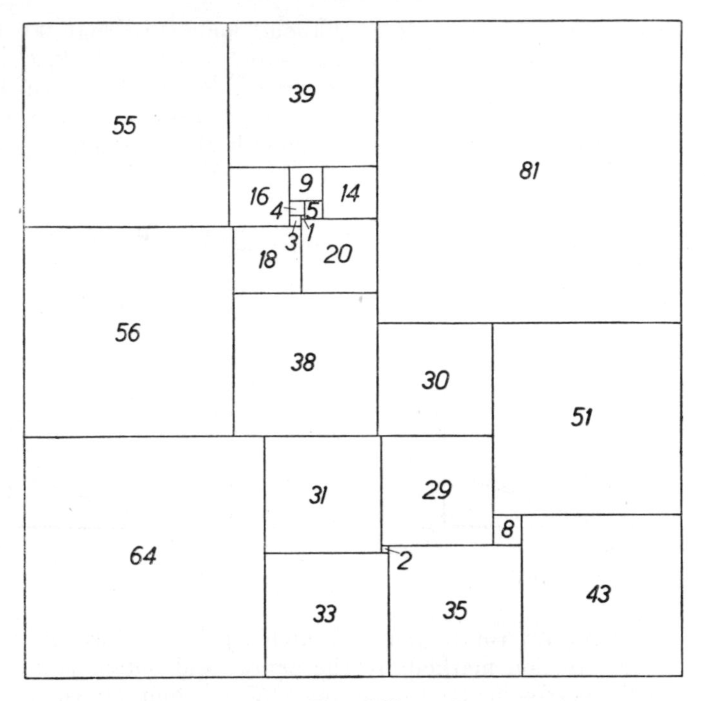

Fig. 162

125. A square *inscribed* in the square $ABCD$ is any square whose vertices lie on the boundary of the square $ABCD$. According to this definition the square $ABCD$ itself belongs to the squares inscribed in $ABCD$.

If the given point K lies on the boundary of the square $ABCD$, the problem is solved immediately. Namely, if point K is one of the vertices of that square, then the only solution is the square $ABCD$ itself; if point K lies between two vertices, then there exist two solutions: the square $ABCD$ and the square whose vertices are obtained by a rotation of point K about the centre of the square $ABCD$ through multiples of a right angle.

We shall now consider the cases where point K is inside the square $ABCD$ (Fig. 163) or outside this square (Fig. 164).

Analysis. Suppose that the given point K lies on side MN or on an extension of side MN of the square $MNPQ$ inscribed in the given square $ABCD$ (Fig. 163 and 164).

The squares $ABCD$ and $MNPQ$ have a common centre O (see problem 76).

Let us rotate the square $MNPQ$ about point O through $90°$ in the direction determined by the succession of points A, B, C, D and marked in the drawing by an arrow. Points M, K will be found after the rotation at points N, L respectively. Since $\sphericalangle KNL = 90°$, point N lies on a circle with diameter KL.

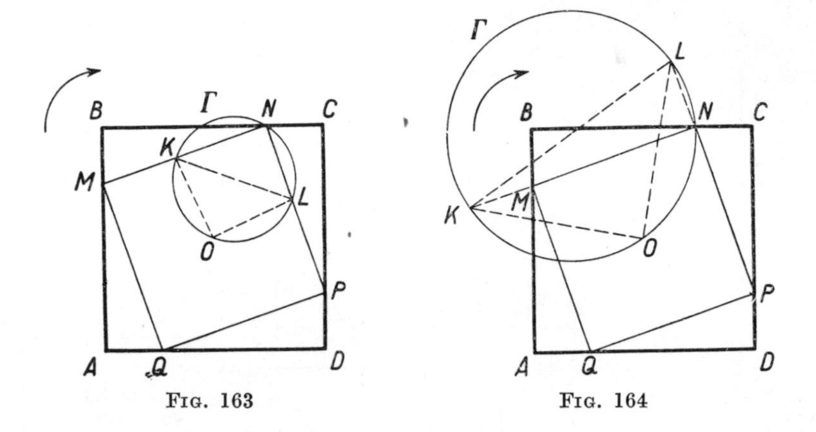

FIG. 163 FIG. 164

Construction. We rotate point K about point O through $90°$ e.g. in the direction marked by the arrow, and obtain point L, lying on the perpendicular to the line OK at point O; then $OL = OK$. We describe a circle Γ with diameter KL.

Suppose that circle Γ passes through point N, lying on the boundary of the square $ABCD$. We construct a square $MNPQ$ inscribed in the square $ABCD$, denoting by M that vertex of the inscribed square which after a rotation about O through $90°$ in the direction of the arrow becomes point N.

We shall show that the square $MNPQ$ satisfies the condition of the problem, i.e. that the line MN passes through point K. We shall consider two cases:

(1) If point K lies inside the square $ABCD$ (Fig. 163), then $\sphericalangle OKM = \sphericalangle OLN$, since angle OLN arises from the rotation of angle OKM and $\sphericalangle OLN + \sphericalangle OKN = 180°$, $\sphericalangle OLN$ and $\sphericalangle OKN$ being opposite angles of a quadrilateral inscribed in a circle. It follows that

$$\sphericalangle OKM + \sphericalangle OKN = 180°.$$

Since the half-lines OM and ON, and thus also the points M and N, lie on opposite sides of the line OK, it follows from the preceding equality that the half-lines KM and KN form one straight line.

(2) If point K lies outside the square $ABCD$ (Fig. 164), then $\sphericalangle OKM = \sphericalangle OLN$, as in case (1), and $\sphericalangle OLN = \sphericalangle OKN$ as inscribed angles subtended by the arc ON.

It follows that

$$\sphericalangle OKM = \sphericalangle OKN.$$

Since the half-lines OM and ON, and thus also the points M and N, now lie on the same side of the line OK, it follows from the preceding equation that the half-lines KM and KN coincide.

Discussion. The problem has as many solutions as there are common points of circle Γ and the boundary of the square $ABCD$.

We shall consider two cases:

(1) Point K, and thus also point L, lies inside the square $ABCD$. In this case the vertices of the square $ABCD$ lie outside the circle Γ, since from each of them the diameter KL of Γ lying inside the square appears at an acute angle.

The segments OK and OL are shorter than half the diagonal of the square $ABCD$, whence the segment KL is shorter than the side of this square. Consequently a circle with diameter KL cannot have common points with two opposite sides of the square $ABCD$, and only the following cases are possible:

(a) circle Γ intersects two adjacent sides of the square, which gives 4 solutions;

(b) circle Γ intersects one side of the square and is tangent to the adjacent side—3 solutions;

(c) circle Γ intersects one side of the square—2 solutions;

(d) circle Γ is tangent to two adjacent sides of the square—2 solutions;

(e) circle Γ is tangent to one side of the square—1 solution;

(f) circle Γ has no common points with the boundary of the square—the problem has no solutions.

(2) Point K, and thus also point L, lies outside the square $ABCD$.

We shall show that in this case circle Γ intersects the perimeter of the square at two points, i.e. that the problem has two solutions.

The arcs OK and OL of the semicircle KOL join the interior point O of the square with its exterior points K and L, whence they must have common points with the boundary of the square. Let N be the first of those points on the arc OK passing from O to K, and N_1 the first of those points on the arc OL passing from O to L (Fig. 165).

We shall show that on circle Γ there are no points common with the boundary of the square except points N and N_1. It suffices to ascertain this for the arc $NKLN_1$, since the arc NON_1

has been chosen in such a way that only the points N and N_1 lie on the boundary of the square.

Since the arc NON_1 is less than the semicircle KOL, the angle NON_1 is obtuse and points N and N_1 do not lie on the same side of the square. Two cases are possible:

(i) Points N and N_1 lie on two adjacent sides of the square $ABCD$, e.g. on the sides AB and AD (Fig. 165). The arc $NKLN_1$ does not intersect either the polygonal line $NBCDN_1$, since it lies on the opposite side of the straight line NN_1, or the polygonal line NAN_1, since $\measuredangle NAN_1 = 90°$, and an angle inscribed in the arc $NKLN_1$ (greater than a semi-circle) is acute, whence point A lies inside the circle Γ. The arc $NKLN_1$, and thus also the circle Γ, has only the points N and N_1 in common with the boundary of the square.

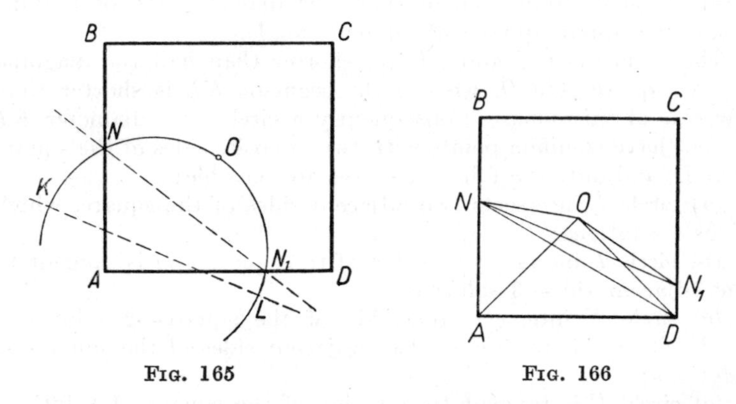

Fig. 165 Fig. 166

(ii) Points N and N_1 lie on two opposite sides AB and CD of the square $ABCD$ (Fig. 166). Suppose that point O is on the same side of the straight line NN_1 as the vertices B and C. The arc $NKLN_1$ is on the opposite side and thus does not intersect the polygonal line $NBCN_1$. We shall prove that this arc does not intersect the polygonal line $NADN_1$ either, showing that the vertices A and D lie inside the circle Γ, whence it follows that the polygonal line $NADN_1$ runs inside circle Γ.

In the triangle NON_1 the angle NON_1 is greater than 90°, whence at least one of the remaining angles is less than 45°. Suppose that $\measuredangle ONN_1 < 45°$. Since $\measuredangle ODN_1 = 45°$ and points N and D lie on the same side of the straight line ON_1, point D lies inside the circle passing through the points N, O, N_1, i.e. inside circle Γ. In that case $\measuredangle ON_1N < \measuredangle ODN$ and, since $\measuredangle ODN < \measuredangle ODA = \measuredangle OAN$, we have $\measuredangle ON_1N < \measuredangle OAN$, whence it follows that point A, lying on the same side of the straight line ON as point N_1, lies inside circle Γ.

We have thus proved that, if the given point K lies outside the given square, the problem always has two solutions. If, moreover, point K lies on an extension of any of the sides of the square $ABCD$, then one of the solutions is the square $ABCD$ itself.

REMARK. In the first part of the above discussion we have found that, according to the position of point K in the square $ABCD$ the number of solutions of the problem may be 4, 3, 2, 1 or 0.

We can investigate this question more thoroughly and determine which number of solutions corresponds to which part of the square.

The most convenient method of doing this will be the method of coordinates. Let us choose a system of orthogonal coordinates as in Fig. 167. In this system let point K have coordinates (x, y). Point L, which has arisen from the rotation of point K clockwise through 90°, has, as can easily be verified, the coordinates $(y, a-x)$ where a denotes the length of the side of the square.

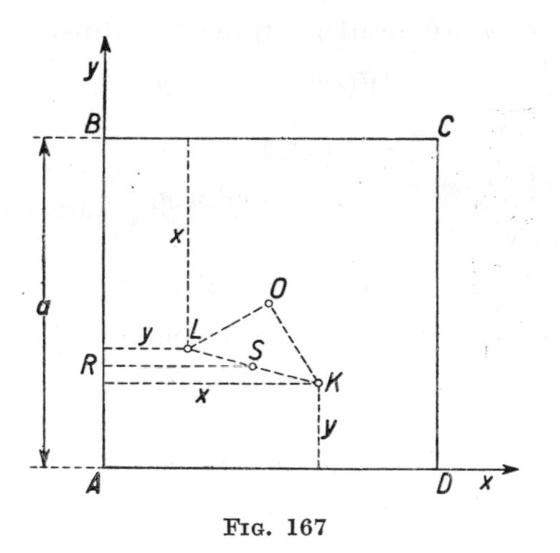

FIG. 167

We shall find out how many points the circle Γ with diameter KL has in common with the segment AB—according to the values of x and y.

The distance RS of the centre S of Γ from the line AB is equal to the arithmetical mean of the distances of points K and L from this line:

$$RS = \frac{x+y}{2}. \tag{1}$$

The radius of circle Γ is equal to

$$SL = \tfrac{1}{2}KL = \tfrac{1}{2}\sqrt{[(x-y)^2+(a-x-y)^2]}. \tag{2}$$

Since points A and B lie, as we have found before, outside circle Γ, the required number of points is 2, 1 or 0, if $RS < SL$, $RS = SL$ or $RS > SL$ respectively.

Using formulas (1) and (2) we obtain the corresponding relations between x, y and a. The equality $RS = SL$ gives the relation

$$(x+y)^2 = (x-y)^2+(a-x-y)^2,$$

whence

$$(x+y)^2-(x-y)^2 = (a-x-y)^2,$$

i.e.

$$4xy = (a-x-y)^2.$$

Extracting the square root from both sides we obtain

$$2\sqrt{(xy)} = |a-x-y|. \tag{3}$$

(i) If $a-x-y \geqslant 0$, relation (3) can be written as

$$2\sqrt{(xy)} = a-x-y,$$

i.e.

$$x+2\sqrt{(xy)}+y = a,$$

$$(\sqrt{x}+\sqrt{y})^2 = a,$$

and ultimately

$$\sqrt{x}+\sqrt{y} = \sqrt{a}. \tag{4}$$

(ii) If $a-x-y < 0$, relation (3) assumes the form

$$2\sqrt{(xy)} = x+y-a,$$

i.e.

$$x-2\sqrt{(xy)}+y = a,$$

$$(\sqrt{x}-\sqrt{y})^2 = a,$$

whence

$$|\sqrt{x}-\sqrt{y}| = \sqrt{a},$$

and finally

$$\sqrt{x} = \sqrt{a}+\sqrt{y} \quad \text{or} \quad \sqrt{y} = \sqrt{a}+\sqrt{x}. \tag{5}$$

A point lying inside the square has coordinates less than a and thus it cannot satisfy either of the equalities (5).

We have obtained the following result: circle Γ has one point in common with the side AB of the square $ABCD$ if x and y satisfy equation (4).

Equation (4) represents the arc BD of a parabola (Fig. 168) whose focus is the centre of the square $ABCD$ and directrix—the line parallel to the diagonal BD passing through point A.

If point K lies in the shaded part of the square, its coordinates satisfy the inequality

$$\sqrt{x}+\sqrt{y}<\sqrt{a},$$

which corresponds to the inequality $RS < SL$. Circle Γ then intersects the side AB at two points.

If point K lies in the unshaded part of the square, its coordinates satisfy the inequality

$$\sqrt{x}+\sqrt{y}>\sqrt{a};$$

then $RS > SL$ and circle Γ has no points in common with the segment AB.

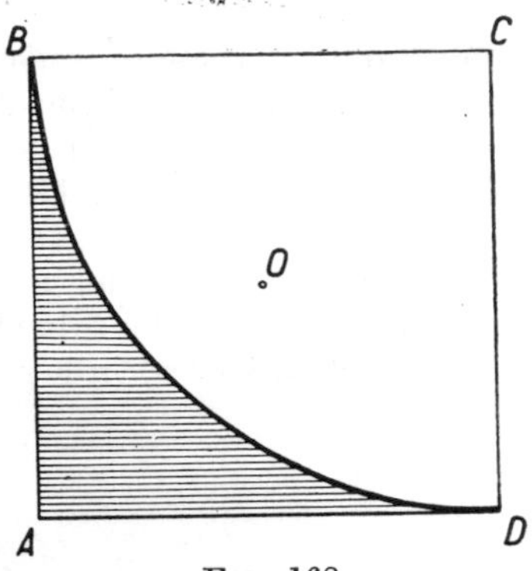

FIG. 168

The case of the remaining sides of the square is similar: the corresponding parabolas are obtained by rotating the parabola of Fig. 168 about point O through 90°, 180° and 270°.

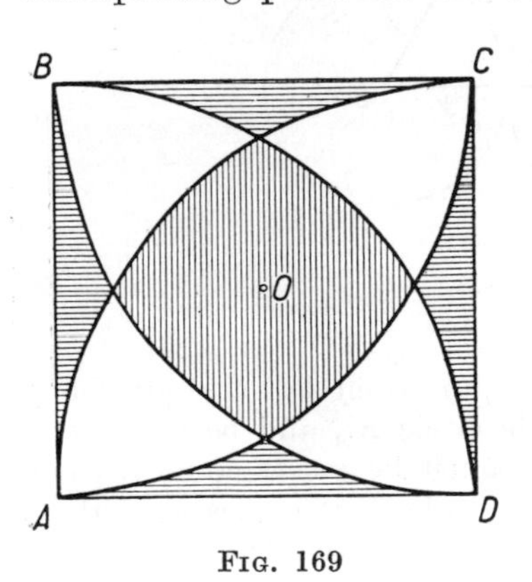

FIG. 169

The ultimate result of the discussion is shown in Fig. 169.

If point K lies in one of the areas shaded with horizontal lines, the problem has 4 solutions; if it lies in the unshaded areas, the problem has 2 solutions; for the points of the area shaded with vertical lines there are no solutions. It is left to the reader as an exercise to find out what numbers of solutions correspond to the lines delimiting these areas.

126. Let us denote by x, y, z, respectively, the radii of the required circles K_A, K_B, K_C with centres A, B, C and let $AB = c$, $BC = a$, $CA = b$.

1. Suppose that the circles of each pair are *externally* tangent. Then the sum of the radii of each pair of circles is equal to the distance of their centres:

$$\begin{aligned} x+y &= c, \\ y+z &= a, \\ z+x &= b. \end{aligned} \tag{1}$$

Hence

$$x = \frac{b+c-a}{2}, \quad y = \frac{c+a-b}{2}, \quad z = \frac{a+b-c}{2}.$$

According to these formulas, it is easy to construct the required segments x, y, z. Since the lengths x, y, z satisfy equations (1), the circles described from points A, B, C by the radii x, y, z are externally tangent (Fig. 170). It will be observed that the same formulas express the lengths of the segments which a circle inscribed in a triangle determines on the sides of the triangle. The problem could thus be solved by constructing in the well-known way a circle inscribed in a given triangle.

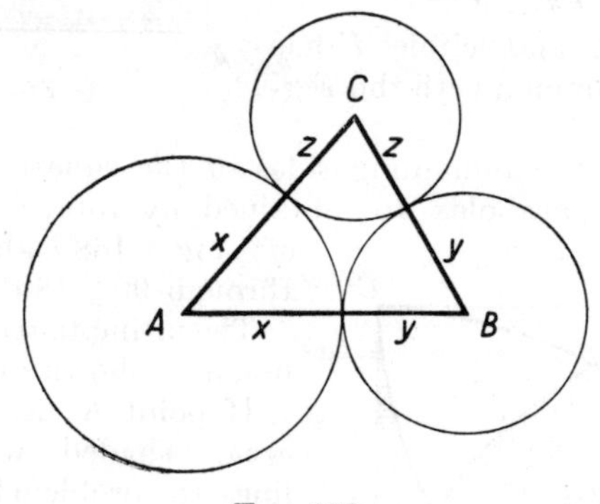

FIG. 170

2. Suppose that two of the required circles are internally tangent, e.g. let circle K_B lie inside circle K_A and be tangent to it at point M. Therefore circle K_C must lie inside circle K_A and be externally tangent to circle K_B. Indeed, if circles K_C and K_A were externally tangent, then the point of contact of circles K_B and K_C would have to be at point M since all the other points of circle K_B lie inside circle K_A. But then both point A and point C would lie on the straight line BM (since the centres of tangent circles and their point of contact are collinear), i.e. points A, B, C would be collinear, contrary to the assumption that they are the vertices of a triangle. In the same way we ascertain that circle K_C must be externally tangent to circle K_B.

Thus the following relations hold:

$$\begin{aligned} x-y &= c, \\ x-z &= b, \\ y+z &= a. \end{aligned} \tag{2}$$

System (2) has the solution:

$$x = \frac{a+b+c}{2},$$

$$y = \frac{a+b-c}{2},$$

$$z = \frac{c+a-b}{2}.$$

If we construct segments x, y, z according to these formulas and describe from points A, B, C with the radii x, y, z circles K_A, K_B, K_C, then circle K_A will be internally tangent to circles K_B and K_C, and circles K_B and K_C will be externally tangent (Fig. 171), since x, y, z satisfy equations (2).

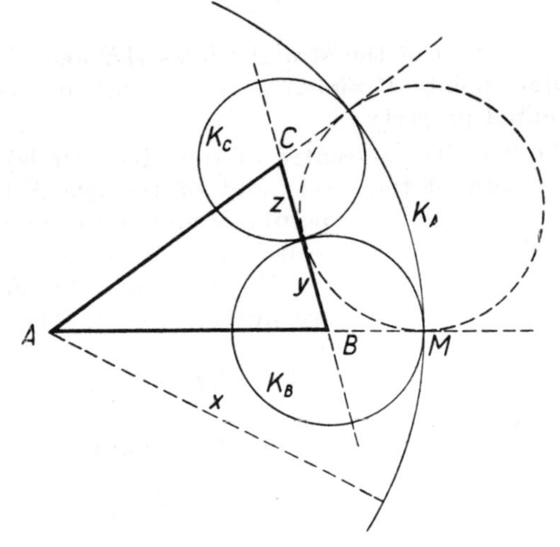

FIG. 171

It will be observed that the above formulas for x, y, z express the lengths of the segments determined on the straight lines AB, AC and BC by an escribed circle of triangle ABC tangent to the side BC; the segments x, y, z could thus be found by constructing that escribed circle.

Besides the above solution we have two analogous solutions, in which circle K_B or circle K_C is internally tangent to the remaining two circles.

The problem thus has four solutions in all.

127. We produce the segment BC both ways and mark off $BE = BC$ and $CF = BC$ (Fig. 172). We join points E and F with point A.

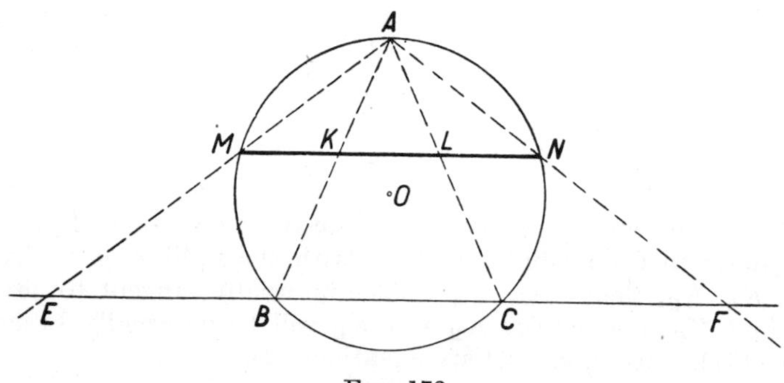

FIG. 172

At the intersection of the straight lines AE and AF with the circle we obtain points M and N; the segment MN is the chord with the required property.

Indeed, the line AO (O—centre of the given circle) is an axis of symmetry both of the circle and of triangle EAF, whence points M and N are symmetric with respect to AO, and MN is parallel to EF. Segments MK, KL and LN are proportional to the equal segments EB, BC and CF, whence $MK = KL = LN$.

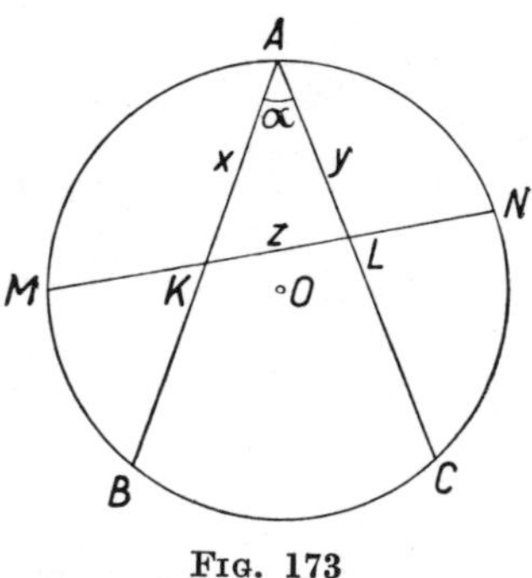

FIG. 173

It remains to consider whether, besides the chord MN (Fig. 172) parallel to BC, there are any chords oblique to BC and divided by the chords AB and AC into three equal parts.

Let MN (Fig. 173) be a chord of the circle such that $MK = KL = LN$. Let us denote $AK = x$, $AL = y$, $KL = z$, $\sphericalangle BAC = \alpha$, $AB = AC = a$.

By the theorem on chords of a circle we have

$$AK\times KB = MK\times KN,$$

i.e.

$$x(a-x) = 2z^2 \tag{1}$$

and analogously

$$y(a-y) = 2z^2, \tag{2}$$

whence

$$x(a-x) = y(a-y),$$

and consequently, after an easy transformation, we obtain

$$(x-y)(x+y-a) = 0. \tag{3}$$

Thus two cases are possible:

(a) $x-y = 0$, i.e. $x = y$; this leads to the solution worked out before, in which we obtained a chord parallel to the straight line BC.

(b) $x+y-a = 0$, i.e.

$$y = a-x. \tag{4}$$

In order to determine the unknown segments x, y, z we thus have two equations (1) and (4); the third equation will be obtained from triangle AKL:

$$z^2 = x^2+y^2-2xy\cos\alpha. \tag{5}$$

From the system of equations (1), (4) and (5) we shall obtain the values of x, y, z.

Eliminating y from equations (4) and (5) and taking into account equation (1), we obtain the equation

$$x(a-x) = 2x^2+2(a-x)^2-4x(a-x)\cos\alpha,$$

which, rearranged, gives

$$(5+4\cos\alpha)x^2-(5+4\cos\alpha)\,ax+2a^2 = 0.$$

Since $5+4\cos\alpha > 0$, we can divide both sides of the equation by $5+4\cos\alpha$ and obtain for the unknown x the equation

$$x^2-ax+\frac{2a^2}{5+4\cos\alpha} = 0. \tag{6}$$

This equation has real roots if

$$\Delta = a^2-\frac{8a^2}{5+4\cos\alpha} \geqslant 0,$$

which, when solved with respect to $\cos\alpha$, gives the condition $\cos\alpha \geqslant \frac{3}{4}$.

If $\cos\alpha = \frac{3}{4}$, the roots of equation (6) are equal. Since the sum of these roots is equal to a, we have $x = \frac{1}{2}a$ and formula (4) gives $y = \frac{1}{2}a$, whence $x = y$, i.e. in this case there are no solutions besides the chord parallel to BC.

Thus, ultimately, chords oblique to BC and satisfying the condition of the problem exist if and only if $\cos\alpha > \frac{3}{4}$. Then there exist two such chords; as can easily be seen, they lie symmetrically with respect to diameter AO. In order to draw them we can solve equation (6) and construct segment x according to the algebraic expression obtained.

EXERCISE. Solve a more general problem, taking in the given circle, instead of chords AB and AC with a common end-point, any two equal chords AB and CD.

128. *Analysis.* Suppose that point C of the arc AB (Fig. 174) satisfies the condition of the problem, i.e. that $AC+CB = a$. On segment AC produced beyond point C let us mark off

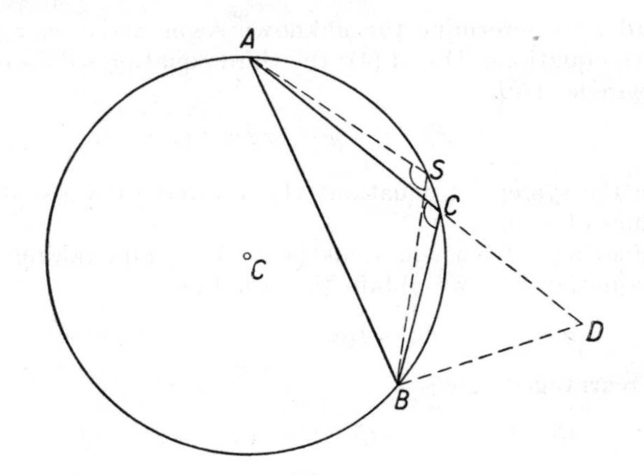

FIG. 174

a segment $CD = CB$. Then $AD = a$ and the angle ADB is equal to half the angle ACB since $\measuredangle ADB = \measuredangle DBC$, whence $\measuredangle ACB = \measuredangle ADB + \measuredangle DBC = 2\times \measuredangle ADB$. Point D thus lies: (1) on the circle with centre A and radius a; (2) on the locus of points from which segment AB appears at an angle equal to $\frac{1}{2}\times \measuredangle ACB$ and which lie on the same side of the line AB as the arc ABC; this locus is an arc of circle $S(SA)$† where

† Circle $S(SA)$ is a circle with centre S and radius SA.

S is the mid-point of the arc ACB ($\sphericalangle ASB = \sphericalangle ACB$, as angles subtended by the same arc).

Construction. From the centre at mid-point S_1 of one of the arcs of the given circle with end-points A and B (Fig. 175) we describe an arc joining points A and B on that side of the

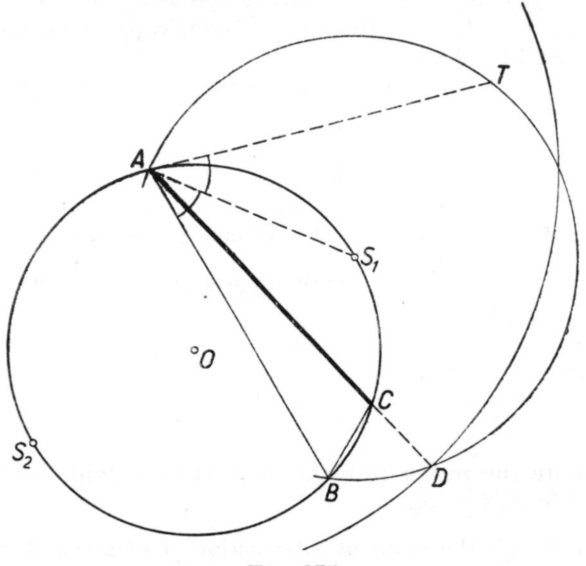

FIG. 175

straight line AB on which lies point S_1. From point A as centre we describe a circle with radius a. If this circle intersects the arc described before at point D and if the segment AD intersects the given circle at point C, then point C satisfies the condition of the problem, i.e. $AC+CB = a$. Indeed, $\sphericalangle ACB = \sphericalangle AS_1B = 2\times\sphericalangle ADB$, whence $\sphericalangle DBC = \sphericalangle ACB - \sphericalangle ADB = \sphericalangle ADB$, which implies that $CD = CB$ and $AC+CB = AC+CD = a$.

The above construction can be executed if the following conditions are satisfied: (1) Circles $S_1(S_1A)$ and $A(a)$ have common points D; this condition is expressed by the inequality $a \leqslant 2S_1A$; (2) segment AD intersects the arc AS_1B; thus point D must lie inside the arc BT of circle $S_1(S_1A)$, determined by the chord AB and the half-line AT tangent to the arc AS_1B at point A. Now $AT = AB$ since $\sphericalangle TAS_1 = \sphericalangle S_1AB$, these angles being respectively equal to the angles inscribed in the given circle and subtended by the arcs AS_1 and S_1B. Condition (2) is thus expressed (condition (1) being satisfied) by the inequality $a > AB$.

Consequently, for the existence on the arc AS_1B of points C with the required property $AC+CB = a$, it is necessary and sufficient that the following inequality be satisfied:

$$AB < a \leqslant 2S_1A.$$

If $a = 2S_1A$, we have one solution—point S_1, and if $a < 2S_1A$, we obtain two points situated symmetrically with respect to the straight line AS_1.

For the arc AS_2B we obtain analogously the condition

$$AB < a \leqslant 2S_2A.$$

Assuming that the condition $a > AB$ is satisfied and that the arc AS_1B is smaller than the arc AS_2B, i.e. that $AS_1 < AS_2$, we can tabulate the results of the discussion as follows:

Case	Corresponding No. of Solutions
$a > 2S_2A$	0
$a = 2S_2A$	1
$2S_1A < a < 2S_2A$	2
$a = 2S_1A$	3
$a < 2S_1A$	4

We invite the reader to make an analogous table for the case of $a > AB$, $S_1A = S_2A$.

129. Through the point of intersection A of given circles with centres O and S we draw a straight line intersecting those circles at points P and Q.

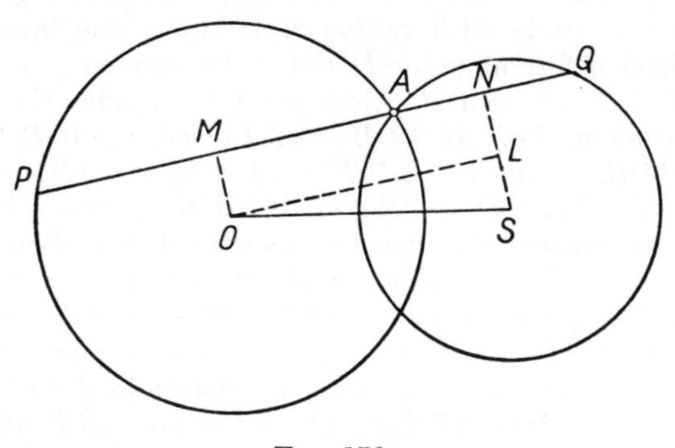

FIG. 176

Points P and Q lie either on opposite sides (Fig. 176) or on the same side of point A (Fig. 177).

In the first case $PQ = AP + AQ$; in the second case the secant PQ can be in such a position that point Q will lie between points A and P (as in Fig. 177) and then $PQ = AP - AQ$, or in such

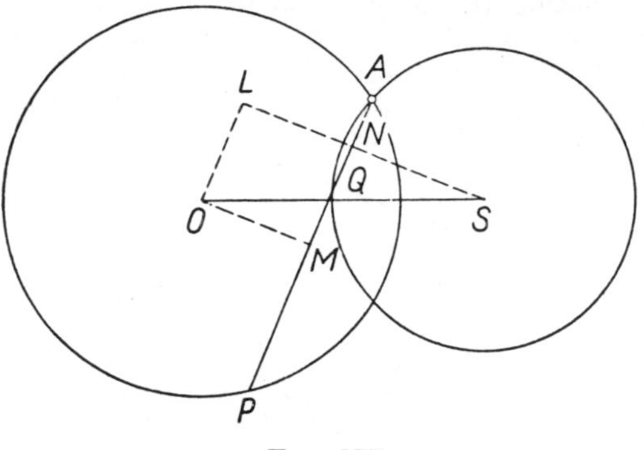

Fig. 177

a position that point P lies between points A and Q and then $PQ = AQ - AP$. In order to make one formula answer both cases we shall use the sign of absolute value:

$$PQ = |AP - AQ|.$$

Let us draw from points O and S the perpendiculars OM and SN to the straight line PQ and observe that

$$AM = \tfrac{1}{2}AP, \quad AN = \tfrac{1}{2}AQ.$$

The length of the segment MN is equal to half the length of the segment PQ. Indeed, in the first case we have

$$MN = AM + AN = \tfrac{1}{2}AP + \tfrac{1}{2}AQ = \tfrac{1}{2}(AP + AQ) = \tfrac{1}{2}PQ,$$

and in the second case

$$MN = |AM - AN| = |\tfrac{1}{2}AP - \tfrac{1}{2}AQ| = \tfrac{1}{2}|AP - AQ| = \tfrac{1}{2}PQ.$$

Let us draw through point O a line parallel to PQ; it will intersect NS at point L. In the right-angled triangle OLS the hypotenuse OS is the distance between the centres of the given circles and $OL = MN = \frac{1}{2}PQ$.

If the length $PQ = d$ is given, we can construct triangle OLS and consequently we can draw the straight line PQ as the line parallel to the line OL passing through point A.

This construction is shown in Fig. 178.

We describe a circle with diameter OS. From point O as centre we describe a circle with radius $\frac{1}{2}d$. If the two circles have a common point, L, then the line parallel to OL passing through point A is the required straight line since $PQ = 2OL = d$, which we ascertain by conducting the above reasoning in the inverse order.

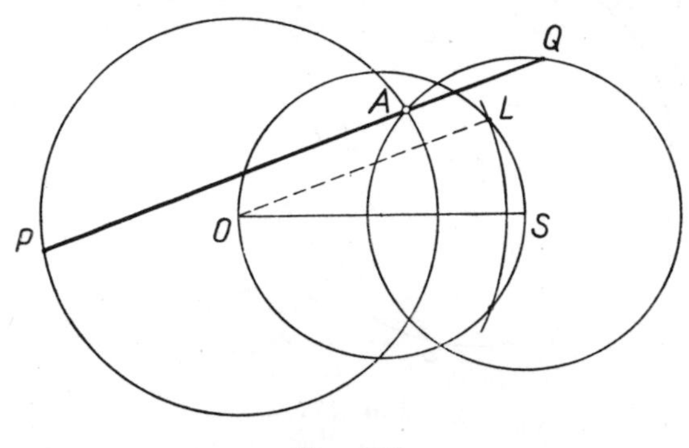

FIG. 178

The condition of the existence of point L is the inequality

$$\tfrac{1}{2}d \leqslant OS, \quad \text{i.e.} \quad d \leqslant 2\times OS.$$

If $d < 2\times OS$, the problem has two solutions: both secants are symmetric with respect to the straight line passing through the points of intersection of the given circles.

If $d = 2\times OS$, there is one solution, namely a straight line parallel to OS.

130. Suppose that $KL = l$ is the required segment of a tangent and that point S is the projection of centre O of the given circle K onto the straight line KL. The segment KL will be called a solution of the first kind (Fig. 179) if points K and L lie on segments MP and MQ; if points K and L lie on these segments produced, we shall say that the segment KL is a solution of the second kind (Fig. 183).

(a) We shall deal first with the determination of the solutions of the first kind: namely we shall find the lengths of segments OK and OL. Let $\sphericalangle PMQ = \varphi$. It will be observed that in triangle KOL we have the side $KL = l$, the altitude $OS = r$ (radius of the circle), $\sphericalangle KOL = \frac{1}{2}\sphericalangle POQ = \frac{1}{2}(180^\circ - \varphi) = 90^\circ - \varphi/2$, the remaining angles being acute as the angles of the right-angled triangles OKS and OLS. Thus if a solution of the problem exists, then segments OK and OL are equal to two sides of an acute-angled triangle whose third side is equal to l, the corresponding altitude is equal to the radius r of the given circle and the opposite angle is $90°-\varphi/2$, where φ is the angle between the given tangents of circle k.

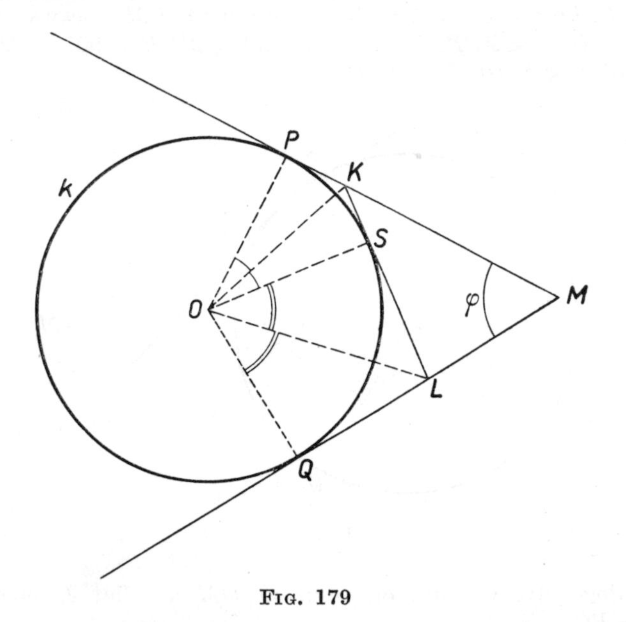

Fig. 179

The construction of such a triangle is as follows: we mark off a segment $AB = l$ (Fig. 180) and on one side of the straight line AB we describe an arc passing through A and B and including the inscribed angle $90°-\varphi$. We then draw on the same side of AB a line parallel to AB at a distance r. The required triangle will be obtained if this parallel line intersects the arc at such a point C that triangle ABC will be acute-angled, i.e. that the foot H of its altitude from C will lie inside the segment AB.

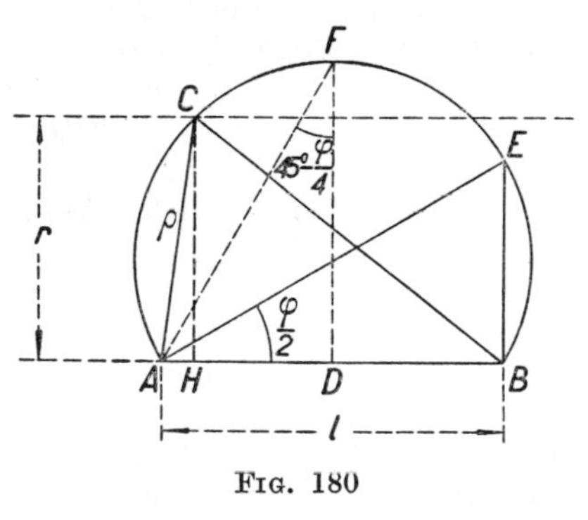

Fig. 180

Suppose that we have constructed the required triangle ABC: we shall investigate the necessary conditions later. The solution of our problem is now obtained immediately. Namely from the centre O of the given circle k (Fig. 181) we describe a circle with radius ϱ equal to segment AC in Fig. 180. This circle will intersect the half-line PM at a point K because $AC > CH$, whence $\varrho > OP$. Point K will lie on the segment PM because $\sphericalangle KOP = \sphericalangle ACH < \sphericalangle ACB = 90° - \varphi/2$ and $\sphericalangle MOP = 90° - \varphi/2$; consequently $\sphericalangle KOP < \sphericalangle MOP$.

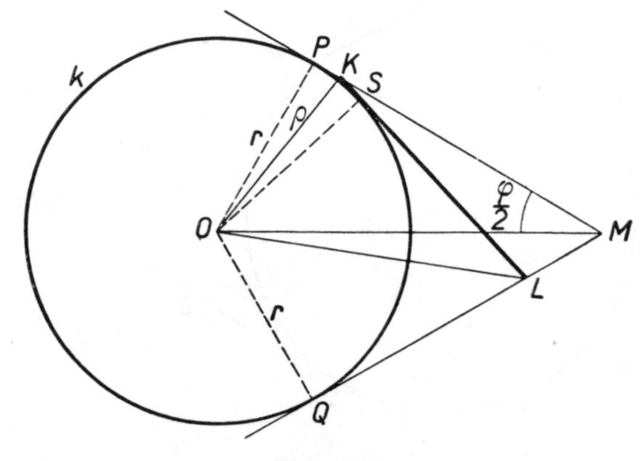

FIG. 181

Analogously, we find on segment QM a point L such that $OL = BC$.

We shall prove that the straight line KL is the solution of the problem, i.e. that it is tangent to the given circle k, and $KL = l$. Indeed, according to the construction we have $\sphericalangle ACH = \sphericalangle KOP$; analogously $\sphericalangle BCH = \sphericalangle LOQ$. Now $\sphericalangle ACH + \sphericalangle BCH = \sphericalangle ACB = 90° - \varphi/2$, whence $\sphericalangle KOP + \sphericalangle LOQ = 90° - \varphi/2$, and consequently $\sphericalangle KOL = \sphericalangle POQ - (\sphericalangle KOP + \sphericalangle LOQ) = 180° - \varphi - (90° - \varphi/2) = 90° - \varphi/2 = \sphericalangle ACB$. Triangles ACB and KOL are thus congruent, which implies that $KL = AB = l$ and the altitude $OS = CH = r$, i.e. the straight line KL is tangent to the given circle.

Let us return to the construction of triangle ABC. Let us draw in Fig. 180 two perpendiculars: $BE \perp AB$ and $DF \perp AB$ (D being the mid-point of segment AB). The required point C exists if and only if $BE < r < DF$.

In order to explain the meaning of these conditions, let us draw a tangent K_1L_1 (Fig. 182) at the point of intersection T_1 of the given circle k with segment OM and denote $MP = m$, $K_1L_1 = t_1$. The right-angled triangles AEB and MOP are similar because

$\sphericalangle BAE = \varphi/2 = \sphericalangle PMO$; consequently $BE : AB = OP : MP$, whence we can see that the condition $BE < OP$ is equivalent to the condition $AB < MP$, i.e. $l < m$. The right-angled triangles ADF and K_1T_1O are also similar because $\sphericalangle AFD = 45° - \varphi/4 = \sphericalangle K_1OT_1$; consequently $FD : AD = OT_1 : K_1T_1$ and the condition $FD \geqslant OT_1$ is equivalent to the condition $AD \geqslant K_1T_1$, i.e. $l \geqslant t_1$.

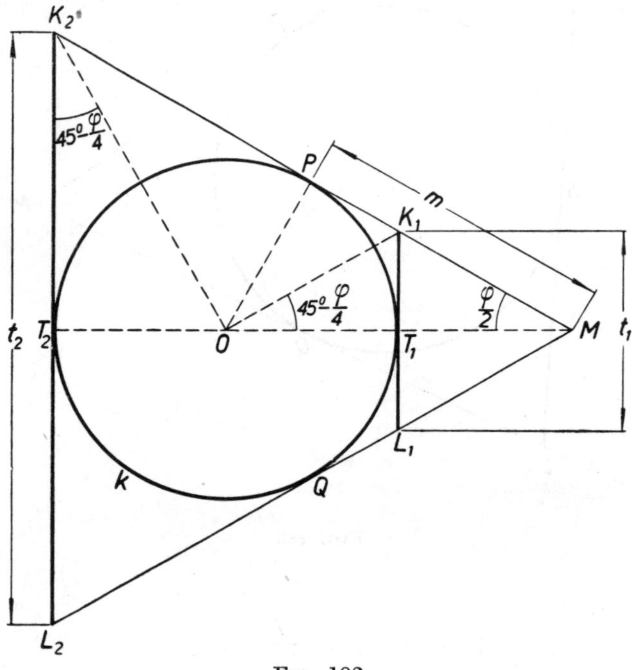

FIG. 182

Finally the condition of the existence of solutions of the problem assumes the form

$$t_1 \leqslant l < m.$$

If this condition is satisfied, we have two solutions symmetrical with respect to the straight line OM, except the case where $l = t_1$, where we have only one solution—the straight line K_1L_1.

(b) The solutions of the second kind are determined in an analogous way. In this case $\sphericalangle KOL = 90° + \varphi/2$, which we shall easily read from Fig. 183. We draw, as before, the segment $AB = l$ and describe an arc AB including the inscribed angle $90° + \varphi/2$.

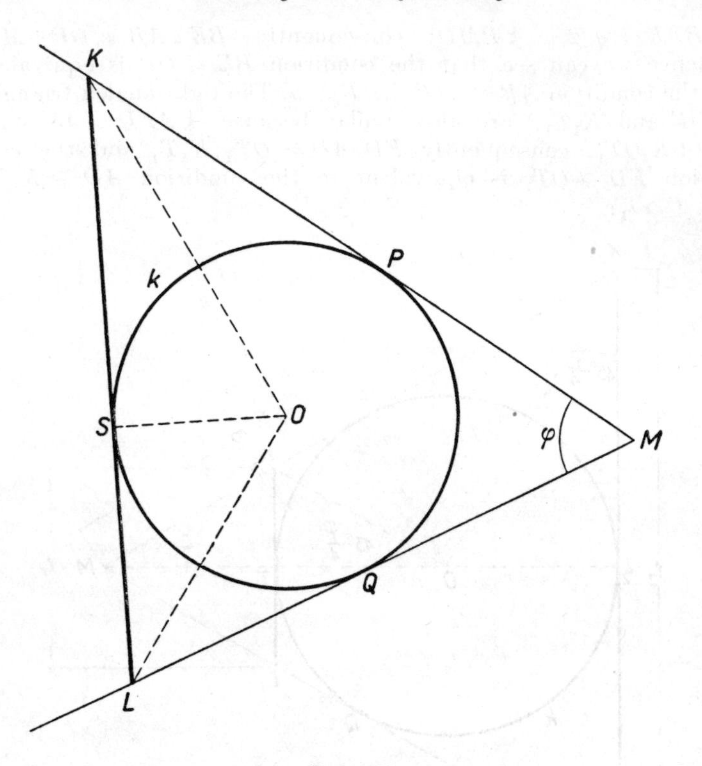

FIG. 183

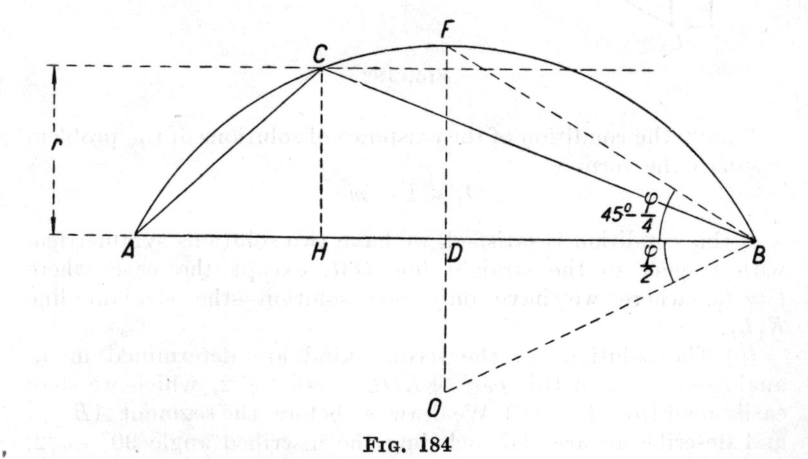

FIG. 184

Then, on the same side of AB, we draw a line parallel to AB at a distance r. Suppose that this parallel intersects the arc at point C (Fig. 184).

On the extensions of segments MP and MQ (Fig. 185) we determine such points K and L that $OK = AC$ and $OL = CB$; such points must exist because $AC > r$ and $CB > r$. The line KL is the required tangent, which we prove in the same way as before, showing that triangles KOL and ACB are congruent.

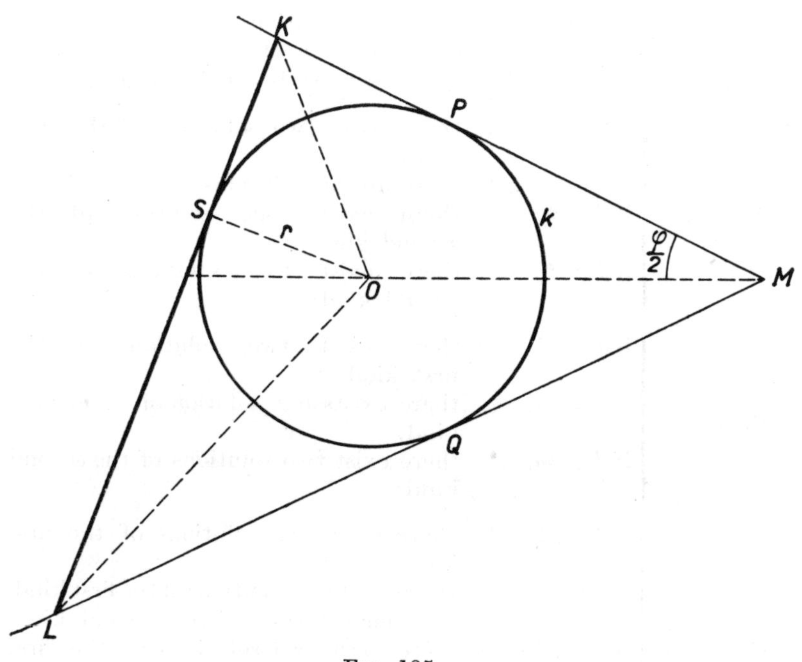

FIG. 185

The above construction can be executed if point C exists, i.e. if $r \leqslant DF$. The meaning of this condition will be clear if we draw the tangent K_2L_2 to the given circle k at point T_2, lying on segment MO produced (Fig. 182), and we shall see that the right-angled triangles BDF and K_2T_2O are similar, whence $DF : DB = T_2O : T_2K_2$. The condition $DF \geqslant r$, i.e. $DF \geqslant T_2O$, is thus equivalent to the condition $DB \geqslant T_2K_2$. This means that the given length l of the segment of the tangent cannot be less than the length of the segment of tangent L_2K_2, which we shall denote by t_2. If $l > t_2$, there are two solutions of the second kind, symmetric with respect to the straight line OM, and if $l = t_2$, the only solution is the tangent L_2K_2.

Let us tabulate the results obtained. It will be observed that the segment $K_1L_1 = t_1$ (Fig. 182) is always smaller than the segment $K_2L_2 = t_2$ and than the segment $MP = m$, where m is equal to half the perimeter of the triangle MK_1L_1, whereas t_2 can be smaller than m, equal to m or greater than m, according to the size of angle φ. In the light of the previous discussion we can establish the following propositions

1. If $l < t_1$, the problem has no solutions.
2. If $l = t_1$, there exists one solution of the first kind.
3. If $l > t_1$, the following cases must be distinguished:

$$\text{A. } m < t_2 \begin{cases} \text{if } t_1 < l < m, & \text{there exist two solutions of the first kind,} \\ \text{if } m \leqslant l < t_2, & \text{there are no solutions,} \\ \text{if } l = t_2, & \text{there exists one solution of the second kind,} \\ \text{if } l > t_2, & \text{there exist two solutions of the second kind;} \end{cases}$$

$$\text{B. } m = t_2 \begin{cases} \text{if } t_1 < l < m, & \text{there exist two solutions of the first kind,} \\ \text{if } l = m, & \text{there exists one solution of the second kind,} \\ \text{if } l > m, & \text{there exist two solutions of the second kind;} \end{cases}$$

$$\text{C. } m > t_2 \begin{cases} \text{if } t_1 < l < t_2, & \text{there exist two solutions of the first kind,} \\ \text{if } l = t_2, & \text{there exist two solutions of the first kind, and one solution of the second kind,} \\ \text{if } t_2 < l < m, & \text{there exist two solutions of the first kind and two solutions of the second kind,} \\ \text{if } l \geqslant m, & \text{there exist two solutions of the second kind.} \end{cases}$$

Figure 186 illustrates the case where $t_2 < l < m$ and the problem has four solutions.

REMARK. Another method of solving the problem will be obtained if, instead of seeking the lengths of OK and OL (Figs. 179 and 183), we seek the lengths of MK and ML.

In the case of a solution of the first kind (Fig. 179) the given circle k is escribed to the triangle MKL, whence the length MP

$= m$ is equal to half the perimeter of that triangle (see the note to problem 84); consequently

$$MK + ML = 2m - l.$$

The required lengths MK and ML will be found by constructing in the well-known way a triangle in which one side is equal to l; the opposite angle is equal to φ and the sum of the remaining two sides is equal to $2m-l$.

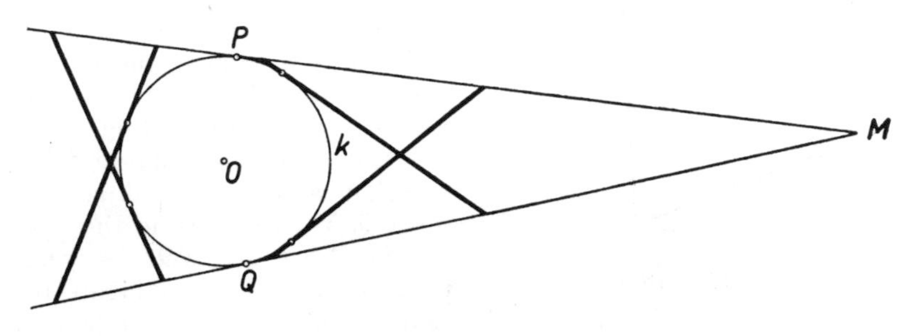

FIG. 186

In the case of a solution of the second kind (Fig. 183) the given circle k is inscribed in the triangle MKL, the perimeter of that triangle is equal to

$$MK+ML+KL = MP+KP+MQ+LQ+KL,$$

and since

$$MP = MQ = m, \quad KP = KS, \quad LQ = LS, \quad KP+LQ = KL = l,$$

we have

$$MK+ML = 2m+l,$$

and the problem is reduced to the construction of a triangle in which one side is equal to l, the opposite angle is equal to φ and the sum of the remaining two sides is equal to $2m+l$.

We leave it to the reader to carry out a detailed discussion (which should of course lead to the same results as in the preceding solution).

131. Since the centre O of the circle lies on the axis of symmetry of each isosceles triangle inscribed in that circle, the problem can be solved only if points M and N lie on opposite sides of point O, which is what will henceforth be assumed. If $MO = NO$, the problem is solved immediately; the vertex of the required triangle lies on the perpendicular bisector of segment MN (two solutions). We shall thus assume that, say, $MO > NO$. Let $\triangle ABC$ be the required triangle (Fig. 187).

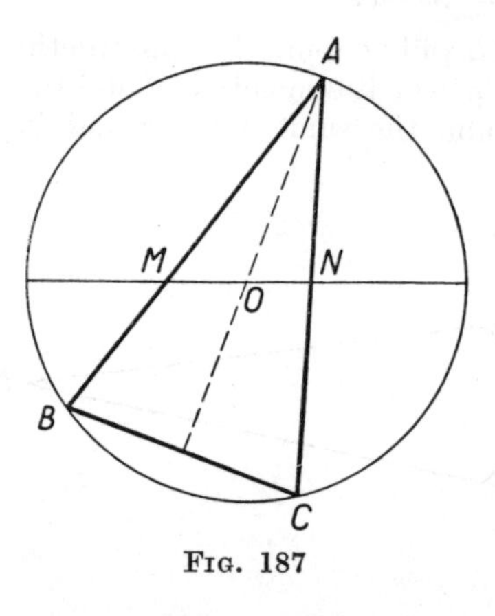

FIG. 187

The straight line AO is the bisector of angle A of the triangle MAN, whence

$$\frac{MA}{NA} = \frac{MO}{NO};$$

thus point A lies on the circle of Apollonius for the segment MN and the ratio MO/NO. The construction is the following (Fig. 188).

We determine on the straight line MN a point P dividing segment MN externally in the ratio equal to MO/NO, i.e. the point harmonically conjugate to point O with respect to points M and N. Accordingly we draw parallel segments $MK = MO$ and $NL = NO$ and find point P of intersection of lines KL and MN. We draw the above-mentioned circle of Apollonius, i.e. the circle with diameter OP; if this circle intersects the given

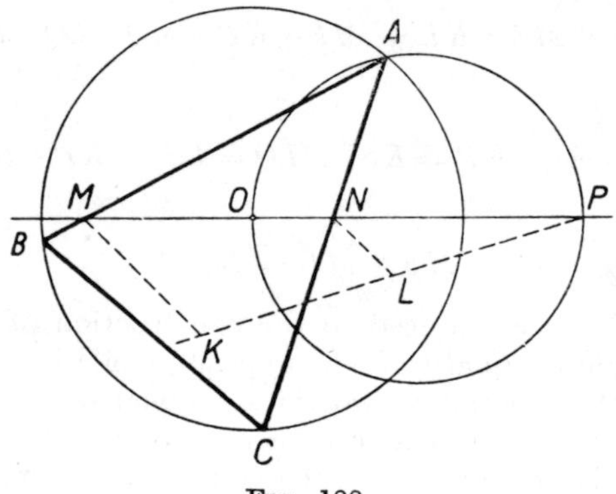

FIG. 188

circle at point A, then A is the vertex of the required triangle. Indeed, since, by the construction, $MA/NA = MO/NO$, the straight line AO is the bisector of angle MAN and the chords AB and AC are equal, being symmetrical with respect to the diameter AO.

The solution exists if the circle of Apollonius intersects the given circle, i.e. if point P lies outside that circle. This condition can be expressed as a relation between the lengths of the segments $MO = a$, $NO = b$ and the radius r of the given circle.

Let $OP = x$; we have the equality $MP/NP = MO/NO$, which can be written as

$$\frac{a+x}{x-b} = \frac{a}{b}, \quad \text{whence} \quad x = \frac{2ab}{a-b}.$$

The condition of solvability of the problem is the inequality $x > r$, i.e. the inequality

$$\frac{2ab}{a-b} > r.$$

If this condition is satisfied, the problem has two solutions, corresponding to two points of intersection of the circle of Apollonius with the given circle; the triangles obtained are symmetric with respect to the straight line MN.

REMARK. In the same way we can solve a more general problem: Given a circle and points M and N lying on a straight line passing through the centre of the circle, inscribe in that circle an isosceles triangle whose equal sides or those sides produced pass through points M and N respectively.

132. *Method I* (by rotation). *Analysis.* Let ABC be an equilateral triangle whose vertices A, B, C lie on given parallel lines a, b, c, respectively (Fig. 189).

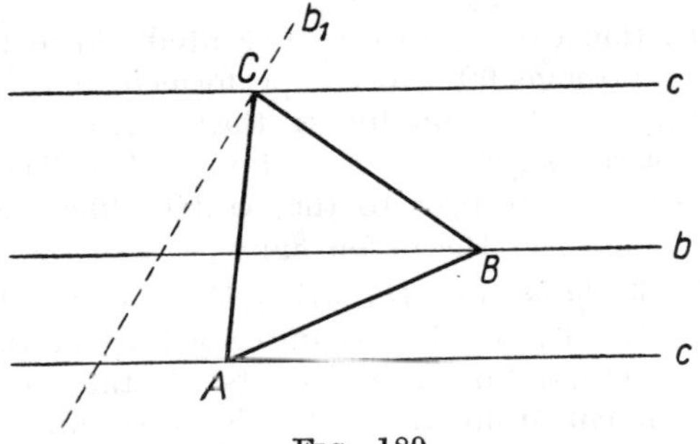

FIG. 189

Let us rotate the whole figure about point A through an angle $BAC = 60°$. After the rotation, point B will be at point C, and the straight line b will assume the position b_1. Given the centre

A and the angle of rotation, we can draw the straight line b_1, whence we will find point C at the intersection of lines b_1 and c.

Construction. We take an arbitrary point A on the straight line a (Fig. 190). We rotate the line b about point A through 60°, drawing $AM \perp a$, $\measuredangle MAN = 60°$ and $b_1 \perp AN$. The line b_1 intersects c at point C. From point A we describe a circle with radius AC. Let B be a point of intersection of this circle with b lying on the other side of the straight line AM with respect to point N. The triangle ABC is equilateral.

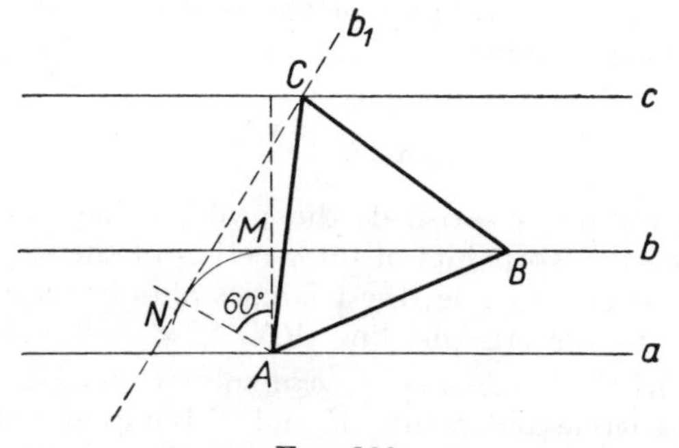

FIG. 190

In order to prove this, let us rotate the whole figure of Fig. 190 through the angle MAN. The straight line b will assume the position b_1 and point B will describe an arc with radius AB and will find itself at the intersection of that arc with the line b_1, i.e. at point C. Consequently $\measuredangle BAC = \measuredangle MAN = 60°$, which shows that triangle ABC is equilateral.

The construction can always be executed. Since the rotation about point A through 60° can be performed in two directions, for an arbitrary point A on line a there exist two equilateral triangles with vertices lying on given lines a, b, c. Those triangles are symmetrical with respect to the straight line AM, which is an axis of symmetry of the given figure.

Method II. Analysis. Let the triangle ABC be the required triangle (Fig. 191). The circle circumscribed about triangle ABC intersects the straight line a at points A and D (why must point D be different from A?). By the theorem on inscribed angles we have

$$\measuredangle ADB = \measuredangle ACB = 60°, \quad \measuredangle BDC = \measuredangle BAC = 60°.$$

Taking an arbitrary point D, we can thus determine points B and C, and then point A.

Construction. From the centre at point D, chosen arbitrarily on a, we describe a circle with an arbitrarily large radius DK (Fig. 192). We mark off $KM = MN = DK$ and draw the

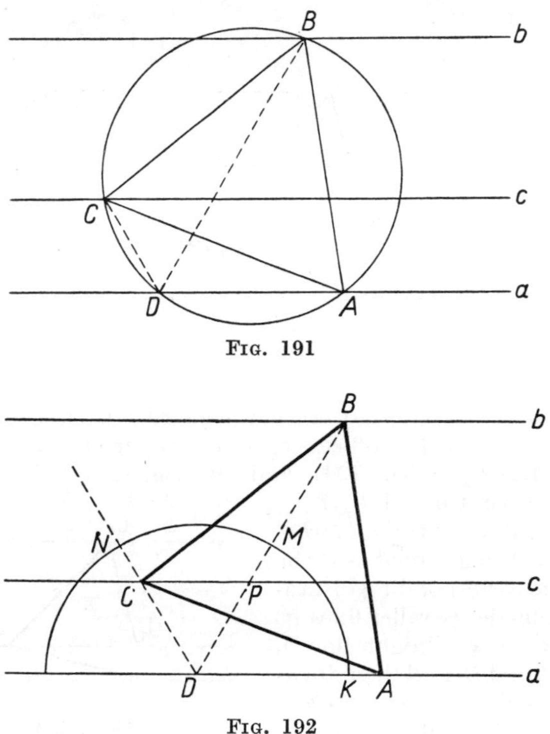

Fig. 191

Fig. 192

straight lines DM and DN, which intersect lines b and c at points B and C. From point C as centre we describe an arc with radius CB as far as the intersection with the half line DK of a at point A.

Triangle ABC is equilateral. In order to make sure of this, let us consider the circle circumscribed about triangle BCD. Since $\measuredangle BDC = 60°$, BC is the side of a regular triangle inscribed in this circle. The third vertex of that triangle must lie on the half-line DK (since $\measuredangle KDB = 60°$) and also on the circle which has been described from centre C. Thus point A is the third vertex.

The construction can always be performed. The circle with centre C and radius CB intersects the half-line DK at one point, since $BC > CP$ and thus also $BC > CD$. For a chosen point D we obtain two symmetrical solutions, since point K can be taken either on one side or on the other side of point D on the straight line a.

Method III (of similarity). *Analysis.* Let ABC be the required triangle (Fig. 193).

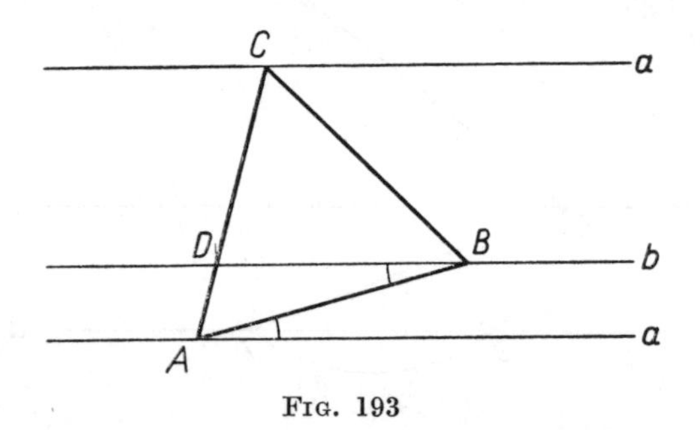

FIG. 193

It will be observed that we can construct a figure similar to the required one. It suffices to draw an arbitrary equilateral triangle $A_1B_1C_1$ (Fig. 194) and determine on side A_1C_1 a point D_1 such that $A_1D_1/D_1C_1 = AD/DC$. But the ratio AD/DC is known, being equal to the ratio of the corresponding distances between the parallel lines a, b, c. Drawing a straight line b_1 through B_1 and D_1 and then drawing $a_1 || b_1$ and $c_1 || b_1$, we obtain a figure which is similar to that of Fig. 193, whence the angles marked in Figs. 193 and 194 are equal. Having the angle between the line AB and the line a, we can draw the triangle ABC.

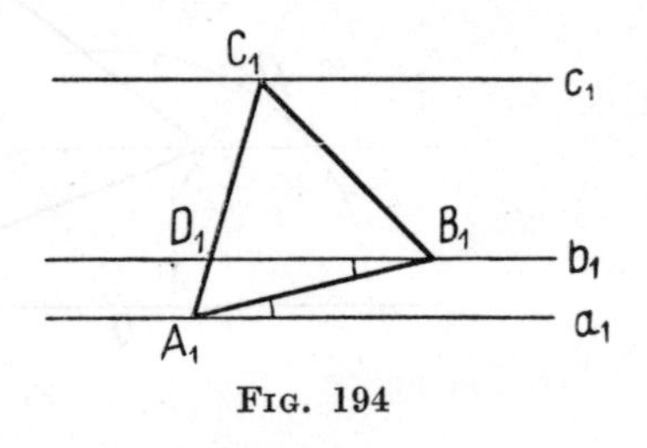

FIG. 194

Construction. Points A_1 and C_1 are best chosen on the straight lines a and c (Fig. 195) since then point D_1 is found immediately at the intersection of the segment A_1C_1 with the straight line b. Constructing the equilateral triangle $A_1B_1C_1$, we obtain the angle $A_1B_1D_1$. Taking an arbitrary point A of a as vertex, we draw an angle equal to the angle $A_1B_1D_1$ with one arm lying on a, and thus obtain point B. Finally we construct $\sphericalangle BAC = 60°$. The triangle ABC is equilateral; we shall prove this fact.

The figure $ABCD$ is similar to the figure $A_1B_1C_1D_1$. Indeed, triangle ABD is similar to triangle $A_1B_1D_1$ (two pairs of angles respectively equal) and point C of the line AD corresponds to point C_1 of the line A_1D_1 in view of the equality of the ratios A_1D_1/D_1C_1 and AD/DC. Triangle ABC is thus equilateral.

The construction can always be executed. Since an angle equal to the angle $A_1B_1D_1$ can be drawn with one arm lying on the straight line a on either side of point A, we obtain two symmetric solutions.

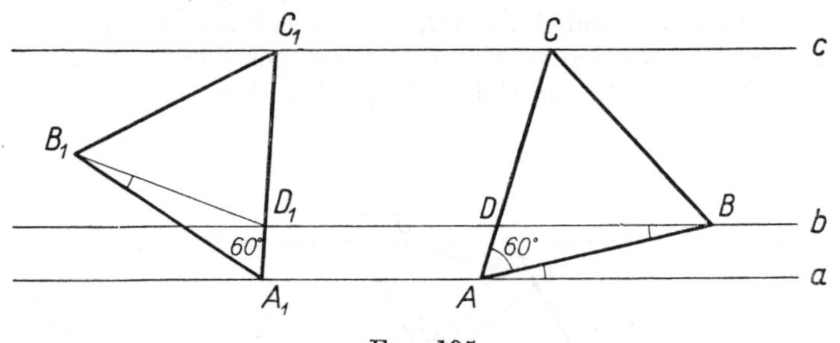

FIG. 195

Method IV (of geometrical loci). *Analysis.* If triangle ABC is the triangle with the required properties, then, translating it in the direction of the given lines a, b, c, we can place vertex A at an arbitrarily chosen point of a. The problem is then reduced to the determination of one of the remaining vertices, e.g. the vertex c.

Point C will be found at the intersection of two loci:

(1) the straight line c,

(2) the locus of the third vertex of an equilateral triangle of which one vertex is at point A and the second vertex, B, moves along the straight line b.

We shall determine the second locus.

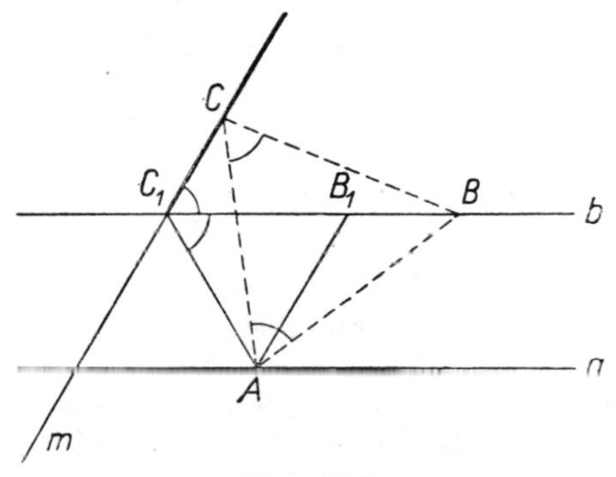

FIG. 196

Since there exist two equilateral triangles with the same base AB, let us first consider those triangles ABC in which points A, B, C follow one another in the positive cyclic order, i.e. in the anti-clockwise direction

Suppose that, in the position AB_1C_1, point C_1 lies on the straight line b and that ABC is any other position of the triangle. We shall prove that point C lies on the straight line m passing through C_1 and parallel to AB_1.

In the proof we shall distinguish three possible positions of B:

(a) point B lies to the right of B_1 (Fig. 196),

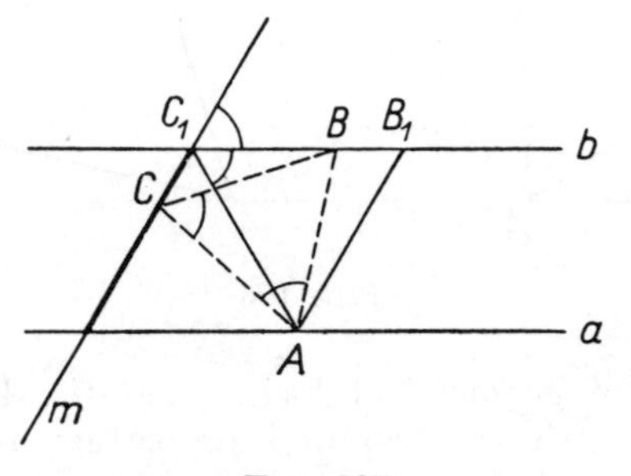

Fig. 197

(b) point B lies between points B_1 and C_1 (Fig. 197),

(c) point B lies to the left of point C_1 (Fig. 198).

In case (a) we have $\sphericalangle AC_1B = \sphericalangle ACB = 60°$; the quadrilateral $ABCC_1$ can be inscribed in a circle, whence $\sphericalangle CC_1B = \sphericalangle CAB = 60°$ and thus indeed $CC_1 || AB_1$. If point B runs over the half-

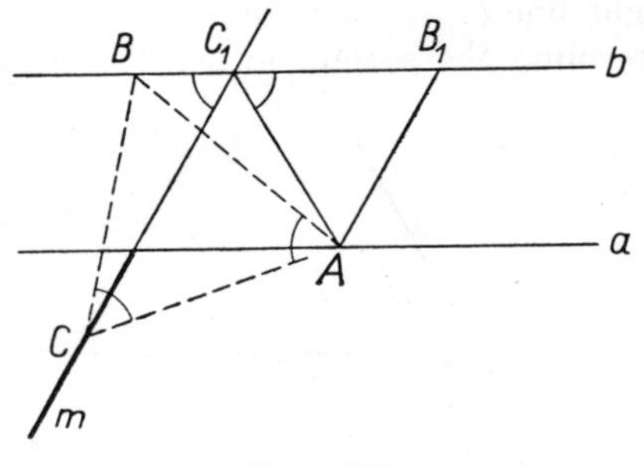

Fig. 198

line of b lying to the right of B_1, point C runs over that half-line of m which lies on the other side of b with respect to the straight line a.

Indeed, if C lies on that half-line and $\sphericalangle CAB = 60°$, then $\sphericalangle CAB = \sphericalangle CC_1B$, whence in quadrilateral $ABCC_1$ we have $\sphericalangle ACB = \sphericalangle AC_1B = 60°$ and triangle ABC is equilateral.

In cases (b) and (c) the reasoning is analogous. In Figs. 197 and 198 equal angles are marked; point C runs over the remaining parts of the straight line m.

Finally, considering triangles ABC with the negative cyclic order of points A, B, C, i.e. triangles symmetrical to the previous triangles with respect to their bases AB, we obtain the second part of the required locus, namely the straight line n passing through B_1 and parallel to AC_1.

Thus the required locus is the pair of straight lines m and n.

Construction. We draw (Fig. 199) the equilateral triangle AB_1C_1 and the straight lines m and n, obtaining at their intersections with the straight line c points C and $\bar{C}$.

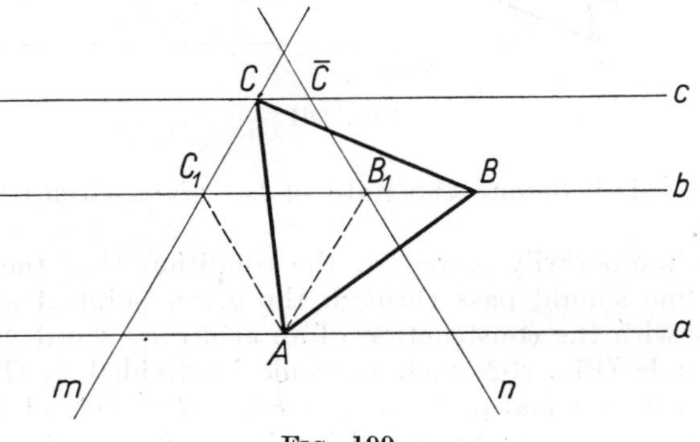

FIG. 199

We construct $\sphericalangle CAB = 60°$. Triangle ABC is the required equilateral triangle; the second solution is the triangle $A\bar{B}\bar{C}$ symmetrical to ABC.

The construction can always be performed, and the problem has two symmetrical solutions.

Method V (algebraical). Introducing the notation indicated in Fig. 200 we find that

$$\tan x = \frac{p\sqrt{3}}{p+2q}.$$

According to this formula we can construct angle x as shown in Fig. 201. The proof that triangle ABC drawn in this way is equilateral is left to the reader as an exercise.

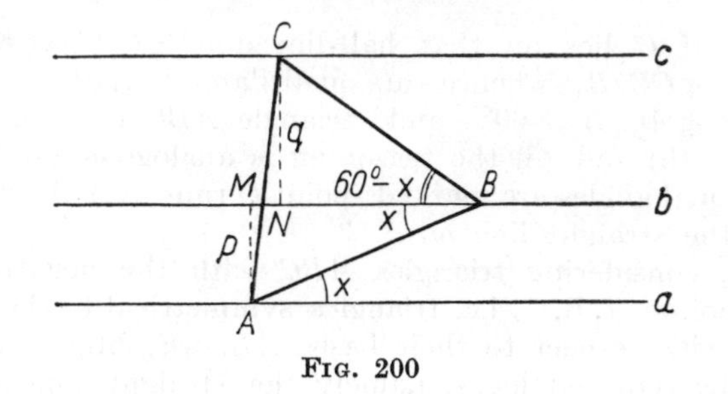

FIG. 200

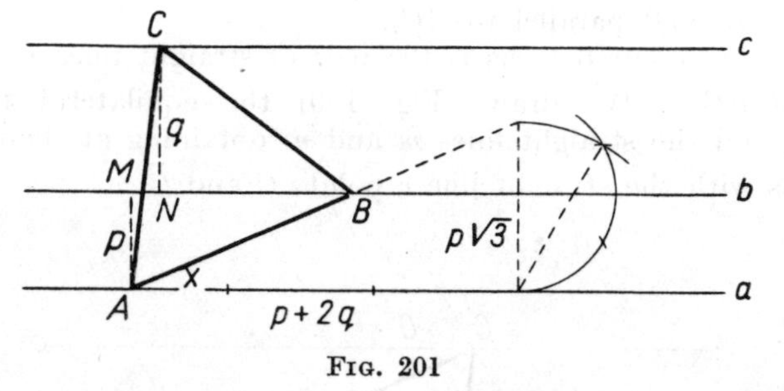

FIG. 201

133. We shall denote the radii of the given circles by R and $r, R > r$.

Let us temporarily disregard the condition that the required straight line should pass through the given point A and let us deal only with the construction of an arbitrary chord MN of the greater circle (Fig. 202) such as would be divided by the smaller circle into three equal parts $MK = KL = LN$. One of the points K, L, M, N can be chosen arbitrarily on its proper circle.

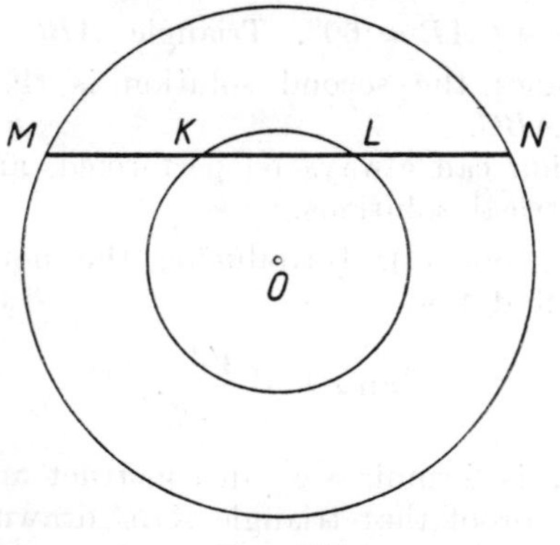

FIG. 202

Several methods of solving this problem can be shown.

Method I (of geometrical loci). *Analysis.* Let us take an arbitrary point M on the greater circle and seek point L. It satisfies two conditions: (1) it lies on the smaller circle and (2) $ML = \frac{2}{3}MN$, which means that point L corresponds to point N in the homothety with centre M and ratio $\frac{2}{3}$. The locus of point L is the circle corresponding to the greater of the given circles in this homothety, i.e. the circle with radius $\frac{2}{3}R$, tangent to the greater circle at point M.

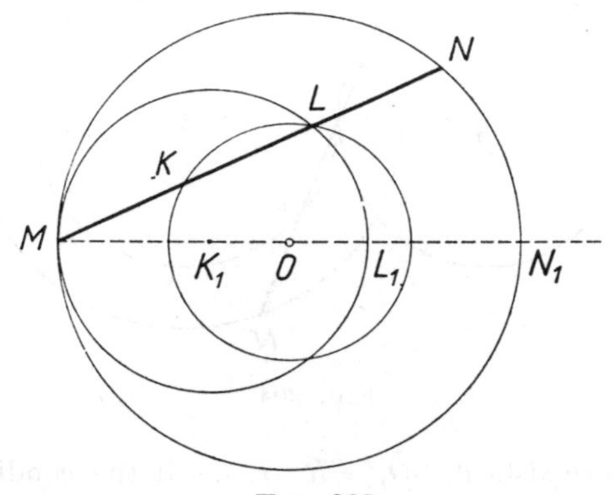

FIG. 203

Construction. We divide the diameter $MN_1 = 2R$ of the greater circle into three equal parts at points K_1 and L_1 (Fig. 203). We describe a circle with centre K_1 and radius $K_1L_1 = \frac{2}{3}R$. If this circle intersects the smaller of the given circles at point L, the straight line ML is the required one, since by our construction $ML = \frac{2}{3}MN$.

The problem has a solution if there exists a common point L of the smaller circle with the circle described, i.e. if point L_1 lies in the smaller circle. This occurs if $OL_1 \leqslant r$, i.e. if $r \geqslant \frac{1}{3}R$. If $r = \frac{1}{3}R$, then of course the diameter of the greater circle is the solution.

Method II (of rotation). *Analysis.* Suppose that the required chord MN is to pass through point K chosen arbitrarily on the smaller circle (Fig. 204)

Let us rotate the smaller circle about the point K through 180°. Point L will then coincide with point M. Thus we determine point M as the point of intersection of the rotated circle with the greater of the given circles.

The *construction* is shown in Fig. 204. We have determined point O_1 symmetric to point O with respect to point K chosen arbitrarily on the smaller circle and we have described from point O_1 as centre a circle with radius $O_1K = r$, obtaining point M at the intersection with the greater circle. The straight line MK is the required one because the segments MK and KL are equal, being symmetric with respect to point K.

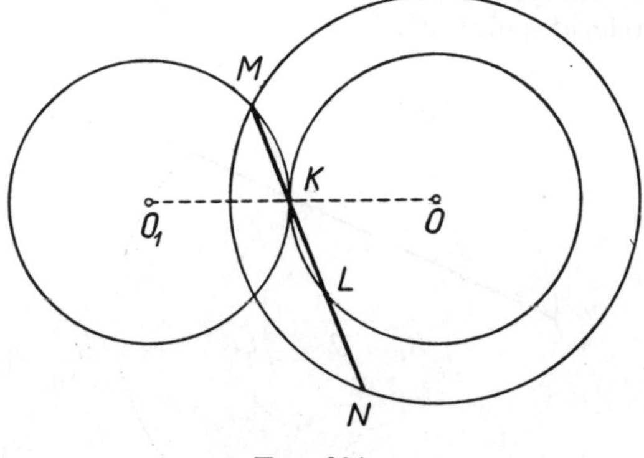

FIG. 204

The solution exists if $OO_1 \geqslant R - r$, i.e. if the condition $r \geqslant \frac{1}{3}R$ is satisfied.

Method III. Analysis. Let MN (Fig. 205) be the required chord. Let us mark off $LP = OL$ on OL produced. The quadrilateral $OKPN$ is a parallelogram because its diagonals bisect each

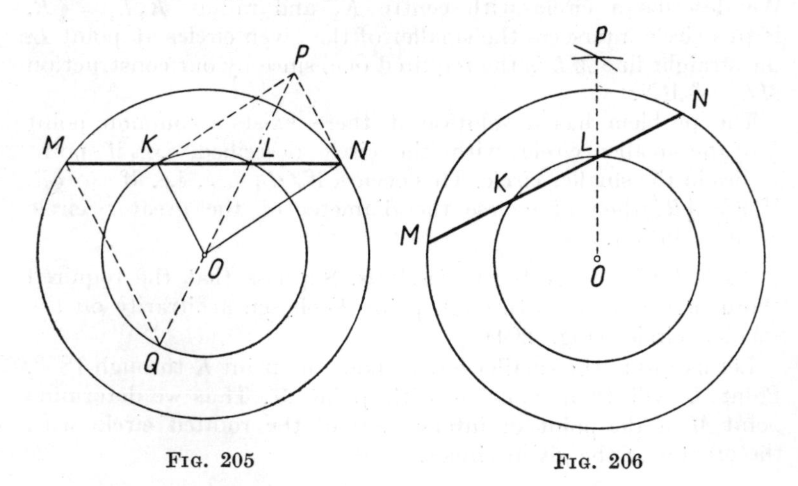

FIG. 205 FIG. 206

other; consequently $KP = ON = R$ and in triangle OKP we know the lengths of all the sides.

Construction. From the centre at an arbitrary point K of the smaller circle we describe a circle with radius R, and from point O—a circle with radius $2r$ (Fig. 206). We join point P, at which these circles intersect, with point O; at the intersection of the straight line OP with the smaller circle we obtain point L.

The solution exists if the above-mentioned circles have a point in common, i.e. if

$$R-2r \leqslant r, \quad \text{i.e.} \quad r \geqslant \frac{R}{3}.$$

Method IV. Analysis. Let us produce LO (Fig. 205) until it intersects the smaller circle at point Q. In triangle LQM the segment OK joins the mid-points of two sides, whence $MQ = 2 \times OK = 2r$. Hence we have the following *construction* (Fig. 207):

From an arbitrary point M of the greater circle we describe a circle with radius $2r$; suppose that it intersects the smaller circle at point Q; we then find point L at the intersection of the straight line QO with the smaller circle.

The condition of existence of the solution—as in method III.

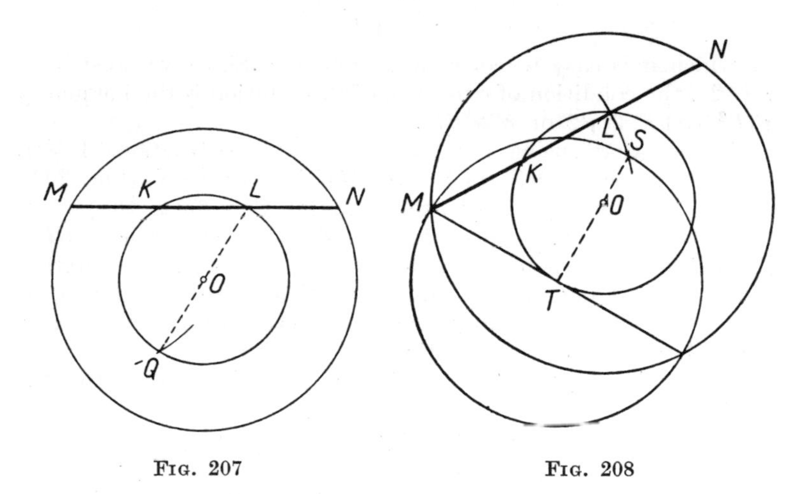

Fig. 207

Fig. 208

Method V. Analysis. If MN is the required chord (Fig. 208) and MT a tangent to the smaller circle with the point of contact T, then

$$MT^2 = MK \times ML = \tfrac{1}{2}ML^2, \quad \text{whence} \quad ML^2 = MT \times 2MT.$$

The length ML can thus be determined as the geometrical mean of the lengths MT and $2MT$.

The *construction* is shown in Fig. 208, where

$$ML = MS = MT \times \sqrt{2}.$$

The condition of existence of the solution is expressed by the inequality

$$MS \leqslant R+r, \quad \text{i.e.} \quad \sqrt{[2(R^2-r^2)]} \leqslant R+r.$$

Transformed, this inequality gives as before:

$$r \geqslant \frac{R}{3}.$$

Method VI (of similarity). This method, less convenient in point of construction than the preceding ones, consists in drawing first an arbitrary figure similar to that of Fig. 202.

The figure obtained should then be transformed by homothety of a suitable ratio.

Method VII (calculatory). Let $KL = x$ (Fig. 202) and let H be the mid-point of segment KL. Applying to triangles OHL and OHN the theorem of Pythagoras, we find the formula

$$x = \frac{\sqrt{(R^2-r^2)}}{\sqrt{2}}$$

by which it is easy to construct segment x. Since we must have $x \leqslant 2r$, the condition of existence of the solution is the inequality $\sqrt{(R^2-r^2)} \leqslant 2r\sqrt{2}$ or $r \geqslant R/3$.

Suppose that, using one of the above constructions I–VII, we draw a secant MN such that $MK = KL = LN$ (Fig. 202). There are infinitely many such secants; they form the set of tangents to a circle with centre O and radius OH, where OH is the distance from point O to the straight line MN. The determination of that one of those secants which passes through point A is reduced to the well-known construction of a tangent from point A to the circle in question.

Since $OH = \frac{1}{4}\sqrt{2} \times \sqrt{(9r^2-R^2)}$, the general result of the discussion is as follows:

(a) if $r > R/3$ and
- $OA > \frac{1}{4}\sqrt{2} \times \sqrt{(9r^2-R^2)}$, there exist two solutions,
- $OA = \frac{1}{4}\sqrt{2} \times \sqrt{(9r^2-R^2)}$, there exists one solution,
- $OA < \frac{1}{4}\sqrt{2} \times \sqrt{(9r^2-R^2)}$, there are no solutions;

(b) if $r = R/3$, there exists one solution—the line OA;
(c) if $r < R/3$, there are no solutions.

134. Let us denote the radii of the given circles by R and r, $R > r$.

Method I (of rotation). *Analysis.* Let $ABCD$ (Fig. 209) be the required square. Let us rotate the smaller circle about point A through $90°$ in the direction indicated by the arrow. Point D will then lie at point B. Thus, if we choose an arbitrary vertex A of the square, then vertex B will be found at the intersection of the rotated circle with the given larger circle.

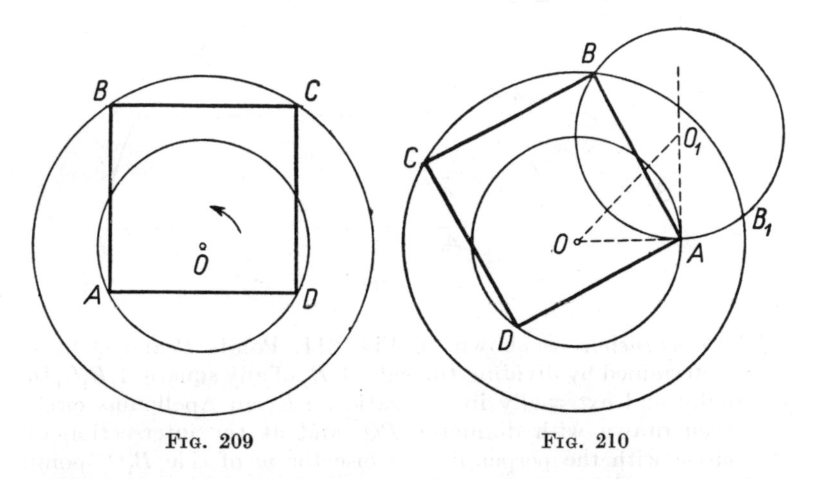

FIG. 209 FIG. 210

The *construction* is shown in Fig. 210. We draw at point A a perpendicular to the straight line OA and mark off on it $AO_1 = AO$; from point O_1 as centre we describe a circle with radius O_1A. If this circle intersects the larger of the given circles at point B, the segment AB will be a side of the required square. Indeed, drawing $AD \perp AB$ we have $AD = AB$, because AD is obtained by the rotation of segment AB through $90°$.

The solution exists if the circle rotated has a point in common with the larger circle, which occurs if

$$OO_1 \geqslant R - r.$$

From the triangle OAO_1 we have $OO_1 = r\sqrt{2}$. Thus the above condition gives the inequality

$$r\sqrt{2} \geqslant R - r, \quad \text{i.e.} \quad r \geqslant R(\sqrt{2} - 1).$$

If $r > R(\sqrt{2} - 1)$, there exist (for a chosen vertex A) four solutions pairwise symmetric with respect to the line OA.

If $r = R(\sqrt{2}-1)$, there exist (for a chosen vertex A) two solutions symmetric with respect to OA.

Method II (of similarity). *Analysis.* We construct a figure (Fig. 211) similar to the given one (Fig. 209), taking an arbitrary square $A_1B_1C_1D_1$ instead to the given square $ABCD$. Point O_1 corresponding to point O will be found by considering (1) that point O_1 lies on the axis of symmetry of the square $A_1B_1C_1D_1$ parallel to the side A_1B_1 and (2) that the ratio $O_1A_1 : O_1B_1$ is equal to the ratio $OA : OB$, i.e. $r : R$.

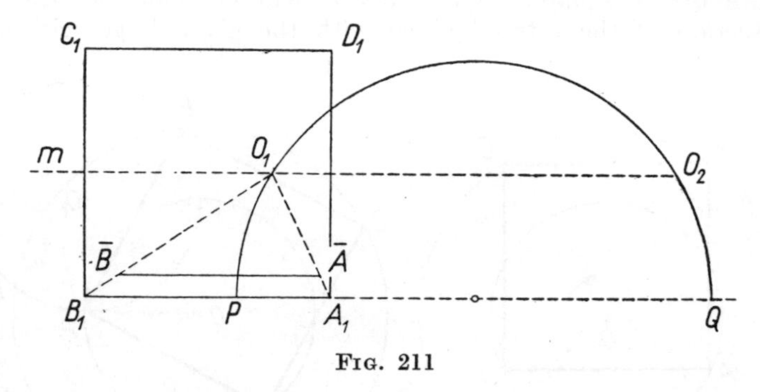

FIG. 211

The *construction* is shown in Fig. 211. Points P and Q have been determined by dividing the side A_1B_1 of any square $A_1B_1C_1D_1$ internally and externally in the ratio $r : R$; an Apollonius circle has been drawn with diameter PQ, and at the intersection of this circle with the perpendicular bisector m of side B_1C_1 point O_1 has been obtained.

If we mark off $O_1\bar{A}_1 = r$ and draw $\bar{A}\bar{B} \| A_1B_1$, then $O_1\bar{A}/O_1\bar{B} = O_1A_1/O_1B_1$ whence $r/O_1\bar{B} = r/R$ and consequently $O_1\bar{B} = R$. The length of $\bar{A}\bar{B}$ is the length of the side of the required square.

The solution exists if the Apollonius circle has a point in common with the line m, which occurs if $PQ \geqslant A_1B_1$.

Now

$$PQ = PA_1 + A_1Q, \quad PA_1 = \frac{r}{R+r} A_1B_1, \quad A_1Q = \frac{r}{R-r} A_1B_1.$$

Thus the above condition expresses the fact that

$$\frac{r}{R+r} + \frac{r}{R-r} \geqslant 1.$$

Transforming this equality, we obtain

$$r^2 + 2rR - R^2 \geqslant 0, \quad (r+R)^2 - 2R^2 \geqslant 0$$

and finally

$$r \geqslant R(\sqrt{2}-1),$$

as before.

Method III (of geometrical loci). *Analysis.* Let us draw a diameter AE of the smaller circle and a diagonal AC of the required square (Fig. 212). Since $\measuredangle ADC = 90°$, point E lies on the half-line CD, whence the angle ACE equals 45°. Consequently if we choose an arbitrary diameter AE, point C will be found at the intersection of two loci: the larger circle and an arc with the chord AE; the centre of that arc lies at the mid-point of the semi-circle with diameter AE.

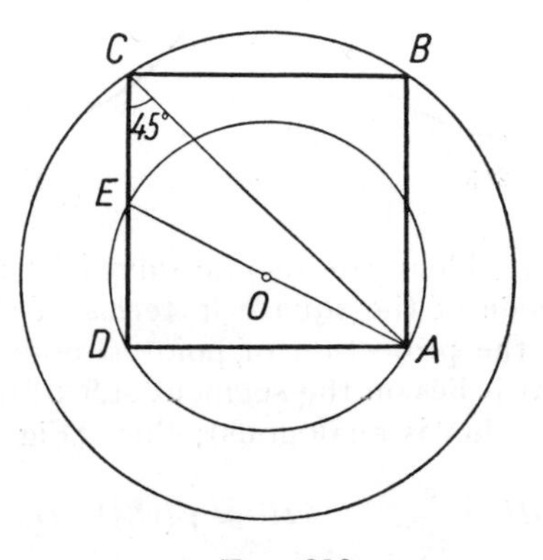

FIG. 212

Construction. We draw an arbitrary diameter AE of the smaller circle (Fig. 213) and a radius OM perpendicular to it. From point M as centre we describe an arc with radius MA as far as the intersection at point C with the larger circle; we then determine on the smaller circle point B at the intersection with the straight line CE and join point B with point A. We draw $AD \| BC$ and join point C with point D. The quadrilateral $ABCD$ is a square.

Indeed, $\measuredangle ACE = \frac{1}{2}\measuredangle AME = 45°$ and $\measuredangle ABC = 90°$ as an inscribed angle subtended by the diameter AE, consequently $AB = BC$ and $\measuredangle BAD = 90°$; the whole figure is symmetric with respect to the perpendicular bisector of segment AB, whence $AD = BC$.

The solution exists if the circle with centre M has a point in common with the given larger circle, which occurs if

$$MA \geqslant MN, \quad \text{i.e. if} \quad r\sqrt{2} \geqslant R-r,$$

which gives—as before—the condition $r \geqslant R(\sqrt{2}-1)$.

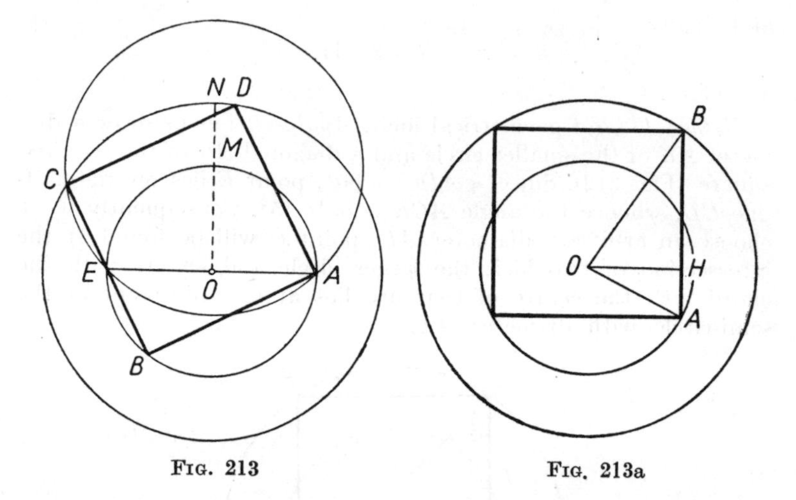

FIG. 213 FIG. 213a

REMARK. The problem can also be solved by finding the length $AB = x$ of the side of the square in terms of the radii R and r.

Let H denote the projection of point O upon the straight line AB. Suppose that it lies on the segment AB (if it lies on AB produced, the calculation is analogous); then (Fig. 213a)

$$OH = \frac{x}{2}, \qquad AH = \sqrt{(OA^2 - OH^2)},$$

$$HB = \sqrt{(OB^2 - OH^2)}, \qquad AB = AH + HB.$$

Consequently

$$x = \sqrt{\left[r^2 - \left(\frac{x}{2}\right)^2\right]} + \sqrt{\left[R^2 - \left(\frac{x}{2}\right)^2\right]}.$$

This equation reduces to a bi-quadratic equation. The construction of segment x according to the formula resulting from it is cumbersome and we shall not discuss it here.

135. *Method I* (of geometrical loci). The problem reduces to finding a point C satisfying the following conditions:

(1) point C lies on the given circle k;

(2) point C is a vertex of a triangle ABC in which vertex A is given, vertex B lies on the given straight line p, $\sphericalangle A = 60°$ and $\sphericalangle B = 90°$.

We shall determine the locus of point C satisfying condition (2). We shall distinguish two cases:

(a) Point A does not lie on the line p. Then there exist on p

two points M and N of the required locus (Fig. 214); they are vertices of triangles ATM and ATN where AT is the distance of point A from p and $\sphericalangle AMT = \sphericalangle ANT = 30°$.

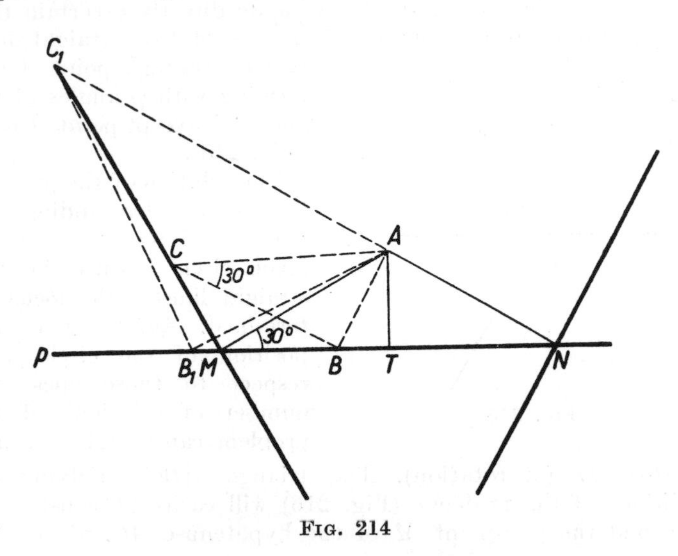

FIG. 214

Let C be a point satisfying condition (2) and let angle ABC have the same orientation as angle ATM.

Point M then lies on the circumcircle of triangle ABC. Indeed, point B coincides with point M, lies on the half-line MT (points M and C then lying on the same side of AB and $\sphericalangle AMB = \sphericalangle ACB = 30°$) or lies on the TM produced beyond point M; points M and C then lie on opposite sides of AB and $\sphericalangle AMB + \sphericalangle ACB = 180°$, since $\sphericalangle AMB = 180° - \sphericalangle AMT = 150°$ and $\sphericalangle ACB = 30°$.

Since AC is the diameter of the circumcircle of triangle ABC, $\sphericalangle AMC = 90°$. We have thus found that all those points of the required locus for which the angle ABC has the same orientation as the angle ATM lie on a perpendicular drawn to AM from point M. Conversely, every point C of that perpendicular belongs to the locus; for, if the circumcircle of triangle AMC passes through point B of the straight line p, the triangle ABC satisfies condition (2) because $\sphericalangle ABC = \sphericalangle AMC = 90°$ and $\sphericalangle ACB = \sphericalangle AMT = 30°$.

Analogously, all points satisfying condition (2) and such that the angle ABC has the same orientation as the angle ATN form a line perpendicular to AN at point N.

We have proved that the required locus consists of two straight lines perpendicular to AM and AN at points M and N, respectively.

(b) Point A lies on p. In this case we directly ascertain that the required locus consists of all points of two straight lines passing through point A and forming with p angles of 60° and 120° except point A itself (Fig. 215).

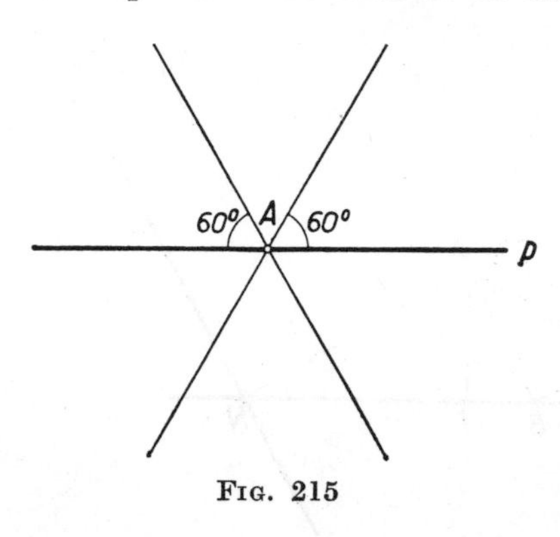

Fig. 215

The solution of the problem is obtained by finding the points of intersection of the given circle k with the two straight lines of the locus determined. According to the position of the circle with respect to those lines, the number of solutions of the problem can be 4, 3, 2, 1 or 0.

Method II (of rotation). The triangle ABC satisfying the conditions of the problem (Fig. 216) will easily be constructed if we find the mid-point M of the hypotenuse AC. Since AM

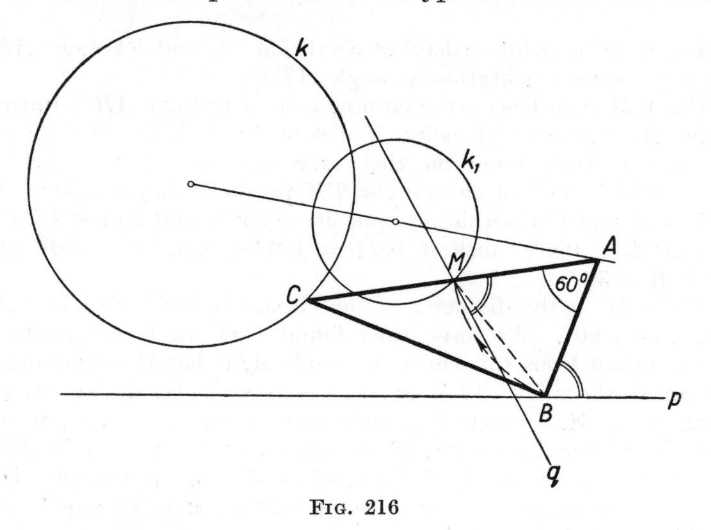

Fig. 216

$= AB$ and $\sphericalangle MAB = 60°$, point M lies on a straight line q which will be obtained by rotating p about point A through an angle of 60°, point B then falling upon point M. On the other hand, considering that $AM = \frac{1}{2}AC$, point M lies on a circle k_1 homothetic (directly) to the given circle k in the ratio $1:2$ with respect to the centre of homothety A. Thus, if the solution of the problem exists, point M is the point of intersection of circle k_1 with q. Conversely, if M is such a point, then point C, homothetic to M in the ratio $1:2$ with respect to point A, and that point B of p which after the rotation falls on point M are vertices of the required triangle.

Since the straight line p can be rotated about point A through an angle of 60° in two directions, which gives two straight lines q_1 and q_2, the problem has as many solutions as there are common points of circle k_1 and the lines q_1 and q_2, i.e. 4, 3, 2, 1 or 0.

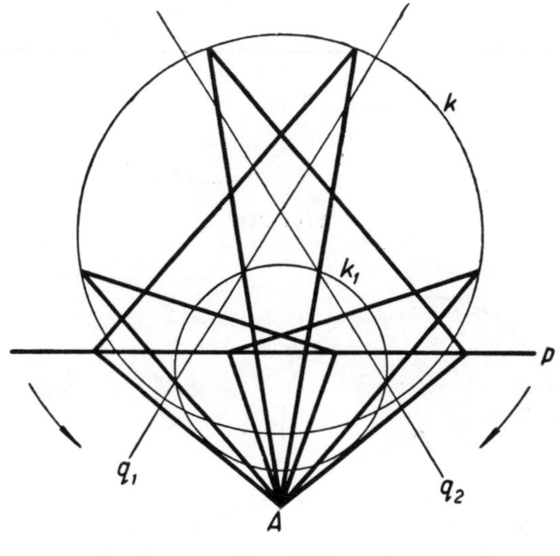

FIG. 217

In Figure 217 point A, line p and circle k have been chosen in such a manner that the problem has four solutions pairwise symmetric.

It will be observed that we can solve the problem in a slightly different way.

(a) Instead of rotating the straight line p, we can rotate the circle k_1 about point A through 60° and thus obtain at the intersection of the circle and line p the vertex B of the required triangle (Fig. 218).

(b) We can also rotate circle k about point A through 60°. At the intersection of the circle with a straight line p_1 homo-

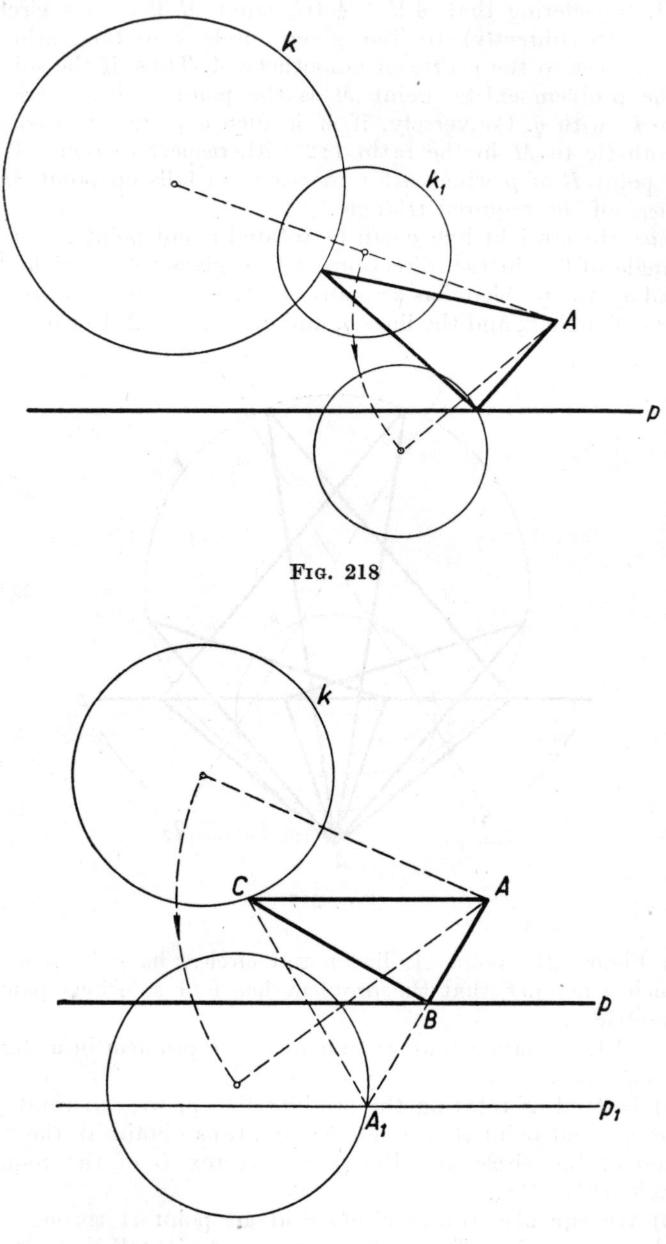

FIG. 218

FIG. 219

thetic to p with respect to point A in the ratio 2:1 we then obtain point A, corresponding in this homothety to point B (Fig. 219).

Finally, it will be observed that our problem is equivalent to the following one: construct an equilateral triangle with a given vertex A (triangle ABM in Fig. 216 or triangle AA_1C in Fig. 219) in which one of the remaining vertices lies on a given straight line and the other on a given circle.

136. *Analysis.* The problem is reduced to determining the mid-point X of the common chord of the given circle and the required one; for, having point X, we can draw in the given circle k a chord CD with mid-point X and describe a circle passing through points A, B, C, D.

The locus of the centres X of chords of circle k having a given length d is a circle k' concentric with k, its radius being $\sqrt{(r^2-d^2/4)}$ where r denotes the radius of circle k. It should be assumed here that $d \leqslant 2r$.

(1) If the given points A and B are equidistant from centre O of circle k (Fig. 220), then the required circle intersects circle

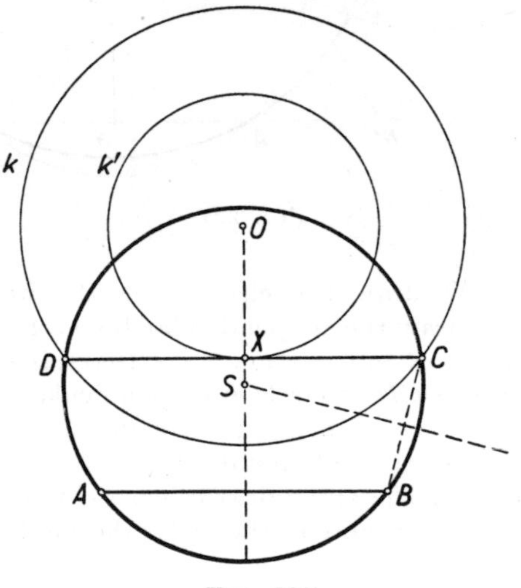

Fig. 220

k at points symmetric with respect to the perpendicular bisector of segment AB, i.e. point X will be found at the intersection of that perpendicular bisector with circle k'.

(2) If A and B are not equidistant from point O (Fig. 221), then all common chords of circle k and of the circles passing through

points A and B lie on straight lines which intersect AB at the same point M (see problem 86).

Point M can be found by drawing any auxilliary circle l passing through points A and B and intersecting circle k. Point X will be found at the intersection of circle k' with a circle with diameter OM which is the locus of the mid-points of chords for secants across M.

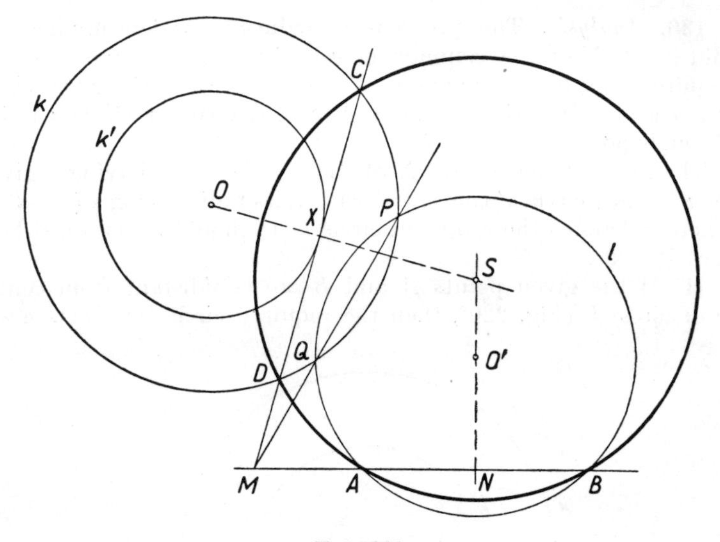

Fig. 221

Construction. We draw the circle k' in the well-known way.

In case (1) we draw the perpendicular bisector of segment AB. Through the point of intersection X of that bisector with circle k' we draw a chord CD of circle k perpendicular to the line OX. Finally we draw a circle passing through points A, B, C; this circle will also pass through point D, which is symmetric to C with respect to OX. The circle drawn has a common chord CD of length d with circle k, and is therefore the circle sought in the problem.

In case (2) we draw an arbitrary circle l passing through points A and B and intersecting circle k at points P and Q. The straight line PQ intersects the straight line AB at point M. We describe a circle with diameter OM. If X is a point of intersection of that circle with circle k', then the straight line MX intersects circle k along a chord CD of length d. We now draw a circle passing through points A, B, C; this circle will pass also through point

D because $MA \times MB = MP \times MQ$, $MP \times MQ = MC \times MD$, whence $MA \times MB = MC \times MD$. It will be the circle sought in the problem.

Discussion. We shall investigate the existence and the number of solutions of the problem according to the choice of the data. We have already observed that the problem can be solved only if the given length d satisfies the condition $d \leqslant 2r$ where r is the radius of the given circle k.

Assuming that this condition is satisfied, we shall investigate all the cases that can occur.

I. The centre O of circle k lies on the perpendicular bisector of segment AB.

(a) $d < 2r$. The perpendicular bisector of AB intersects circle k' at two points X_1 and X_2, and thus there exist two chords, C_1D_1 and C_2D_2 of circle k of length d and the direction of the straight line AB.

If AB does not coincide with any of the lines C_1D_1 and C_2D_2, i.e. if the distance of AB from point O is not equal to $\sqrt{(r^2-d^2/4)}$, the problem has two solutions.

If AB coincides with, for example, C_1D_1 but points A and B do not lie on circle k, i.e. do not coincide with points C_1 and D_1, the problem has only one solution, namely the circle passing through points A, B, C_2, D_2.

Finally, if points A and B coincide with points C_1 and D_1, every circle passing through points A and B except circle k itself is a solution of the problem.

(b) $d = 2r$. In this case circle k' is reduced to point O; point X also coincides with point O, and in the circle k there exists one chord CD of length d and the direction of the straight line AB.

If the straight line AB does not pass through point O, then the problem has one solution.

If the straight line AB passes through point O but points A and B do not lie on circle k, the problem has no solutions.

Finally, if the segment AB is a diameter of circle k, then every circle passing through points A and B except circle k is a solution of the problem.

II. The centre of circle k does not lie on the perpendicular bisector of segment AB.

(a) $d < 2r$. If the straight line AB lies outside circle k', i.e. if its distance from point O is greater than $\sqrt{(r^2-d^2/4)}$, then point M also lies outside circle k'. The circle with diameter OM intersects circle k' at two points, X_1 and X_2. The straight lines MX_1 and MX_2 determine in circle k the chords C_1D_1 and C_2D_2 of length d. The problem has two solutions: the circle passing through points A, B, C_1, D_1 and the circle passing through points A, B, C_2 D_2.

If the straight line AB is tangent to circle k' at point T and point M is different from point T, then the circle with diameter OM intersects circle k' at two points, T and X. In this case there is only one solution: the circle passing through points A, B and through points C, D at which MX intersects circle k.

If AB is tangent to circle k' and point M coincides with the point of contact T, there is no solution. This case occurs if point T lies inside segment AB and the geometric mean of segments AT and BT is equal to $\frac{1}{2}d$.

Finally, if AB intersects circle k', the problem has two solutions, one solution or no solution according to whether point M lies outside circle k', on that circle or inside it.

(b) $d = 2r$. In this case point X coincides with point O.

If the straight line AB does not pass through point O, then the straight line MO intersects circle k at two points, C and D. The problem has one solution: the circle passing through points A, B, C, D.

If AB passes through point O and point M is different from point O, then MO intersects circle k at points C and D of the line AB and the problem has no solution.

Finally, if point M of the line AB coincides with point O, then every circle passing through points A and B is a solution of the problem. This case occurs if point O lies inside segment AB and radius r of circle k is the geometric mean of segments OA and OB.

Remark. In the above problem number d, as the length of a segment, is a positive number. However, we can consider the "limiting" case of this problem with $d = 0$, i.e. the following problem:

Through two given points A and B draw a circle tangent to the given circle k.

The method of solution remains the same—the only difference being the identity of circle k' with circle k.

The discussion of the possible number of solutions can easily be derived from the discussion carried out above for the case $d < 2r$.

This problem has been discussed together with problem 86.

137. To begin with, it will be observed that, if points M and N lie on a given circle k, every point C of the circle except points M and N satisfies the conditions of the problem: points A and B coincide with points M and N and triangle ABC coincides with triangle MNC. In the sequel we shall disregard this case and assume that at least one of the points M and N lies away from circle k.

If C is the required point, then for the similar triangles ABC and MNC, having equal angles at the vertex C, one of the following cases must occur:

I. $\sphericalangle A = \sphericalangle M$, $\sphericalangle B = \sphericalangle N$ (solution of the first kind),

II. $\sphericalangle A = \sphericalangle N$, $\sphericalangle B = \sphericalangle M$ (solution of the second kind).

If the triangles are isosceles, the two cases occur simultaneously.

I. *Seeking solutions of the first kind.*

In solutions of the first kind, the sides AB and MN of triangles ABC and MNC are parallel, whence the points M and N either both lie on the segments AC and BC or both lie on those segments produced. Thus our problem can have a solution of the first kind only if points M and N either both lie inside the given circle k or both lie outside that circle.

We shall give two methods of obtaining such solutions.

Method I. If point C is a point with the required property (Fig. 222), then triangles ABC and MNC are homothetic with

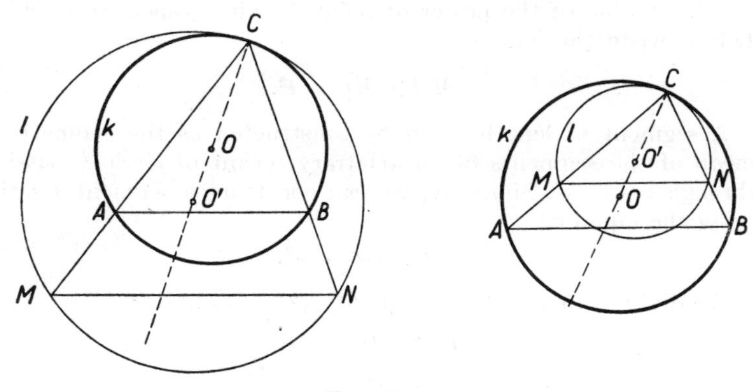

Fig. 222

respect to point C. In this homothety, to circle k, passing through the centre of homothety C and through points A and B, corresponds a circle l, tangent to k at point C and passing through points M and N homothetic to A and B.

The problem is thus reduced to drawing a circle passing through given points M, N and tangent to the given circle k.

The construction of such a circle was explained in problem 86 and again in problem 136.

If we draw circle l, then point C at which it is tangent to circle k will be the required point. Indeed, the straight lines MC and NC intersect circle k, homothetic to circle l from centre C, at points A and B, corresponding to points M and N. The triangles ABC and MNC are homothetic, and we have $\sphericalangle A = \sphericalangle M$ and $\sphericalangle B = \sphericalangle N$.

The problem has as many solutions as there are circles passing through points M and N and tangent to circle k. Consequently (see remark 2 to problem 86):

(1) If points M and N both lie inside circle k, or if they both lie outside circle k but the straight line MN is not tangent to that circle, the problem has two solutions.

(2) If points M and N lie outside circle k and the straight line MN is tangent to that circle, the problem has one solution.

Method II. From the similarity of triangles MNC and ABC (Fig. 222) it follows that

$$\frac{MA}{NB} = \frac{MC}{NC}. \tag{1}$$

It will be observed that the product $MA \times MC$ is equal to the absolute value of the power of point M with respect to circle k. Let us write the equality

$$MA \times MC = t^2. \tag{2}$$

A segment of length t can be constructed as the geometrical mean of the segments of an arbitrary secant of circle k passing through point M. Similarly, we can construct a segment s satisfying the equality

$$NB \times NC = s^2. \tag{3}$$

We divide equality (2) by equality (3) and obtain:

$$\frac{MA \times MC}{NB \times NC} = \frac{t^2}{s^2}.$$

Taking into account equality (1), we obtain

$$\left(\frac{MC}{NC}\right)^2 = \left(\frac{t}{s}\right)^2,$$

whence

$$\frac{MC}{NC} = \frac{t}{s}. \tag{4}$$

Equality (4) states that the ratio of the segments MC and NC is equal to the ratio of the segments t and s, which we are able to construct. Point C will thus be found at the intersection of circle k with the Apollonius circle for segment MN and for the ratio $t:s$.

To perform the construction according to the above plan presents no difficulties; however, it is less convenient than that of method I.

If the straight line MN is tangent to circle k, the construction is simpler because segments t and s are then given; this is shown in Fig. 223.

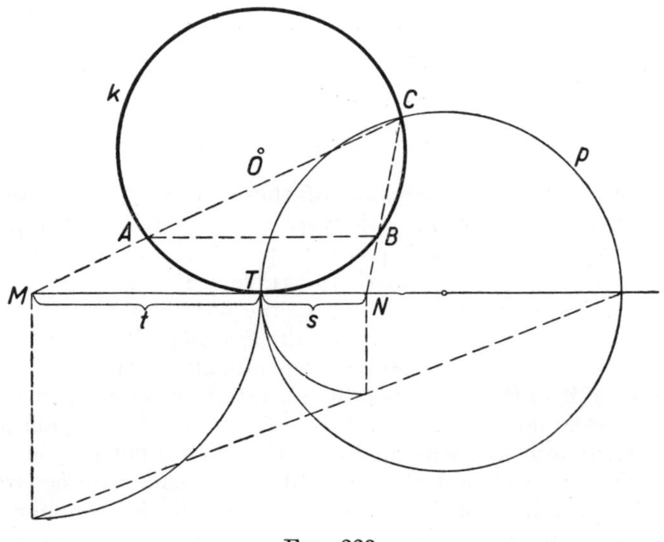

Fig. 223

In this case the Apollonius circle p intersects circle k at two points, one of them being the point of contact T of circle k with MN and the other the required point C. Thus the problem always has one solution.

From the solution by method I we know that if MN is not tangent to circle k, the problem has two solutions; consequently, also in this case the Apollonius circle always intersects circle k.

II. *Seeking solutions of the second kind.*

Suppose that in the triangles ABC and MNC (Fig. 224) we have

$$\measuredangle A = \measuredangle N.$$

Method I. From the similarity of triangles ABC and MNC it follows that

$$\frac{CA}{CN} = \frac{CB}{CM},$$

whence

$$CA \times CM = CB \times CN.$$

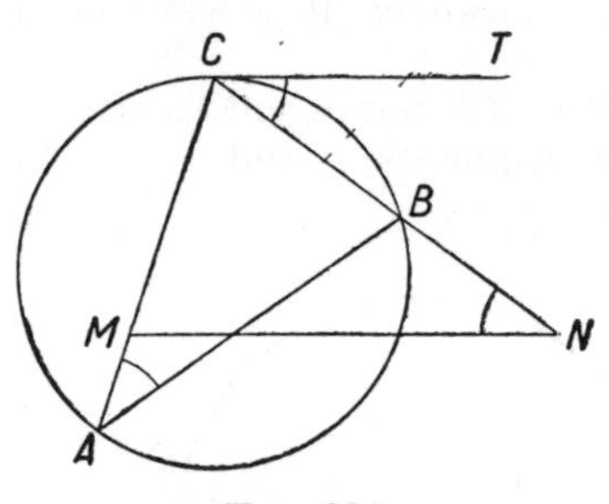

FIG. 224

Denoting the common value of these products by r^2 we can see that in an inversion with respect to a circle with centre C and radius r point M corresponds to point A, and point B corresponds to point N (see remark to problem 85).

The centre of inversion C lies on the axis of symmetry of the figure formed by circle k and straight line MN. The construction is thus reduced to drawing a perpendicular from centre O of circle k to MN. If that perpendicular intersects circle k at a point C which does not lie on MN, that point gives the solution of the problem. Indeed, the straight line CM, which is not perpendicular to CO, intersects circle k also at point A, and similarly the straight line CN has, besides C, one more point, B, in common with circle k. In an inversion with centre C which transforms point A into point M, to circle k corresponds the straight line passing through point M and perpendicular to CO, i.e. the line MN. Consequently, in that inversion, point N corresponds to point B, whence we have

$$CA \times CM = CB \times CN,$$

which gives

$$\frac{CA}{CN} = \frac{CB}{CM}.$$

Thus triangles ABC and MNC are similar and $\sphericalangle A = \sphericalangle N$.

The construction is always possible. The perpendicular from point O to MN intersects circle k at two points. One of them may happen to lie on MN; this occurs if MN is tangent to k; in that case the problem has one solution. If MN is not tangent to k, the problem has two solutions.

Method II. The same construction as in method I can be obtained in a different way without the use of transformation by inversion.

Let CT be a tangent to circle k at the required point C (Fig. 225). By the well-known theorem on the angle between a tangent and a chord we have

$$\sphericalangle BCT = \sphericalangle A,$$

and since

$$\sphericalangle A = \sphericalangle N,$$

we have

$$\sphericalangle BCT = \sphericalangle N.$$

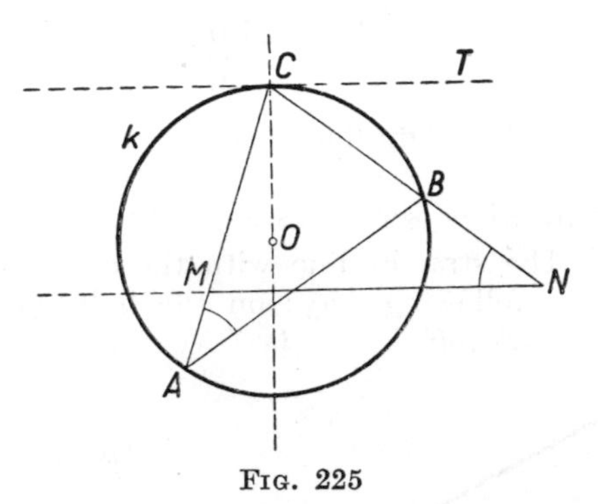

FIG. 225

This implies that the straight lines CT and MN are parallel.

Thus the construction is reduced to drawing a tangent to circle k parallel to MN. The point of contact C will be found, as in method I, at the intersection of circle k with a perpendicular to MN drawn from point O.

138. *Analysis*. Let us choose point M, say, on the side AB of the given triangle produced beyond point B (Fig. 226). Suppose that the straight line m is the solution of the problem, i.e. that $AN = BP$.

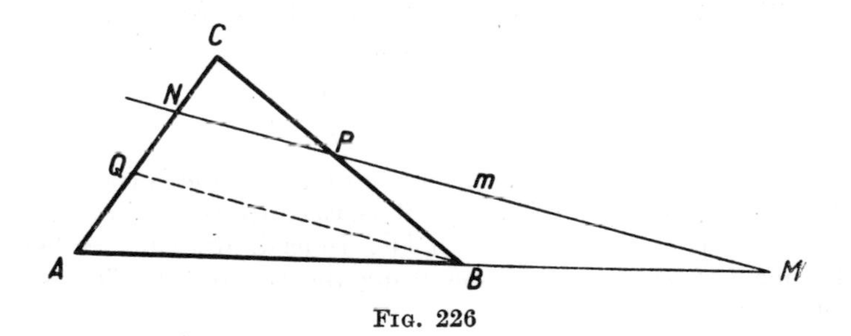

FIG. 226

Let us draw through point B a line parallel to m; it will intersect the side AC of the triangle at a point Q. Applying the theorem of Thales to the straight lines AB and AC, intersected by the parallel lines QB and m, and to the lines AC and BC, intersected by the same parallel lines, we obtain

$$\frac{AM}{BM} = \frac{AN}{QN} \quad \text{and} \quad \frac{BP}{QN} = \frac{PC}{NC}.$$

Since by hypothesis we have $BP = AN$, these properties imply that

$$\frac{PC}{NC} = \frac{AM}{BM}.$$

Thus the required straight line m cuts off from the given triangle a triangle PCN in which the ratio of the sides PC and NC is equal to the ratio of the given segments AM and BM.

Construction. The straight line with the above properties will be drawn in the following way: on the half-lines CA and CB (Fig. 227) we mark off segments $CN_1 = BM$ and $CP_1 = AM$

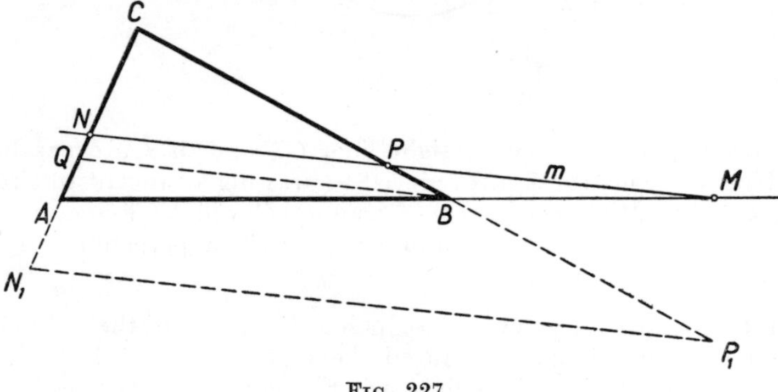

FIG. 227

respectively. Through point M we draw a straight line m parallel to P_1N_1; the line m intersects BC and AC at points P and N, and the following proportions hold:

$$\frac{PC}{NC} = \frac{P_1C}{N_1C} = \frac{AM}{BM}.$$

We must investigate whether the straight line m is the solution of the problem. To begin with, it is easy to ascertain that the required equality $AN = BP$ holds. Indeed, drawing as before the line BQ parallel to m and using the theorem of Thales, we obtain the equalities

$$\frac{AM}{BM} = \frac{AN}{QN}, \qquad \frac{PC}{NC} = \frac{BP}{QN}.$$

The left-hand sides of these equalities are equal by the construction; consequently

$$\frac{AN}{QN} = \frac{BP}{QN} \quad \text{whence} \quad AN = BP.$$

The fulfilment of condition $AN = BP$ does not yet ensure that the straight line m is the required one, since it is necessary for points N and P to lie within the sides AC and BC (as shown in Fig. 227) and not beyond them. Drawing the figure with the data changed we would find out that point N may happen to lie on the extension of segment AC beyond point A or beyond point C.

We must therefore establish the necessary and sufficient conditions for the point N determined by the preceding construction to lie on the segment AC (point P will then lie on the segment BC).

If point N lies on AC, then point Q, which belongs to the side AN of the triangle AMN, also lies on AC, whence we have the following order of points on that segment: A, Q, N, C.

Hence we shall draw two conclusions:

(1) Since $AC/AN = (AQ+QC)/(AQ+QN)$ and $QC > QN$, we have $AC/AN < QC/QN$†, which, in view of the equalities $QC/QN = BC/BP$ and $BP = AN$, gives $AC/AN < BC/AN$ and finally

$$AC < BC. \tag{1}$$

(2) Since, by the construction, $BM/AM = CN_1/CP_1 = CQ/CB$, and $CQ < AC$, we have

$$\frac{BM}{AM} < \frac{AC}{CB}. \tag{2}$$

Inequalities (1) and (2), which we have inferred from the assumption that the line m is a solution of the problem, are thus the *necessary* conditions of the existence of the solution. We can express them in a simple way as follows: The given point M must lie on the base produced beyond the end-point of the *larger* of the remaining sides of the triangle and must divide the base externally in a ratio less than the ratio of the smaller side to the larger one.

---

† We have used here the following theorem of artihmetic: if $a > b > 0$ and $c > 0$, then $(a+c)/(b+c) < a/b$.

We shall show that these necessary conditions are also *sufficient*. Indeed, by the construction, point Q lies on the segment CN_1, and $BM/AM = CN_1/CP_1 = CQ/CB$; thus if condition (2) is satisfied, then

$$\frac{CQ}{CB} < \frac{AC}{CB}, \quad \text{whence} \quad CQ < AC,$$

i.e. point Q lies on the segment AC. In that case point N lies on the half-line QC, because Q lies between A and N.

On the other hand, by the construction we have

$$\frac{AN}{QN} = \frac{AM}{BM} = \frac{CP_1}{CN_1} = \frac{CB}{CQ};$$

thus, if condition (1) is satisfied, then

$$\frac{AN}{QN} > \frac{AC}{QC}, \quad \text{whence} \quad \frac{AQ+QN}{QN} > \frac{AQ+QC}{QC};$$

this inequality implies that

$$\frac{AQ}{QN} > \frac{AQ}{QC}, \quad \text{whence} \quad QN < QC.$$

Consequently point N lies on the segment QC, and thus on AC.

We have obtained the following result: In the case of $AM > BM$, the problem can be solved if and only if inequalities (1) and (2) are satisfied. There is only one solution.

REMARK 1. The necessity of condition (1) can be ascertained in the following simple way: Let $BP = AN$ (Fig. 228). Let us draw $PK || AN$; then $AN > PK$, whence $BP > PK$ and thus also $BC > AC$.

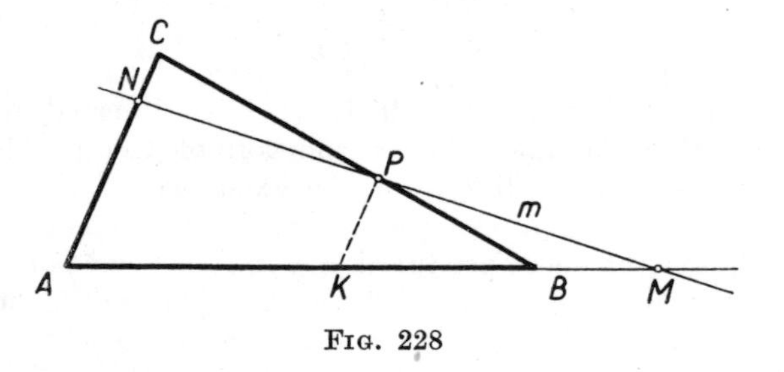

FIG. 228

REMARK 2. Condition (2) can be interpreted as follows: Let $AC'B$ (Fig. 229) be a triangle symmetric to triangle ACB. If $AC < BC$, then $AC' > BC'$ and the bisector of the exterior angle at vertex C' in the triangle $AC'B$ intersects segment AB produced beyond point B at a point M_0. By the theorem on the bisector of an angle in a triangle we have

$$\frac{BM_0}{AM_0}=\frac{BC'}{AC'}=\frac{AC}{BC}.$$

Condition (2) can thus be replaced by the condition

$$\frac{BM}{AM}<\frac{BM_0}{AM_0},$$

which means that point M should lie between points B and M_0.

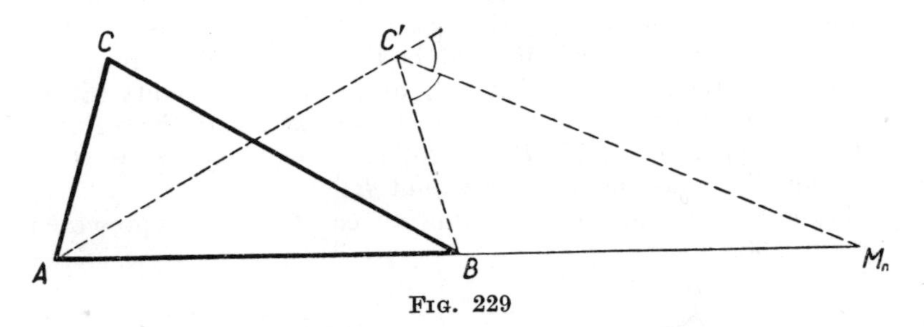

FIG. 229

REMARK 3. The solution of the problem can be made very simple and clear if we make use of the theorem of Menelaus. (See problem 77.)

If the straight line m intersects the directed straight lines AB, BC, CA at points M, P, N respectively, the following equality holds:

$$\frac{AM}{MP}\times\frac{BP}{PC}\times\frac{CN}{NA}=-1. \qquad (\alpha)$$

If m is the solution of our problem, then $BP = NA$ and equality (α) assumes the form

$$\frac{AM}{MB}\times\frac{CN}{PC}=-1. \qquad (\beta)$$

Let us introduce the notation: $AB = c$, $BC = a$, $CA = b$, $BM = p$, $CN = x$; then $AM = c+p$, $PC = BC-BP = BC-$ $-NA = a-(b-x)$ and equality (β) gives

$$\frac{c+p}{-p}\times\frac{x}{a-b+x}=-1, \qquad (\beta')$$

whence

$$x=\frac{(a-b)p}{c}. \qquad (\gamma)$$

Since $0 < CN < CA$, i.e. $0 < x < b$, formula (γ) implies the inequalities

$$b < a, \tag{δ}$$

$$p < \frac{bc}{a-b}. \tag{ε}$$

Inequality (δ) is identical with inequality (1), and inequality (ε) is equivalent to the inequality $p/(c+p) < b/a$, i.e. to inequality (2). We have obtained the same necessary conditions of the existence of the solution as before.

Suppose that conditions (δ) and (ε) are satisfied. If we determine x from formula (γ) and mark off $CN = x$ on the half-line CA, then the straight line MN will be the solution of the problem. Indeed, we then have equality (β'), and thus also equality (β), in which P denotes the point of intersection of MN with the segment BC. Since points M, N, P are collinear, equality (α) is true; and equalities (α) and (β) imply that $BP = NA$.

Formula (γ) leads to the following construction, represented in Fig. 230.

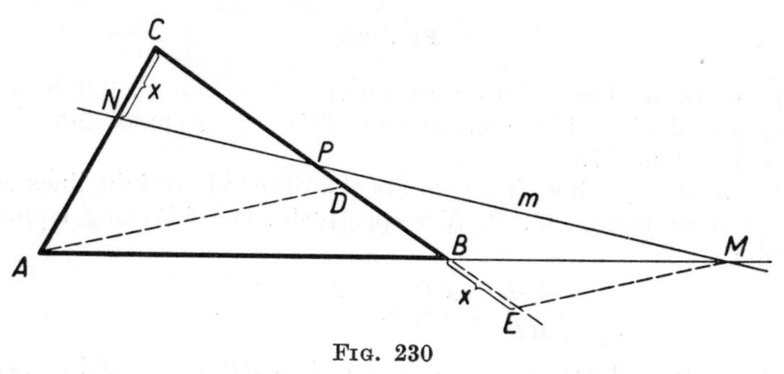

FIG. 230

We mark off $CD = CA$, draw $ME \| AD$ and mark off $CN = EB$; the straight line MN is the required one.

139. (a) Suppose that the parallel straight lines a, b, c, passing through points A, B, C, respectively, satisfy the condition that the distances between the neighbouring parallel lines should be equal. Then that one of the lines a, b, c which lies between the other two is equidistant from them. Suppose that b is the line in question. In that case, points A and C are equidistant from line b and lie on opposite sides of that line; consequently b intersects the segment AC at its mid-point M. The fact that points A, B, C are not collinear implies that point M is different from point B.

Hence the construction: we draw a straight line b through point B and mid-point M of the segment AC and then draw through points A and C straight lines a and c parallel to b (Fig. 231).

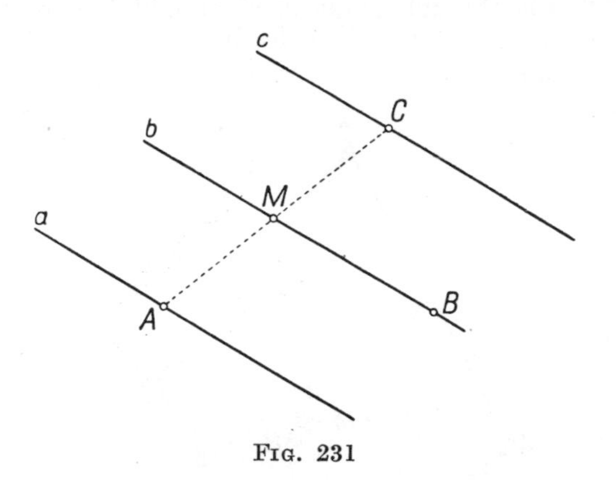

Fig. 231

The parallel lines a, b, c determined in this way give the solution of the problem because the points A and C, and thus also the lines a and c, are equidistant from line b and lie on opposite sides of that line.

The above solution has been found by assuming that line b lies between lines a and c; since that "interior" line can equally well be a or c, the problem has three solutions.

(b) Suppose that the planes α, β, γ, δ, passing through points A, B, C, D respectively and parallel, satisfy the condition that the distances between the neighbouring planes should be equal. Let those planes lie in the order α, β, γ, δ, i.e. let plane β be equidistant from planes α and γ, and plane γ equidistant from planes β and δ.

In that case points A and C are equidistant from plane β and lie on its opposite sides, whence plane β passes through the mid-point M of the segment AC. Similarly plane γ passes through the mid-point N of the segment BD. The fact that points A, B, C, D are not coplanar implies that point M is different from point B and point N is different from point C.

Hence we derive the following construction. We join point B with the mid-point M of segment AC and point C with the mid-point N of segment BD (Fig. 232 represents a parallel projection of the figure). The straight lines BM and CN are skew; for, if they lay in one plane, then points A, B, C, D would—contrary to our assumption—lie in the same plane. We know from solid geometry that through two skew lines BM and CN two and only two parallel planes β and γ can be drawn.

Accordingly, we draw through point M a straight line m parallel to CN and through point N a straight line n parallel to BM; plane β is then determined by lines m and BM and plane γ—by lines n and CN.

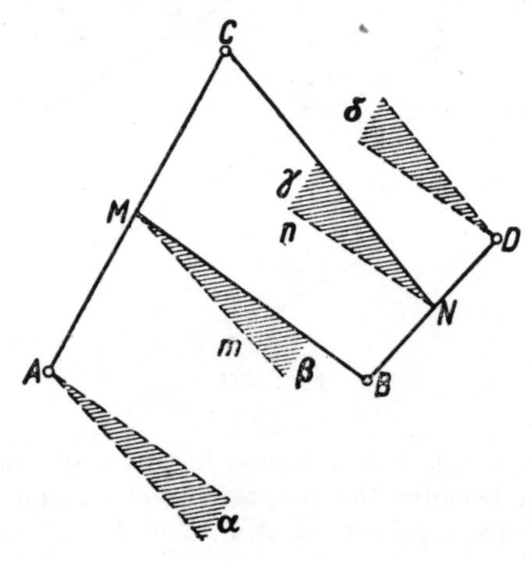

FIG. 232

Finally, through points A and D we draw planes α and δ parallel to planes β and γ; we can determine them, as shown in Fig. 232, by drawing through the points A and D straight lines parallel to lines BM and CN respectively.

The planes α, β, γ, δ determined in this way give the solution of the problem, since the points A and C, and thus also the planes α and γ, are equidistant from plane β, and similarly planes β and δ are equidistant from plane γ and lie on opposite sides of that plane.

We have obtained the above solution assuming that the required planes lie in the order α, β, γ, δ. For different successions of the required planes we shall find—in the same way—other solutions of the problem. The number of all the possible successions, in other words: of permutations of letters α, β, γ, δ, is $4!$, i.e. 24. It will be observed, however, that two "inverse" permutations, such as $\alpha, \beta, \gamma, \delta$ and $\delta, \gamma, \beta, \alpha$ for instance, give the same solution. Consequently, the problem has $\frac{24}{2} = 12$ solutions, corresponding to the permutations:

$$\alpha\beta\gamma\delta,\ \alpha\beta\delta\gamma,\ \alpha\gamma\beta\delta,\ \alpha\gamma\delta\beta,\ \alpha\delta\beta\gamma,\ \alpha\delta\gamma\beta,$$
$$\beta\alpha\gamma\delta,\ \beta\alpha\delta\gamma,\ \beta\gamma\alpha\delta,\ \beta\delta\alpha\gamma,\ \gamma\alpha\beta\delta,\ \gamma\beta\alpha\delta.$$

140. *Method I.* Let B', C', D' be the orthogonal projections of points B, C, D respectively upon a plane α passing through point A. The projection of the mid-point M of segment AC is the mid-point M' of segment AC' and the projection of the mid-point N of segment BD is the mid-point N' of segment $B'D'$. The plane quadrilateral $AB'C'D'$, which is the projection of the quadrilateral $ABCD$ upon the plane α, is a parallelogram if and only if points M' and N' coincide. Then the projecting straight lines MM' and NN' also coincide, whence the direction of projecting is the direction of the straight line MN (points M and N are different, points A, B, C, D not being coplanar). The required plane is the plane passing through point A and perpendicular to MN (Fig. 233). The problem always has one and only one solution.

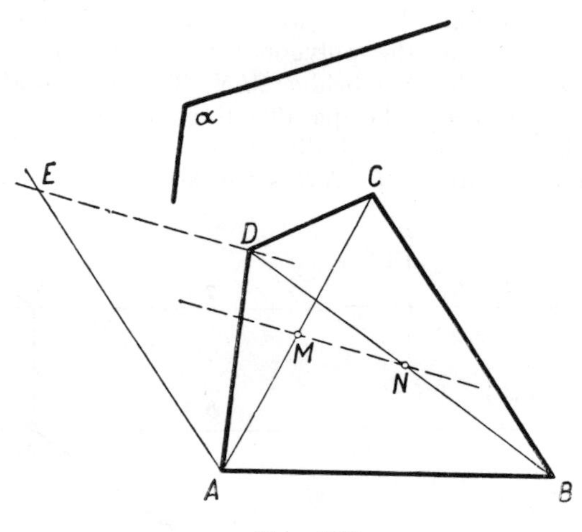

FIG. 233

Method II. Suppose, as in method I, that the quadrilateral $AB'C'D'$ is the orthogonal projection of quadrilateral $ABCD$ upon a plane α passing through point A. Let us translate the segment BC until it assumes the position AE (Fig. 233). Then the projections AE' and $B'C'$ of segments AE and BC upon the plane α are equal and parallel. The quadrilateral $AB'C'D'$ is a parallelogram if and only if the segments AD' and $B'C'$ are equal and parallel, which occurs if and only if segments AE' and AD' are equal and identically directed, i.e. if they coincide. The projection E' of point E thus coincides with the projection D' of point D, which means that the direction of projecting is the direction of the straight line DE (points D and E are different, points A, B, C, D not being coplanar).

Thus the only solution of the problem is the plane passing through point A and perpendicular to DE.

REMARK. The above implies that the straight lines MN and DE mentioned in methods I and II are parallel. This can easily be proved directly. Point M, as the mid-point of the diagonal AC of the parallelogram $ABCE$, is also the mid-point of the diagonal BE of that parallelogram; the straight line MN, passing through the mid-points M and N of the sides BE and BD of triangle BDE, is parallel to the side DE of that triangle.

§ 9. Maxima and Minima

141. Let the straight lines a and b (Fig. 234) represent the bank of the river and the polygonal line $AMNB$ the passage from A to B over the foot-bridge MN. The length MN, equal to the distance between the parallel lines a and b, is constant; thus the problem consists in finding a position of the foot-bridge MN for which the sum $AM+NB$ is the least.

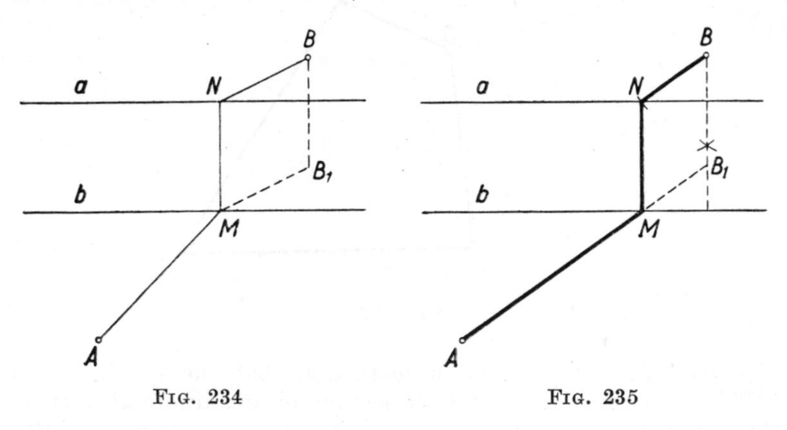

FIG. 234 FIG. 235

Let us translate the segment NB to the position MB_1. Since $AM+NB = AM+MB_1$, the problem is reduced to determining a point M for which the sum $AM+MB_1$ has its minimum. Point B_1 is known, since $BB_1 || NM$ and $BB_1 = NM$; B_1 and A lie on opposite sides of line b. The minimum length $AM+MB_1$ occurs if points A, M and B_1 are collinear.

Hence follows the construction shown in Fig. 235. The problem always has one and only one solution.

Exercise. Using the above method, i.e. that of *translation*, solve the following more general problem.

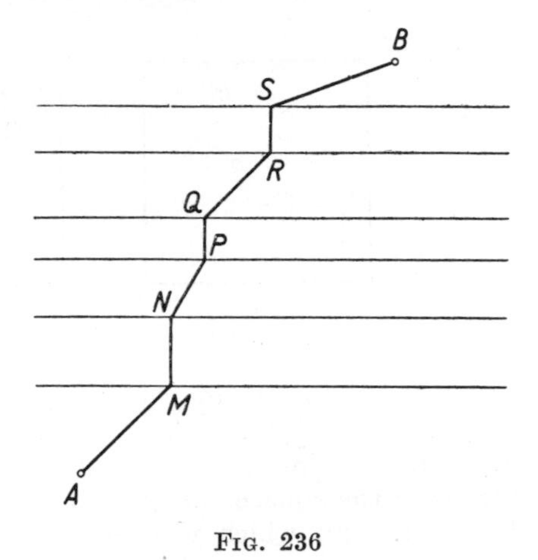

Fig. 236

Over several parallel tracks (Fig. 236) *build bridges* MN, PQ, RS, ... *in such a way as to obtain the shortest passage from* A *to* B.

142. *Answer.* The required straight line is a diagonal of a parallelogram with centre at the given point M and two sides lying on the arms of the given angle.

143. (1) Let M be a point of the square $ABCD$ such that $MA > \frac{1}{2}\sqrt{5}$. Let us draw from point A as centre a circle with radius $\frac{1}{2}\sqrt{5}$. Since $1 < \frac{1}{2}\sqrt{5} < \sqrt{2}$, points B and D lie inside that circle and point C lies outside it, whence the circle intersects the sides BC and DC at certain points S and T. From triangle ABS we find $BS = \sqrt{[(\frac{1}{2}\sqrt{5})^2-1]} = \frac{1}{2}$, i.e. S is the mid-point of BC and likewise T is the mid-point of CD. Point M is in that part of the square (shaded in Fig. 237) which lies in the common part of the right-angled triangles BCT and CDS, and, since the hypotenuse of a right-angled triangle is longer than any other segment lying in that triangle, we have $MB < BT$, $MC < BT$, $MD < DS$, i.e. each of the segments MB, MC, MD is less than $\frac{1}{2}\sqrt{5}$ in length.

(2) Suppose that point M of the given square satisfies the conditions $MA > 1$ and $MB > 1$. Point M then lies outside the circles with radius 1, described from points A and B as centres, i.e. it lies in the domain (shaded in Fig. 238) contained in the common part of triangles BCD and ACD, i.e. in the triangle

COD. Each of the segments *MC* and *MD* is shorter than the hypotenuse *CD* of this triangle, i.e. than 1.

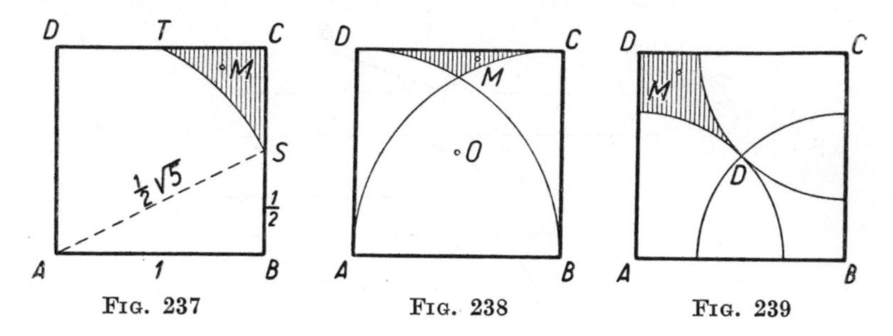

FIG. 237 FIG. 238 FIG. 239

(3) Suppose that for point M of the square we have $MA > \frac{1}{2}\sqrt{2}$, $MB > \frac{1}{2}\sqrt{2}$, $MC > \frac{1}{2}\sqrt{2}$. Point M then lies outside the circles described from points A, B, C as centres and passing through the centre O of the square, i.e. it lies in a part of the square (shaded in Fig. 329) which is contained in the square with the diagonal OD; consequently $MD < OD$, i.e. $MD < \frac{1}{2}\sqrt{2}$.

REMARK. According to the above, number $\frac{1}{2}\sqrt{5}$ is such that at most one of the distances of point M from the vertices of the square is greater than $\frac{1}{2}\sqrt{5}$. Every number greater than $\frac{1}{2}\sqrt{5}$ has this property of course, but is does not apply to any number less than $\frac{1}{2}\sqrt{5}$, which can be seen for example from the fact that the distance of point S in Fig. 237 from two vertices of the square, A and D, is equal to $\frac{1}{2}\sqrt{5}$.

The distances of point C from the remaining three vertices, A, B, C, are not less than 1; and the distances of point O (Fig. 239) from all the four vertices A, B, C, D are equal to $\frac{1}{2}\sqrt{2}$.

We can thus formulate the following theorem.

Numbers $\frac{1}{2}\sqrt{5}$, 1, $\frac{1}{2}\sqrt{2}$ *are the least numbers* k_1, k_2, k_3 *such that in a square with side* 1 *at most one of the distances of a point of the square from its vertices is greater than* k_1, *at most two of those distances are greater than* k_2 *and at most three are greater than* k_3.

Analogous numbers can be found for other figures.

We submit the following exercises to the reader:

(a) Find the least number k_1 such that in a regular hexagon with side 1 at most one of the distances of a point of the hexagon from its vertices is greater than k_1.

(b) Show that for a regular polygon with $2n+1$ sides the least number k_1 mentioned above is equal to the diameter of the polygon. (The definition of the diameter of a figure is given in the remark to problem 151.)

144. We assume that the metal plate is of uniform thickness; the weight of a part of the plate is thus proportional to the area of the plane figure represented by that part. The problem is reduced to showing that, if we cut the triangle along a straight line passing through its centre of gravity, i.e. through the point of intersection of its medians, then each of the parts of the triangle has an area equal to at least $\frac{4}{9}$ of the area of the whole triangle.

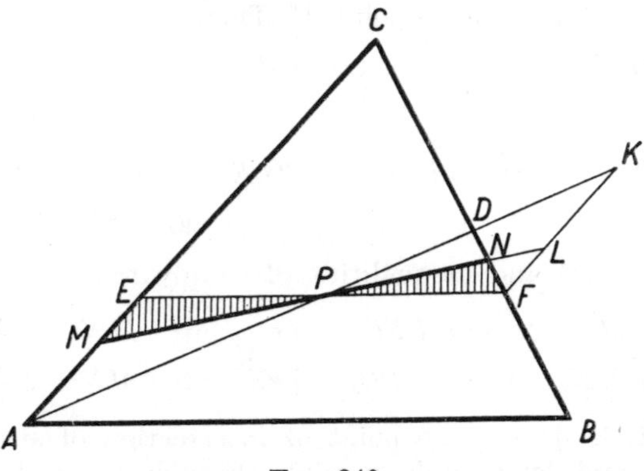

FIG. 240

Let P be the centre of gravity of triangle ABC with area S (Fig. 240). If we cut the triangle along one of the medians, e.g. along AD, we shall divide it into two triangles with areas equal to $\frac{1}{2}S$. Cutting the triangle ABC along a segment passing through point P and parallel to one of the sides of the triangle, e.g. along the segment EF parallel to side AB, we shall divide it into a triangle EFC and a trapezium $ABFE$. The triangle EFC is similar to triangle ABC in the ratio $EC/AC = \frac{2}{3}$. Since the ratio of the area of two similar figures is equal to the square of the ratio of similarity, the area of triangle EFC is equal to $\frac{4}{9}S$ and the area of trapezium $ABFE$ is equal to $\frac{5}{9}S$.

We have thus found that the assertion of the theorem holds in the two cases we have examined. Let us draw through point P any straight line not passing through any of the vertices of the triangle and not parallel to any of its sides. This line will intersect two sides of the triangle, e.g. side AC at point M and side BC at point N. Points M and N lie on opposite sides of the line EF; for instance let point M lie on the segment AE and point N on the segment FD. Our theorem will be proved if we show that

$$\tfrac{4}{9}S < \text{area } MNC < \tfrac{5}{9}S.$$

Accordingly, it will be observed that

$$\begin{aligned} \text{area } MNC &= \text{area } EFC + \text{area } MPE - \text{area } NPF, \\ \text{area } MNC &= \text{area } ADC + \text{area } PND - \text{area } PMA. \end{aligned} \tag{1}$$

Let K be the point symmetric to point A with respect to point P. Point K lies on the segment PD produced beyond point D, since $PK = AP = 2PD$. The triangle PFK is symmetric to triangle PEA with respect to point P. Let L be the point of segment FK symmetric to point M. Then

$$\begin{aligned} \text{area } MPE - \text{area } NPF &= \text{area } LPF - \text{area } NPF \\ &= \text{area } LNF > 0, \\ \text{area } PMA - \text{area } PND &= \text{area } PLK - \text{area } PND \\ &= \text{area } LKDN > 0. \end{aligned} \tag{2}$$

Equations (1) and inequalities (2) imply that

$$\text{area } MNC > \text{area } EFC, \quad \text{i.e.} \quad \text{area } MNC > \tfrac{4}{9}S;$$
$$\text{area } MNC < \text{area } ADC, \quad \text{i.e.} \quad \text{area } MNC < \tfrac{5}{9}S.$$

145. Let M denote the point of intersection of straight lines AB and p and let a circle passing through point A and B intersect the line p at points C and D.

The chord CD is the sum of segments CM and MD, whose product is equal, by the theorem on intersecting chords of a circle, to the product of the given segments AM and MB.

We know from arithmetic that the sum of positive numbers having a given product is least when the numbers are equal†. The chord CD is thus shortest when $CM = MD$, i.e. when point M is the mid-point of that chord. Consequently, the centre of the required circle lies on a perpendicular drawn through point M to the straight line p. Since the centre of that circle must also lie on the perpendicular bisector of the chord AB, we determine it as the point of intersection of the two straight lines mentioned. The problem always has one and only one solution.

146. We shall denote by α a half-plane with edge m passing through point A and by β a similar half-plane passing through B (Fig. 241).

† This can be proved for instance in the following way. Let $a+b = s$, $ab = p$; then $(a-b)^2 = (a+b)^2-4ab = s^2-4p$, whence $s^2-4p \geqslant 0$; consequently (with a constant p) s is least when $s^2-4p = 0$ and then $(a-b)^2 = 0$, i.e. $a = b$. See also problem 62, remark 3.

Let us rotate half-plane α about the line m in such a way as to make it coincide with the extension of half-plane β; point A will then lie at a certain point A_1, coplanar with point B and straight line m, point A_1 and B lying on opposite sides of m. Let C be the point of intersection of segment A_1B with m. If M is an arbitrary point of m, then

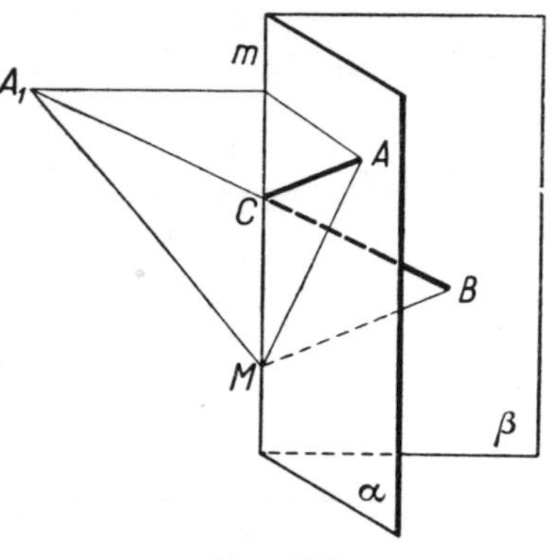

Fig. 241

$$A_1M+MB \geqslant A_1B,$$

i.e. $$A_1M+MB \geqslant A_1C+CB.$$

Now $A_1M = AM$ since A_1M arises by the rotation of segment AM, and similarly $A_1C = AC$. The preceding inequality thus gives

$$AM+MB \geqslant AC+CB.$$

Consequently the sum of the segments $AM+MB$ has the least value when point M coincides with point C.

147. *Method I.* Let S be the mid-point of segment AB, and M an arbitrary point of straight line p. Then

$$MS^2 = \tfrac{1}{2}AM^2+\tfrac{1}{2}BM^2-\tfrac{1}{4}AB^2.$$

Indeed, if M lies away from the straight line AB, this equality is the well-known formula for the square of a median of a triangle (see problem 68): if M lies on AB, the verification of the equality is immediate (see problem 66).

Consequently

$$AM^2+BM^2 = 2MS^2+\tfrac{1}{2}AB^2.$$

Hence we can see that the sum AM^2+BM^2 is least when the segment MS is shortest. Thus the required point M is the projection of point S upon the straight line p. The problem always has one and only one solution.

Method II. Let P (Fig. 242) be an arbitrary point of the straight line p and A' and B' the projections of points A and B on that line; let $AA' = a$, $BB' = b$, $A'B' = c$ and let x denote the relative measure of the vector $A'P$ on the axis $A'B'$. From triangles $AA'P$ and $BB'P$ we have

$$AP^2 = a^2+x^2, \qquad BP^2 = b^2+(c-x)^2.$$

Consequently

$$AP^2+BP^2 = 2x^2-2cx+a^2+b^2+c^2.$$

We see that AP^2+BP^2 is a quadratic function of variable x which has its minimum for $x = c/2$, i.e. when point P is the projection of the mid-point of segment AB upon the straight line p.

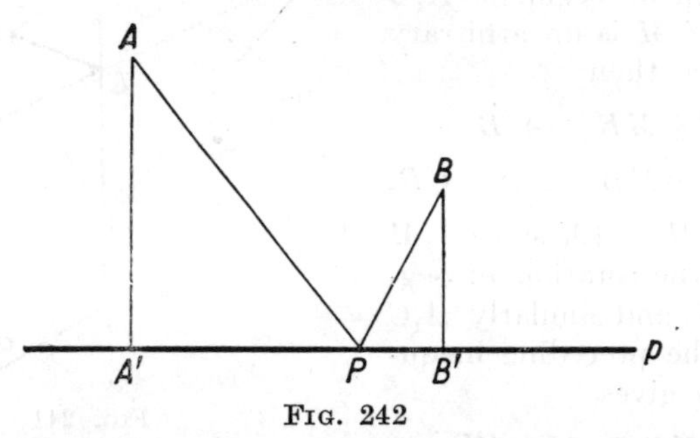

FIG. 242

148. The centres A, B, C, D of the given spheres are the vertices of a regular tetrahedron of edge $2r$. Let O be the centre and R the radius of the sphere circumscribed on the tetrahedron $ABCD$ (Fig. 243).

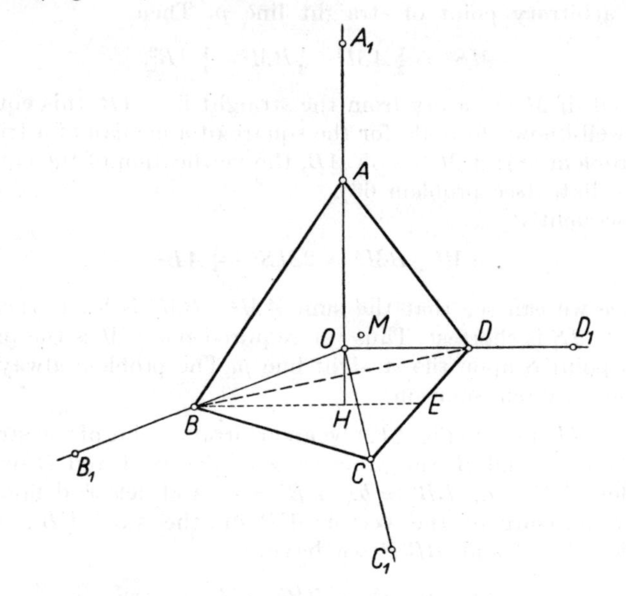

FIG. 243

Let us consider a sphere K with centre O and radius $\varrho = R+r$. It comprises all the given spheres and is internally tangent to them at points A_1, B_1, C_1, D_1 lying on the extensions of segments OA, OB, OC, OD respectively at a distance ϱ from point O.

We shall prove that sphere K is the least sphere comprising the given spheres.

To begin with, it will be observed that no sphere with centre O and radius less than ϱ comprises the given spheres because it does not contain points A_1, B_1, C_1, D_1. Suppose next that a certain sphere Q with centre M different from O and radius ϱ_1 contains all the given spheres; we shall show that $\varrho_1 > \varrho$.

Let us describe from points A_1, B_1, C_1, D_1 as centres spheres K_1, K_2, K_3, K_4 with radius ϱ. The surfaces of those spheres, and thus also the spheres themselves, have only one point O in common because O is the only point equidistant from points A_1, B_1, C_1, D_1. Consequently point M must lie outside at least one of the spheres K_1, K_2, K_3, K_4, e.g. outside the sphere K_1. Then $MA_1 > \varrho$; and since point A_1 lies inside sphere Q, we have $\varrho_1 \geqslant MA_1$, whence $\varrho_1 > \varrho$.

The radius ϱ of sphere K will easily be computed if we consider that the centre O lies on the altitude AH of the tetrahedron $ABCD$ and that the segment BH is the radius of the circle circumscribed on the equilateral triangle BCD of side $2r$, whence

$$BH = \frac{2r}{\sqrt{3}}.$$

Let $\sphericalangle OAB = x$. In triangle AOB we have

$$\cos x = \frac{AB}{2AO} = \frac{r}{R},$$

whence

$$R = \frac{r}{\cos x}.$$

From triangle ABH we have

$$\cos x = \frac{AH}{AB} = \frac{\sqrt{(AB^2 - BH^2)}}{AB} = \frac{1}{2r}\sqrt{\left(4r^2 - \frac{4r^2}{3}\right)} = \sqrt{\frac{2}{3}}.$$

Hence

$$R = r\sqrt{\tfrac{3}{2}},$$

and thus

$$\varrho = R+r = r(1+\sqrt{\tfrac{3}{2}}).$$

149. All triangles with the given base AB (Fig. 244) and the given area P have equal altitudes, whence their vertices lie on two straight lines parallel to the straight line AB and lying symmetrically on both sides of it. Since symmetrical triangles have equal perimeters, it is sufficient to prove the theorem for the triangles lying on one side of AB.

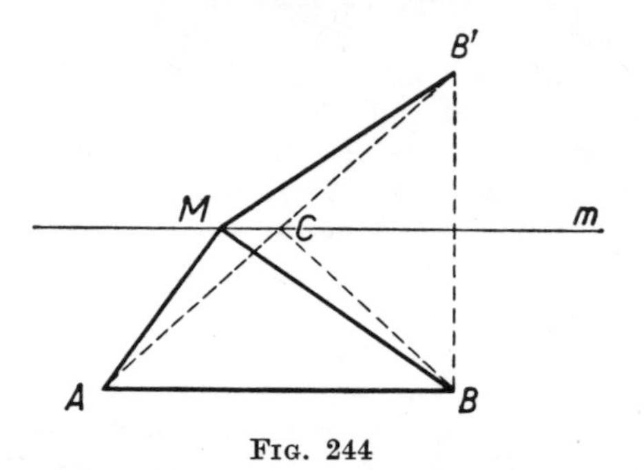

FIG. 244

Let point M move along a straight line m parallel to AB. The perimeter $AM+MB+AB$ of triangle AMB will be least when the sum $AM+MB$ has the least value. Let B' denote the point symmetric to point B with respect to m; then $AM+MB = AM+MB'$. Now $AM+MB' \geqslant AB'$, the equality holding only if point M coincides with point C, at which the straight line AB' intersects the straight line m. Consequently, of all the triangles with base AB whose vertices lie on the straight line m the triangle ABC has the least perimeter. It is an isosceles triangle; indeed, since $CB = CB'$ (as symmetric segments) and $AC = CB'$ (the straight line m being parallel to side AB and passing through the mid-point of the side BB' of the triangle ABB'), we have $AC = CB$.

REMARK. From the above theorem we can easily draw the following conclusion:

(α) *Of all triangles with a given base and a given perimeter the isosceles triangle has the greatest area* (the greatest altitude).

Let ABC be an isosceles triangle ($AC = BC$) and let AMB be any triangle with the same perimeter. Let h_C denote the altitude of the first triangle and h_M the altitude of the second triangle. We are to prove that $h_C \geqslant h_M$. Now, if the inequality $h_C < h_M$ hold, we could construct an isosceles triangle $AC'B$ with altitude $h_{C'} = h_M$; the perimeter of that triangle would be greater than the perimeter of triangle ACB, which has a smaller altitude; consequently, it would also be greater than the perimeter of the triangle AMB. But the triangles $AC'B$ and AMB have equal bases and equal altitudes, whence their areas are equal; this contradicts the theorem of problem 5 and thus proves the validity of theorem (α).

The reader is invited to prove theorem (α) without referring to problem 149 (the formula of Heron should be used) and to show that the theorem of problem 149 is a conclusion from theorem (α). The two theorems are *equivalent.*

150. Let triangle ABC be inscribed in a circle $O(R)$† and circumscribed on a circle $S(r)$ (Fig. 245). Suppose that $AC \neq BC$ and let us consider an isosceles triangle ABC_1 inscribed in the circle $O(R)$, point C_1 lying on the same side of the straight line AB as point C.

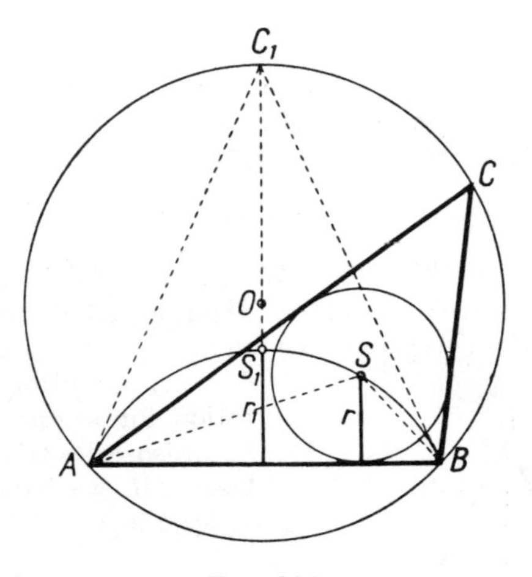

Fig. 245

We shall show that radius r_1 of the circle inscribed in triangle ABC_1 is greater than the radius r of the circle inscribed in triangle ABC. Indeed, when point C describes the arc BC_1A, the centre S of the circle inscribed in triangle ABC describes an arc of a circle passing through points A and B (see problem 113); the distance of point S from AB, i.e. the radius r of the circle inscribed in the triangle ABC, will be greatest when point S lies on the perpendicular bisector of segment AB, i.e. at the centre S_1 of the circle inscribed in the triangle ABC_1; consequently $r_1 > r$.

The inequality $r_1 > r$ can also be proved by computation, for example in the following way: Let A, B, C denote the angles

† The symbol $O(R)$ denotes a circle with centre O and radius R.

of triangle ABC; let us denote the side AB by c and find the area of triangle ABS:

$$\text{area } ABS = \frac{c^2 \sin \frac{A}{2} \sin \frac{B}{2}}{2 \sin \frac{A+B}{2}} = \frac{c^2}{4} \times \frac{\cos \frac{A-B}{2} - \cos \frac{A+B}{2}}{\sin \frac{A+B}{2}}$$

$$= \frac{c^2}{4} \times \frac{\cos \frac{A-B}{2} - \sin \frac{C}{2}}{\cos \frac{C}{2}}.$$

When point C describes the arc BC_1A, angle C does not change its magnitude; the area ABS has the greatest value when $\cos (A-B)/2 = 1$, i.e. when $A = B$.

Consequently

$$\text{area } ABS < \text{area } ABS_1, \quad \text{whence} \quad r < r_1.$$

It follows from this inequality that if the inequality $R \geqslant 2r$ holds for an isosceles triangle, then it holds for any triangle. It is thus sufficient to prove this inequality for an isosceles triangle.

Now, adopting the usual notation for segments and angles in an isosceles triangle ABC with base AB, we have the formulas (Fig. 246):

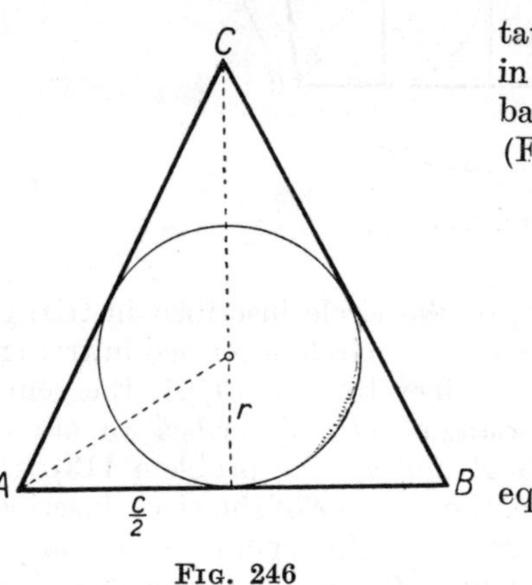

Fig. 246

$$r = \frac{c}{2} \tan \frac{A}{2},$$

$$R = \frac{c}{2 \sin C} = \frac{c}{2 \sin 2A}.$$

The inequality $R \geqslant 2r$ is thus equivalent to the inequality

$$\frac{1}{2 \sin 2A} \geqslant \tan \frac{A}{2},$$

which in turn is equivalent to the inequality

$$1 - 2 \sin 2A \tan \frac{A}{2} \geqslant 0.$$

Let $\tan(A/2) = m$; then

$$\sin A = \frac{2m}{1+m^2}, \qquad \cos A = \frac{1-m^2}{1+m^2}, \qquad \sin 2A = \frac{4m(1-m^2)}{(1+m^2)^2}.$$

The above inequality assumes the form

$$1 - \frac{8m^2(1-m^2)}{(1+m^2)^2} \geqslant 0.$$

This inequality is equivalent to the inequality

$$(1+m^2)^2 - 8m^2(1-m^2) \geqslant 0,$$

which, when rearranged, gives the inequality

$$9m^4 - 6m^2 + 1 \geqslant 0,$$

i.e. the inequality

$$(3m^2-1)^2 \geqslant 0.$$

Since the last inequality is always true, the inequality $R \geqslant 2r$, which is equivalent to it, is also true.

REMARK 1. We shall consider when the equality

$$R = 2r$$

occurs.

The answer to this question can easily be deduced from the preceding reasoning. The inequality $r < r_1$ proved at the beginning implies that the equality $R = 2r$ can occur only in an isosceles triangle; if the triangle is isosceles, this equality, as shown in the preceding argument, leads to the equality

$$(3m^2-1)^2 = 0,$$

whence we obtain

$$m = \frac{1}{\sqrt{3}}, \quad \text{i.e.} \quad \tan\frac{A}{2} = \frac{1}{\sqrt{3}},$$

which means that $A = 60°$, i.e. that the triangle is *equilateral*.

REMARK 2. The inequality $R \geqslant 2r$ is an immediate consequence of the following theorem of Euler (1747):

In every triangle the distance d of the centre of the circumcircle $O(R)$ from the centre of the inscribed circle $S(r)$ is expressed by the formula

$$d^2 = R(R-2r).$$

This theorem can be proved in the following way (Fig. 247).

In the circle $O(R)$ circumscribed about triangle ABC we draw a diameter KL through points O and S, and a chord CM through points S and C. Then

$$KS \times SL = CS \times SM. \tag{1}$$

We shall express both sides of equality (1) in terms of R, r and d:

$$KS \times SL = (KO+OS)(OL-OS) = (R+d)(R-d) = R^2-d^2. \tag{2}$$

From triangle CSP, where $SP \perp BC$, we have

$$CS = \frac{SP}{\sin \sphericalangle SCP} = \frac{r}{\sin \frac{C}{2}}. \tag{3}$$

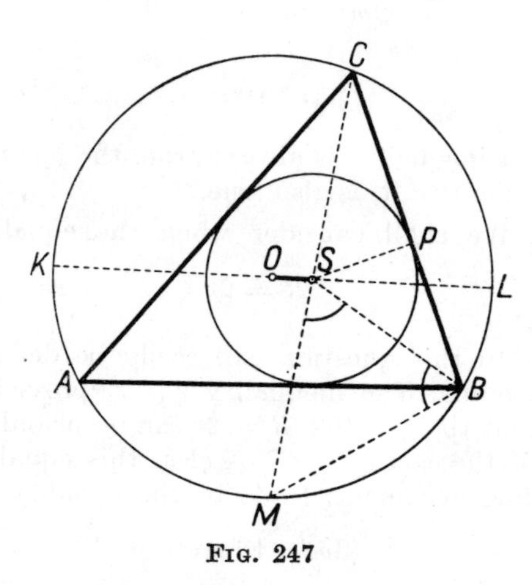

Fig. 247

The segment SM is equal to the segment BM. Indeed, since CS and BS are the bisectors of angles C and B of triangle ABC, we have

$$\sphericalangle MSB = \sphericalangle SCB + \sphericalangle SBC = \frac{C}{2} + \frac{B}{2},$$

$$\sphericalangle SBM = \sphericalangle SBA + \sphericalangle ABM = \sphericalangle SBA + \sphericalangle ACM = \frac{B}{2} + \frac{C}{2},$$

whence $\sphericalangle MSB = \sphericalangle SBM$ and $SM = MB$. From triangle BCM we find

$$BM = 2R \sin \sphericalangle BCM = 2R \sin \frac{C}{2},$$

and thus also

$$SM = 2R\sin\frac{C}{2}. \tag{4}$$

Substituting the expressions from formulas (2), (3), (4) into formula (1), we obtain

$$R^2 - d^2 = \frac{r}{\sin\frac{C}{2}} \times 2R\sin\frac{C}{2},$$

and hence *Euler's formula*

$$d^2 = R(R-2r).$$

In the above proof it is possible to dispense with the use of trigonometry: if we draw the diameter MD of circle $O(R)$, then the similarity of triangles CSP and DMB gives us at once

$$CS \times SM = CS \times BM = MD \times SP = 2Rr.$$

In an analogous way we can prove the following theorem: *The distance d_a of the centre of circle $O(R)$ circumscribed about a triangle from the centre of the escribed circle $S_a(r_a)$ tangent to side a is expressed by the formula*

$$d_a^2 = R(R+2r_a).$$

151. Let us consider an arbitrary triangle ABC and let AB be a side which is not shorter than either of the remaining sides, i.e. $AC \leqslant AB$ and $BC \leqslant AB$. Assume that $AB \leqslant a$. Let us describe circles with radius equal to the segment AB (Fig. 248) from points A and B as centres.

The triangle ABC lies in one of the domains delimited by the segment AB and by two arcs of those circles, e.g. in the domain ABD. Thus it lies in the circumcircle of the regular triangle ABD. The radius r of this circle, i.e. the radius of the circle passing through the vertices of the equilateral triangle with side AB, is $AB/\sqrt{3}$.

Since $AB \leqslant a$, we have $r \leqslant a/\sqrt{3}$. Consequently every triangle with sides not longer than a is contained in a circle with radius $a/\sqrt{3}$. It is not every triangle of this kind, however, that will fit into a circle with radius R less than $a/\sqrt{3}$. For instance, an equilateral triangle with side a will not go into such a circle; this is shown in Fig. 249. Hence the conclusion:

The circle with radius $a/\sqrt{3}$ is the smallest circle in which every triangle with sides not longer than a can be enclosed.

REMARK. Let F be an arbitrary geometrical figure, and let A and B be points of that figure which lie farthest from each other. The length d of the segment AB is called the *diameter of figure* F. Thus for instance the diameter of a circle with radius r is equal to $2r$, the diameter of a rectangle is the length of its diagonal, the diameter of a triangle is the length of its longest side.

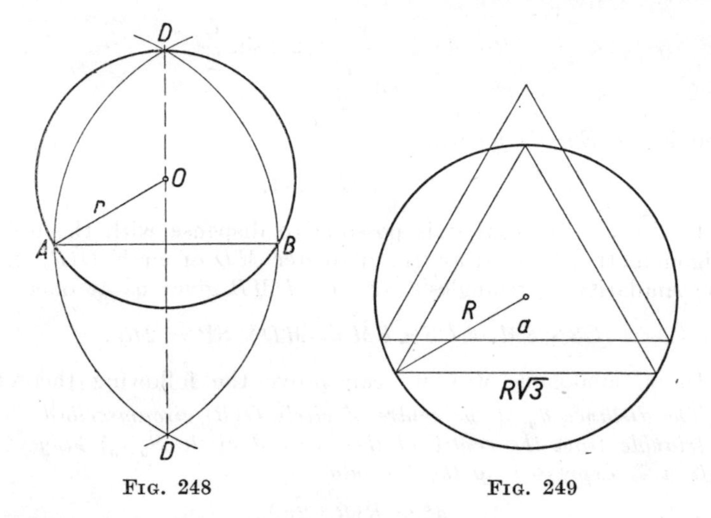

FIG. 248 FIG. 249

The theorem proved above is a particular case of the following theorem:

The circle with radius $d/\sqrt{3}$ is the smallest circle in which every plane figure with diameter not longer than d can be enclosed.

An analogous theorem holds in space:

The sphere with radius $d\sqrt{\frac{3}{8}}$ is the smallest sphere in which every solid figure with diameter not greater than d can be enclosed.

The proofs of these theorems are much more difficult than the proof given above and are beyond the scope of this book.

152. The nut can be unscrewed if and only if the following two conditions are satisfied:

(1) The square Q of the nut can be enclosed by hexagon S of the spanner.

(2) When being turned, the spanner strikes against the nut, which occurs if the greatest distance of two points of the nut, i.e. the diagonal of square Q, is greater than the "least width" of the aperture of the spanner, i.e. the distance between opposite sides in hexagon S.

The above conditions must be expressed as relations between the lengths a and b.

Let b_0 denote the side of the greatest square that can be enclosed by hexagon S. Condition (1) is then expressed by the inequality

$$b \leqslant b_0.$$

Condition (2) has the form

$$b\sqrt{2} > a\sqrt{3}, \quad \text{i.e.} \quad b > a\frac{\sqrt{3}}{\sqrt{2}},$$

since the distance between opposite sides in a regular hexagon of side a is $a\sqrt{3}$.

Joining the two inequalities in one formula we obtain the condition

$$a\frac{\sqrt{3}}{\sqrt{2}} < b \leqslant b_0. \tag{1}$$

We must find the length b_0 in terms of a.

Accordingly, we shall prove that the greatest square that can be enclosed by hexagon S is equal to the square $ABCD$ (Fig. 250) with sides parallel, respectively, to two axes of symmetry of the hexagon and vertices lying on the sides of the hexagon.

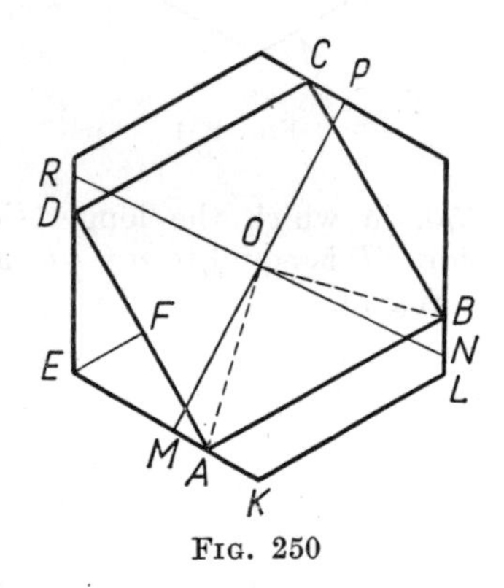

Fig. 250

Let Q be any square contained in the hexagon S. We are to prove that Q is not greater than the square $ABCD$.

Two cases are possible:

Case 1. The centre of square Q lies at the centre O of the hexagon. The diagonals of square Q lie on perpendicular lines MP and NR, cutting the hexagon into four parts. Some of those parts contain whole sides of the hexagon (why?), e.g. let a side KL of the hexagon be inside angle MON.

Without loss of generality we can assume that the side AB of the square $ABCD$ is—as shown in Fig. 250—parallel to the side KL of the hexagon, since the square $ABCD$ can be suitably rotated.

Then either the segments OM and ON coincide with the segments OA and OB, whence $OM = OA$, or one of them, say OM, lies—as in Fig. 250—outside angle AOB, the segment KM being equal to at most half the side of the hexagon, and $OM < OA$. In both cases we have $OM \leqslant OA$.

Now the diagonal of the square Q is not greater than the segment PM, and we have $PM = 2\times OM \leqslant 2\times OA$, i.e. $PM \leqslant AC$. It follows that square Q is not greater than the square $ABCD$.

Case 2. The centre of square Q lies at a point O_1 different from point O (Fig. 251); the diagonals of square Q lie on perpendicular lines M_1P_1 and N_1R_1.

Let us draw through point O straight lines $MP \parallel M_1P_1$ and $NR \parallel N_1R_1$. As has been proved in case 1, one of the segments MP and NR is not greater than the diagonal AC of the square $ABCD$ (not shown in Fig. 251). For example, let $MP \leqslant AC$. Now $M_1P_1 \leqslant MP$, whence $M_1P_1 \leqslant AC$.

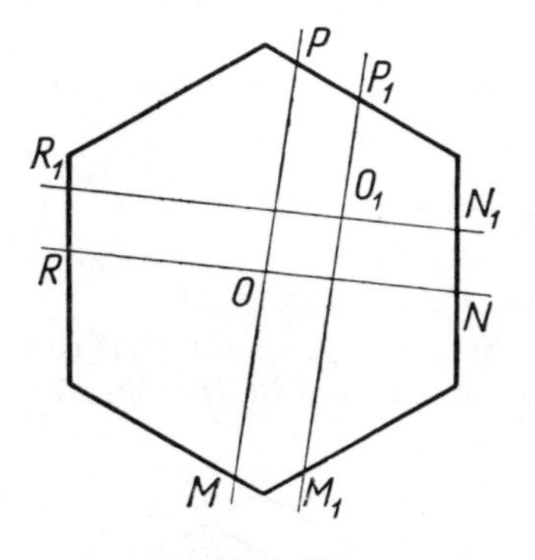

FIG. 251

Since the diagonal of square Q is not greater than the segment M_1P_1, it is not greater than the segment AC, i.e. square Q is not greater than the square $ABCD$.

It follows that the length b_0 in formula (1) is equal to the length of the side of the square $ABCD$. We can find it, for instance, from the right-angled triangle DEF in Fig. 250, in which the longer side DF is equal to $\frac{1}{2}b_0$, the shorter side EF is equal to $a-\frac{1}{2}b_0$ and the acute angles are 60° and 30°.

We obtain

$$\tfrac{1}{2}b_0 = (a-\tfrac{1}{2}b_0)\sqrt{3},$$

whence

$$b_0 = \frac{2a\sqrt{3}}{\sqrt{3}+1}, \quad \text{i.e.} \quad b_0 = (3-\sqrt{3})\,a.$$

Thus the answer to the question which has been asked is: the nut can be unscrewed if and only if the lengths a and b satisfy the condition

$$\frac{a\sqrt{3}}{\sqrt{2}} < b \leqslant (3-\sqrt{3})\,a.$$

153. We shall assume that $b > 0$; let the reader himself formulate the answer to the question asked if $b = 0$, i.e. if the messenger is on the road.

Method I. Let M denote the point at which the messenger finds himself, S the point of the meeting, t the time which will elapse between the initial moment and the moment of the meeting and x the velocity of the messenger. Applying the Cosine Rule to triangle MOS, in which $OS = vt$, $MS = xt$, $OM = a$, we obtain

$$x^2t^2 = a^2+v^2t^2-2avt \cos \alpha,$$

where α denotes angle MOS. Hence

$$x^2 = \frac{a^2}{t^2}-2av \cos \alpha \times \frac{1}{t}+v^2.$$

Let us write $1/t = s$; then

$$x^2 = a^2s^2-2av \cos \alpha \times s+v^2 = (as-v \cos \alpha)^2+v^2-v^2 \cos^2 \alpha,$$

or, more briefly,

$$x^2 = (as-v \cos \alpha)^2+v^2 \sin^2 \alpha. \tag{1}$$

We seek a *positive* value of s for which the positive quantity x, and thus also x^2, has the least value. We must distinguish two cases here:

Case 1: $\cos \alpha > 0$, i.e. α is an acute angle. It follows from formula (1) that x has the least value $x_{\min}$ if $as-v \cos \alpha = 0$, whence

$$s = \frac{v \cos \alpha}{a}.$$

Then

$$x^2_{\min} = v^2 \sin^2\alpha, \qquad \text{and thus} \qquad x_{\min} = v \sin \alpha.$$

Case 2: $\cos \alpha \leqslant 0$, i.e. α is a right angle or an obtuse one. In this case the required minimum does not exist since $as-v \cos \alpha > 0$, and thus also x^2 is the smaller the nearer s is to zero, i.e. the greater is t. As t increases indefinitely, s tends to zero and x, as shown by formula (1), tends to v.

We shall explain these results with the aid of a drawing. If $\alpha < 90°$ (Fig. 252), the minimum velocity of the messenger is equal to $v \sin \alpha = vb/a$; the meeting will take place at the moment when $1/t = (v \cos \alpha)/a$. Then

$$MS = v \sin \alpha \frac{a}{v \cos \alpha} = a \tan \alpha,$$

which means that $\sphericalangle OMS = 90°$; the messenger should run along a perpendicular to OM.

If $\alpha \geqslant 90°$ (Fig. 253), the messenger must cover a longer route than the cyclist, and thus he can overtake him only if his velocity is greater than the velocity v of the cyclist; the necessary surplus of velocity, however, will be the less the greater is the angle

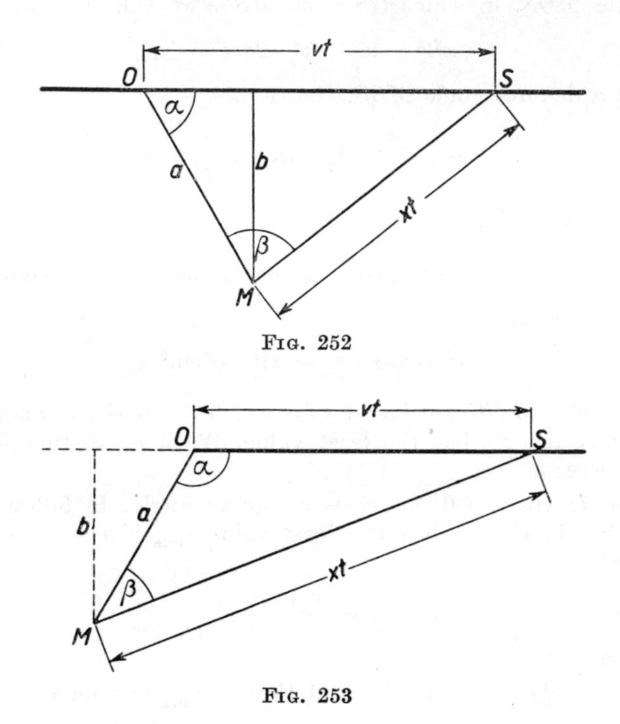

FIG. 252

FIG. 253

$\sphericalangle OMS = \beta$, and can be arbitrarily small if the cyclist rides in a direction which forms with OM an angle sufficiently near $180°-\alpha$.

Method II. Adopting the same notation as before, we have

$$\frac{x}{v} = \frac{xt}{vt} = \frac{MS}{OS},$$

whence by the Sine Rule (Fig. 252)

$$\frac{x}{v} = \frac{\sin\alpha}{\sin\beta} \quad \text{and} \quad x = \frac{\sin\alpha}{\sin\beta}\times v.$$

This equality implies that x assumes the least value when $\sin\beta$ is greatest. If $\alpha < 90°$, this occurs for $\beta = 90°$, whence

$$x_{\min} = v \sin \alpha = \frac{vb}{a}.$$

If $\alpha \geqslant 90°$ (Fig. 253), then angle β is acute; a greatest value of β does not exist, the velocity x is the smaller the nearer the angle β is to $180° - \alpha$. As angle β increases and tends to $180° - \alpha$, velocity x decreases and tends to v.

REMARK. In the above solution we can dispense with the use of trigonometry, reasoning as follows.

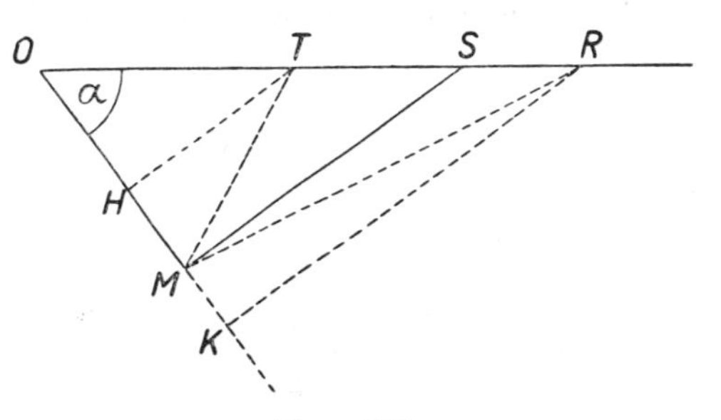

FIG. 254

If $\alpha < 90°$ (Fig. 254) and $MS \perp OM$, we draw $TH \perp OM$ and $RK \perp OM$. Then

$$\frac{MS}{OS} = \frac{HT}{OT} < \frac{MT}{OT}, \qquad \frac{MS}{OS} = \frac{KR}{OR} < \frac{MR}{OR},$$

whence at point S of the road the ratio of the distances from points M and O is smallest.

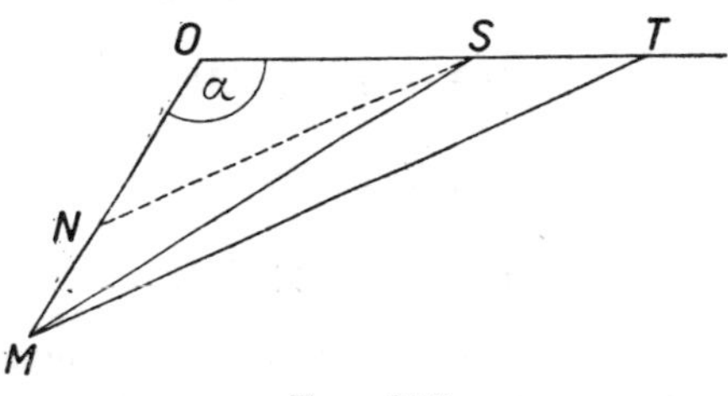

FIG. 255

If $\alpha \geqslant 90°$ and point T lies farther from point O than point S (Fig. 255), then, drawing NS parallel to MT, we have

$$\frac{MT}{OT} = \frac{NS}{OS} < \frac{MS}{OS},$$

whence it is obvious that the required minimum does not exist.

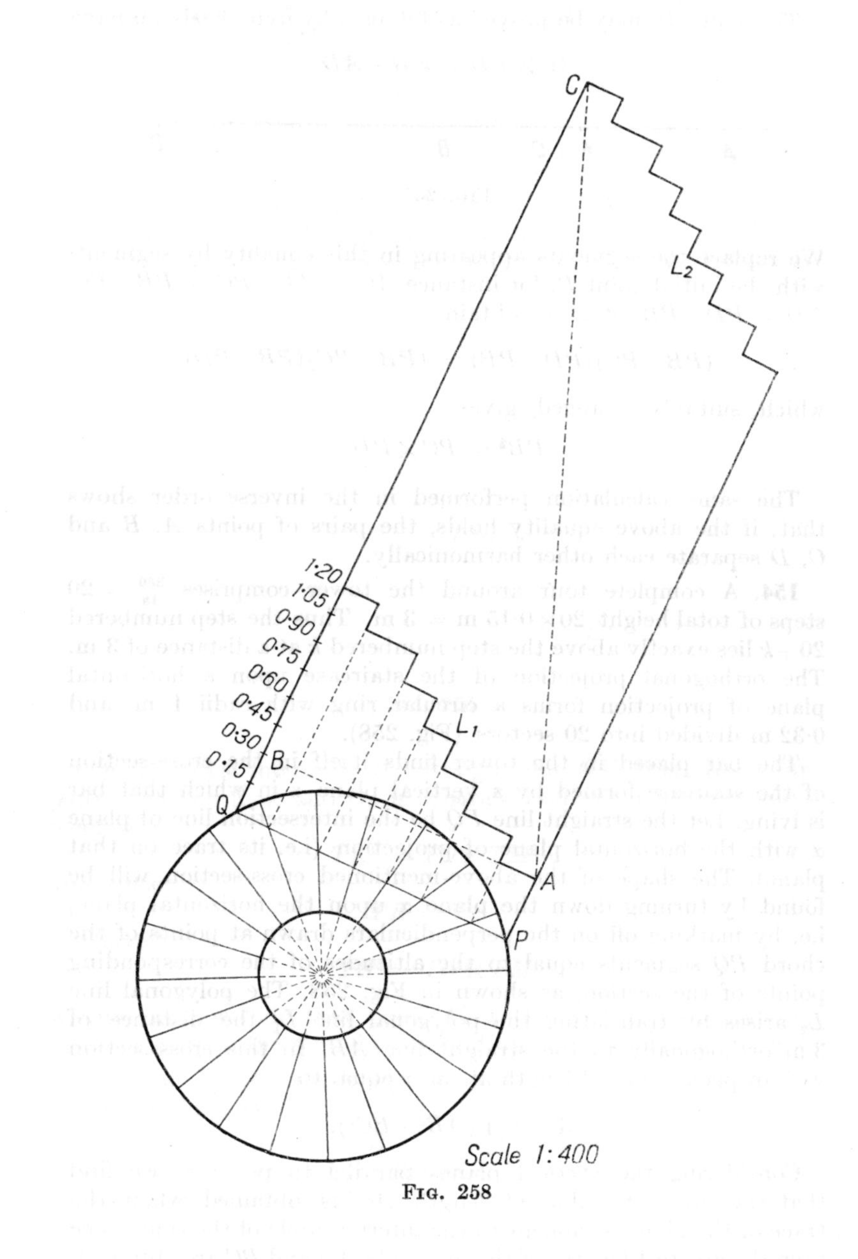

Fig. 258

Thus we must find the length of the longest bar that can be carried up the stairs in such a manner that all the time it remains tangent to the column.

A cross-section of the staircase formed by a plane tangent to the column is drawn in the same way as the section in Fig. 258, the difference being that the straight line AB is tangent to the smaller circle (Fig. 259).

The length of the chord AB is found from the formula $AB^2 = 4 \times (OA^2 - OM^2) = 4 - (0{\cdot}64)^2$, and the central angle AOB from the formula $\cos \sphericalangle \frac{1}{2}AOB = OM/OA = 0{\cdot}32$, whence $\sphericalangle AOB = 142°40'$. The angle AOB includes approximately $7\frac{8}{9}$ sectors corresponding to the steps of the staircase.

The section drawn in Fig. 259 in a continuous line passes through the end-point A of the edge of one of the steps, whence in the lower part of the section there are 7 points of intersection of the plane of the vertical section with the edges of the steps; they are the points $1, 2, \ldots, 7$, belonging to the edges with the projections $OA_1, OA_2, \ldots, OA_7$. In the upper part of the section we have, analogously, the points $1', 2', \ldots, 7'$. We find $BD = 7 \times 0{\cdot}15$ m $= 1{\cdot}05$ m, $BC = BD + DC = 1{\cdot}05$ m $+ 3$ m $= 4{\cdot}045$ m.

In a section of this kind we can place a bar of length at most equal to

$$AC = \sqrt{(AB^2 + BC^2)} \approx 4{\cdot}47 \text{ m}.$$

Let the plane of the vertical section rotate about the axis of the tower. Then the section will undergo changes and all its possible shapes will have appeared during the rotation from 0° to 18°; in further rotation the same sections will appear after every 18°, each time moved 0·15 m upwards.

For illustration by means of a drawing it is convenient to assume that the plane ABC of the section is motionless and *the tower rotates*, e.g. in the direction marked with an arrow in Fig. 259. During a rotation from 0° to 18° point A_1 runs over the arc A_1A and point A_8 runs over the arc A_8A_7. Let us consider the successive stages of that rotation.

(1) While point A_8 describes the arc A_8B, the bar originally placed in the position AC may remain in this position, i.e. have the maximum length 4·47 m which has been calculated above. It will be observed that the position of the bar along the segment AC is not disturbed during the rotation by any step edge; in a carefully executed drawing it is clearly seen that points $1, 2, \ldots, 7$ always lie on one side and points $1', 2', \ldots, 7'$ on the other side of AC. We can verify the fact by calculation, showing that the

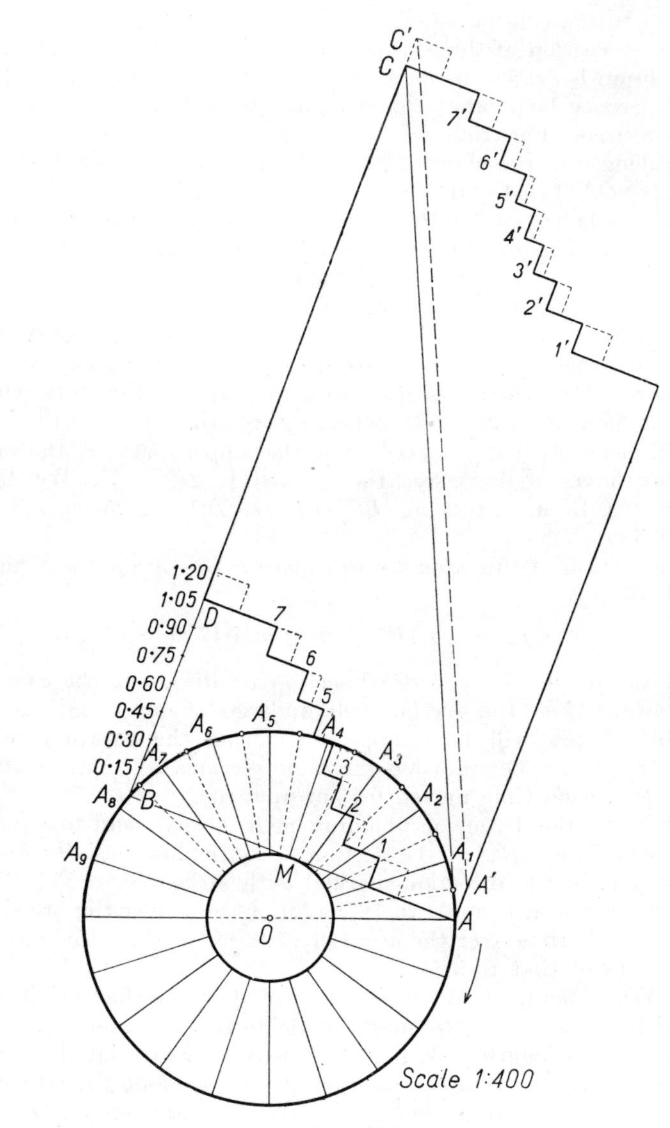
C'
C
7'
6'
5'
4'
3'
2'
1'
1·20
1·05
D
0·90
0·75
0·60
0·45
0·30
0·15
7
6
5
4
3
2
1
A₁
A₂
A₃
A₄
A₅
A₆
A₇
A₈
A₉
B
M
O
A'
A
Scale 1:400

FIG. 259

angles $BA1$, $BA2$, ..., $BA7$ remain smaller than angle BAC and the angles $BC1'$, $BC2'$, ..., $BC7'$ are greater than angle BCA.

(2) When the rotating point A_8 has overtaken point B and runs over the arc BA_7, the situation changes; plane ABC intersects in addition to the preceding step edges also those edges whose projection is the radius OA_8. In the upper part of the section on the left the eighth recess is formed, and the bar can be gradually moved in the plane ABC from the position AC upwards to the position $A'C'$, i.e. it can be raised the height of one step, namely 0·15 m. In Fig. 259 the corresponding shape of the section is outlined with a dotted line.

(3) When point A_8 assumes the position of A_7, we obtain a section of the same shape as the initial one, but it is raised 0·15 m; the bar retains position $A'C'$, in which it will remain until point A_9 reaches point B, after which it will again be possible to move the bar upwards and so forth.

It follows that a bar 4·47 m in length can be carried up the stairs; a longer bar could not be carried upwards since it could not be contained in certain sections, e.g. in the initial section.

§ 10. Trigonometrical Transformations

155. The proof of the theorem will be obtained by transforming the equation

$$\cos 3A + \cos 3B + \cos 3C = 1. \tag{1}$$

In order to perform the transformation properly we should consider what is aimed at. We can reason here in various ways.

Method I. To prove that one of the angles A, B, C is equal to 120° it suffices to prove that one of the differences $1-\cos 3A$, $1-\cos 3B$, $1-\cos 3C$ is equal to zero, which occurs if and only if

$$(1-\cos 3A)(1-\cos 3B)(1-\cos 3C) = 0. \tag{2}$$

Our aim is to infer equality (2) from equality (1). To make the calculation easier we substitute in equality (1) the value $C = 180°-(A+B)$, whence $\cos 3C = -\cos(3A+3B)$, and consequently

$$\cos 3A + \cos 3B - \cos(3A+3B) = 1 \tag{3}$$

or

$$\cos 3A + \cos 3B - \cos 3A \cos 3B + \sin 3A \sin 3B - 1 = 0. \tag{4}$$

Since we want to obtain a relation containing only the cosines of angles, we write equality (4) as

$$\sin 3A \sin 3B = 1-\cos 3A-\cos 3B+\cos 3A \cos 3B,$$

or briefly

$$\sin 3A \sin 3B = (1-\cos 3A)(1-\cos 3B),$$

and then square both sides:

$$\sin^2 3A \sin^2 3B = (1-\cos 3A)^2(1-\cos 3B)^2.$$

We obtain

$$(1-\cos^2 3A)(1-\cos^2 3B)-(1-\cos 3A)^2(1-\cos 3B)^2 = 0,$$

$$(1-\cos 3A)(1-\cos 3B)[(1+\cos 3A)(1+\cos 3B)-$$
$$-(1-\cos 3A)^2(1-\cos 3B)^2] = 0,$$

$$(1-\cos 3A)(1-\cos 3B)(\cos 3A+\cos 3B) = 0,$$

and since by hypothesis (1) we have

$$\cos 3A+\cos 3B = 1-\cos 3C,$$

we finally obtain the required equality (2).

Method II. It is sufficient to prove that one of the numbers $\sin\frac{3}{2}A$, $\sin\frac{3}{2}B$, $\sin\frac{3}{2}C$ is equal to zero, since then one of the angles $\frac{3}{2}A$, $\frac{3}{2}B$, $\frac{3}{2}C$ is equal to 180°.

Accordingly, we must deduce from equation (1) the equation

$$\sin\tfrac{3}{2}A \sin\tfrac{3}{2}B \sin\tfrac{3}{2}C = 0. \tag{5}$$

We transform equation (1) as follows:

$$1-\cos 3A-(\cos 3B+\cos 3C) = 0,$$

$$2\sin^2\tfrac{3}{2}A-2\cos\tfrac{3}{2}(B+C)\cos\tfrac{3}{2}(B-C) = 0,$$

$$2\sin^2\tfrac{3}{2}A+2\sin\tfrac{3}{2}A\cos\tfrac{3}{2}(B-C) = 0,$$

$$2\sin\tfrac{3}{2}A\,[\sin\tfrac{3}{2}A+\cos\tfrac{3}{2}(B-C)] = 0,$$

$$2\sin\tfrac{3}{2}A[-\cos\tfrac{3}{2}(B+C)+\cos\tfrac{3}{2}(B-C)] = 0,$$

$$4\sin\tfrac{3}{2}A\sin\tfrac{3}{2}B\sin\tfrac{3}{2}C = 0,$$

whence $A = 120°$ or $B = 120°$ or $C = 120°$.

Method III. We can also introduce tangents into our calculation and proceed to show that equation (1) implies the equation

$$\tan\tfrac{3}{2}A\tan\tfrac{3}{2}B\tan\tfrac{3}{2}C = 0.$$

This method requires slightly longer transformations, and we must show first that the symbols $\tan\frac{3}{2}A$, $\tan\frac{3}{2}B$, $\tan\frac{3}{2}C$ have a numerical sense, i.e. that none of the angles $3A$, $3B$, $3C$ is equal to 180°, which can be inferred from equation (1).

REMARK. The inverse theorem also holds: if one of the angles A, B, C of a triangle is equal to 120°, then

$$\cos 3A+\cos 3B+\cos 3C = 1.$$

We leave the proof to the reader as an exercise.

156. Suppose that none of the angles of the convex quadrilateral $ABCD$ is right. We are to prove that

$$\frac{\tan A+\tan B+\tan C+\tan D}{\tan A \tan B \tan C \tan D} = \cot A+\cot B+\cot C+\cot D. \quad (1)$$

We know that

$$(A+B)+(C+D) = 360°,$$

whence

$$\tan(A+B)+\tan(C+D) = 0. \quad (2)$$

Hence

$$\frac{\tan A+\tan B}{1-\tan A \tan B}+\frac{\tan C+\tan D}{1-\tan C \tan D} = 0.$$

Multiplying both sides of this equation by

$$(1-\tan A \tan B)(1-\tan C \tan D)$$

we obtain

$$(\tan A+\tan B)(1-\tan C \tan D)+ +(\tan C+\tan D)(1-\tan A \tan B) = 0$$

or, after an easy transformation,

$$\tan A+\tan B+\tan C+\tan D$$
$$= \tan B \tan C \tan D+\tan A \tan C \tan D+\tan A \tan B \tan D+ +\tan A \tan B \tan C.$$

Dividing both sides of this equation by the product

$$\tan A \tan B \tan C \tan D,$$

we obtain the required equation (1).

However, there is a gap in the above proof, because equality (2) requires the assumption that $A+B \neq 90°$ and $A+B \neq 270°$. Thus it is necessary to consider also the case where

$$A+B = 90°, \quad \text{and thus} \quad C+D = 270°. \quad (3)$$

(The case where $A+B = 270°$ and $C+D = 90°$ need not be considered separately, since it is obtained from the preceding one by changing the letters.)

Now equation (3) and the fact that each angle of a convex quadrilateral is contained between 0° and 180° imply the inequalities

$$0° < A < 90°, \quad 90° < C < 180°. \tag{4}$$

From inequalities (4) we infer that

$$90° < A+C < 270°.$$

We can thus deduce from the equation $A+C = 360°-(B+D)$ that

$$\tan(A+C)+\tan(B+D) = 0, \tag{2a}$$

and obtain equation (1) from equation (2a) in the same way as before from equation (2).

REMARK 1. In the preceding proof we derived inequalities (4) using the assumption that the quadrilateral is *convex*. The question arises whether formula (1) is also valid for a concave polygon if none of its angles is right. We find that it is really so; our previous reasoning, however, must be supplemented.

Suppose that the quadrilateral $ABCD$ is *concave*. As in a convex quadrilateral, we have the equality

$$A+B+C+D = 360°,$$

since the diagonal drawn from the vertex of the concave angle divides the quadrilateral into two triangles.

If the quadrilateral has two angles whose sum is equal neither to 90° nor to 270°, then equation (1) will be proved in exactly the same way as for a convex quadrilateral.

But there are concave quadrilaterals in which every two angles total 90° or 270°. Let A denote the concave angle of a quadrilateral; the above-mentioned case occurs if

$$B+C = 90°, \quad C+D = 90°, \quad D+B = 90°,$$

i.e. if

$$B = C = D = 45°, \quad A = 225°.$$

Our preceding argument cannot be applied to a quadrilateral of this kind (Fig. 260). But formula (1) is valid also in this case because all the tangents appearing in it are then equal to 1 and the formula expresses the equality $4 = 4$.

REMARK 2. In the case where one or more angles of the quadrilateral are equal to 90° formula (1) becomes meaningless because there is no tangent of a right angle. However, if we perform the division on the left-hand side of formula (1) and introduce the cotangents of the angles instead of their tangents, we shall obtain the following formula, valid for all quadrilaterals:

$$\cot A \cot B \cot C + \cot A \cot B \cot D + \cot A \cot C \cot D + \\ + \cot B \cot C \cot D = \cot A + \cot B + \cot C + \cot D. \quad (1^*)$$

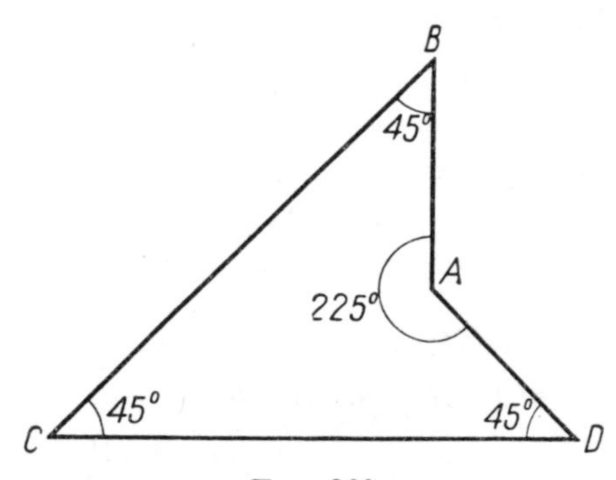

FIG. 260

The proof of formula (1*) can be carried out in exactly the same way as the preceding proof of formula (1); the reader is invited to verify this for himself.

If one of the angles of the quadrilateral, say D, is equal to 90°, formula (1*) gives

$$\cot A \cot B \cot C = \cot A + \cot B + \cot C.$$

If $D = 90°$ and $C = 90°$, formula (1*) is reduced to the equality $\cot A + \cot B = 0$; if $D = C = B = 90°$, formula (1*) is reduced to $\cot A = 0$.

REMARK 3. Inverting problem 156 we can ask what algebraic relation holds between angles A, B, C, D satisfying equality (1) (or (1*). One can easily prove (we leave this to the reader as an exercise) that it is the relation

$$A + B + C + D = k \times 180°,$$

where k is an arbitrary integer.

157. *Method I.* The problem can be solved in a very simple way if we notice that the equation $x = (2k+1) \times 180°$ is equivalent to the equation $\cos(x/2) = 0$. A necessary and sufficient condition for one of the angles $A+B+C$, $A+B-C$, $A-B+C$,

$A-B-C$ to be equal to an odd multiple of 180° can thus be expressed by the following equation:

$$\cos\frac{A+B+C}{2}\cos\frac{A+B-C}{2}\cos\frac{A-B+C}{2}\cos\frac{A-B-C}{2}=0. \qquad (1)$$

The problem is thus reduced to showing that equality (1) is equivalent to the equality

$$\cos^2 A+\cos^2 B+\cos^2 C+2\cos A\cos B\cos C-1=0. \qquad (2)$$

We shall attain this by transforming the left-hand side of equality (1) with the use of well-known formulas for the sums and products of trigonometric functions:

$$\cos\frac{A+B+C}{2}\cos\frac{A+B-C}{2}\cos\frac{A-B+C}{2}\cos\frac{A-B-C}{2}$$

$$=\tfrac{1}{4}[\cos(A+B)+\cos C]\times[\cos(A-B)+\cos C]$$

$$=\tfrac{1}{4}[\cos(A+B)\cos(A-B)+\cos(A+B)\cos C+$$

$$+\cos(A-B)\cos C+\cos^2 C]$$

$$=\tfrac{1}{4}(\cos^2 A\cos^2 B-\sin^2 A\sin^2 B+2\cos A\cos B\cos C+\cos^2 C)$$

$$=\tfrac{1}{4}(\cos^2 A+\cos^2 B+\cos^2 C+2\cos A\cos B\cos C-1).$$

We have found that the left-hand side of equation (1) is identically equal to $\frac{1}{4}$ of the left-hand side of equation (2). Thus the two equations are equivalent, whence it follows that equation (2) holds if and only if one of the angles $A+B+C$, $A+B-C$, $A-B+C$, $A-B-C$ is an odd multiple of 180°.

Method II. Suppose that one of the angles $A+B+C$, $A+B-C$, $A-B+C$, $A-B-C$ is equal to an odd multiple of 180°. For example let

$$A+B+C=(2k+1)\times 180°. \qquad (3)$$

Then $C=(2k+1)\times 180°-(A+B)$ and thus

$$\cos C=-\cos(A+B)=\sin A\sin B-\cos A\cos B.$$

We substitute this value of $\cos C$ in the left-hand side of (2):

$$\cos^2 A+\cos^2 B+\cos^2 C+2\cos A\cos B\cos C$$

$$=\cos^2 A+\cos^2 B+(\sin A\sin B-\cos A\cos B)^2+$$

$$+2\cos A\cos B(\sin A\sin B-\cos A\cos B)$$

$$=\cos^2 A+\cos^2 B+\sin^2 A\sin^2 B-\cos^2 A\cos^2 B$$

$$=\cos^2 A+\cos^2 B+(1-\cos^2 A)(1-\cos^2 B)-\cos^2 A\cos^2 B=1.$$

We have shown that if the angle $A+B+C$ satisfies equation (3), then it satisfies also equation (2). Similarly, we shall verify this for each of the remaining angles, $A+B-C$, $A-B+C$ and $A-B-C$.

It remains to prove the inverse theorem. Suppose that equality (2) holds. Regarding this equality as an equation, say with $\cos C$ as the unknown, let us write it as

$$\cos^2 C+2\cos A\cos B\cos C+(\cos^2 A+\cos^2 B-1)=0$$

and let us find $\cos C$ according to the formula for the roots of a quadratic equation:

$$\begin{aligned}\cos C &= -\cos A\cos B\pm\sqrt{[\cos^2 A\cos^2 B-(\cos^2 A+\cos^2 B-1)]}\\ &= -\cos A\cos B\pm\sqrt{[(1-\cos^2 A)(1-\cos^2 B)]}\\ &= -\cos A\cos B\pm\sin A\sin B = -\cos(A\pm B)\\ &= \cos[180°+(A\pm B)].\end{aligned}$$

Consequently $C=2k\times 180°\pm[180°+(A\pm B)]$, whence

$$C=(2k+1)\times 180°+A\pm B \quad \text{or} \quad C=(2k-1)\times 180°-A\pm B.$$

This means that one of the angles $A+B+C$, $A+B-C$, $A-B+C$, $A-B-C$ is an odd multiple of 180°.

158. From the equation

$$\cot A+\frac{\cos B}{\sin A\cos C}=\cot B+\frac{\cos A}{\sin B\cos C} \tag{1}$$

successively follow the equations

$$\cos A\sin B\cos C+\sin B\cos B=\sin A\cos B\cos C+\sin A\cos A,$$

$$(\cos A\sin B-\sin A\cos B)\cos C=\sin A\cos A-\sin B\cos B,$$

$$\sin(B-A)\cos C=\tfrac{1}{2}(\sin 2A-\sin 2B),$$

$$\sin(B-A)\cos C=\cos(A+B)\sin(A-B),$$

$$\sin(A-B)[\cos(A+B)+\cos C]=0,$$

$$\sin(A-B)\cos\frac{A+B+C}{2}\cos\frac{A+B-C}{2}=0. \tag{2}$$

Equation (2) implies the alternative

$$\sin(A-B)=0 \quad \text{or} \quad \cos\frac{A+B+C}{2}=0$$

$$\text{or} \quad \cos\frac{A+B-C}{2}=0,$$

equivalent to the alternative

$$A-B = k\pi \quad \text{or} \quad \frac{A+B+C}{2} = \frac{\pi}{2}+m\pi$$

$$\text{or} \quad \frac{A+B-C}{2} = \frac{\pi}{2}+n\pi,$$

where k, m, n denote integers.

We have obtained the following result: If equation (1) holds, then

$$A-B = k\pi \quad \text{or} \quad A+B+C = (2m+1)\pi$$

$$\text{or} \quad A+B-C = (2n+1)\pi, \tag{3}$$

where k, m, n are integers.

It will be observed that relations (3) are not mutually exclusive; e.g. if $A = B = \pi/2$, $C = 0$, all three relations hold.

REMARK. As shown in the previous calculation, equality (1) is equivalent to the equality

$$\frac{\sin(A-B)\cos\dfrac{A+B+C}{2}\cos\dfrac{A+B-C}{2}}{\sin A \sin B \cos C} = 0. \tag{1^a}$$

Consequently, the following inversion of problem 158 is true: If alternative (3) occurs and $A \neq 0$, $B \neq 0$, $C \neq \frac{1}{2}\pi+k\pi$, then we have equality (1^a), and thus also equality (1).

159. *Method I.* The equality

$$\tan\alpha+\tan\beta+\tan\gamma = \tan\alpha\tan\beta\tan\gamma \tag{1}$$

implies that

$$\tan\alpha+\tan\beta = \tan\gamma(\tan\alpha\tan\beta-1).$$

If $\tan\alpha\tan\beta \neq 1$, we can divide both sides of the above equality by $1-\tan\alpha\tan\beta$:

$$\frac{\tan\alpha+\tan\beta}{1-\tan\alpha\tan\beta} = -\tan\gamma,$$

whence

$$\tan(\alpha+\beta) = \tan(-\gamma),$$

and consequently

$$\alpha+\beta = -\gamma+k\pi,$$

i.e.

$$\alpha+\beta+\gamma = k\pi \quad (k\text{—an arbitrary integer}). \tag{2}$$

The above calculation has been preformed under the assumption that $\tan\alpha\tan\beta \neq 1$. Now this condition is always satisfied if equality (1) is satisfied; indeed, if we had $\tan\alpha\tan\beta = 1$, then equality (1) would give $\tan\alpha+\tan\beta = 0$, which is impossible, since two real numbers whose product is equal to 1 have the same sign and thus their sum cannot be equal to zero.

We have obtained the following result: if α, β, γ satisfy equality (1), then the algebraic relation (2) holds between them.

Method II. Equality (1) can be written as

$$\frac{\sin\alpha}{\cos\alpha}+\frac{\sin\beta}{\cos\beta}+\frac{\sin\gamma}{\cos\gamma} = \frac{\sin\alpha\sin\beta\sin\gamma}{\cos\alpha\cos\beta\cos\gamma}.$$

It follows that

$$\sin\alpha\cos\beta\cos\gamma+\cos\alpha\sin\beta\cos\gamma+$$
$$+\cos\alpha\cos\beta\sin\gamma-\sin\alpha\sin\beta\sin\gamma = 0;$$

after a suitable grouping of the terms, we have

$$\sin(\alpha+\beta)\cos\gamma+\cos(\alpha+\beta)\sin\gamma = 0,$$

i.e.

$$\sin(\alpha+\beta+\gamma) = 0.$$

We conclude from this equality that

$$\alpha+\beta+\gamma = k\pi \qquad (k\text{—an arbitrary integer}).$$

REMARK. It is easy to verify the inverse theorem:

If $\alpha+\beta+\gamma = k\pi$ (k—an integer) *but none of the numbers* α, β, γ *is equal to* $\frac{1}{2}\pi+m\pi$ (m—an integer) *then equality* (2) *holds.*

160. *Method I*. We shall prove a "stronger" theorem:

If $0° < x_i < 180°$ *for* $i = 1, 2, \ldots, n$, $n \geqslant 2$, *then*

$$|\sin(x_1+x_2+\ldots+x_n)| < \sin x_1+\sin x_2+\ldots+\sin x_n.^{\dagger} \qquad (1)$$

In the proof we shall make use of the well-known properties of the absolute value,

$$|a+b| \leqslant |a|+|b|, \qquad |ab| = |a|\times|b|,$$

and of the fact that if $0° < x < 180°$, then $|\sin x| = \sin x > 0$, $|\cos x| < 1$, and that $|\cos x| \leqslant 1$ for every x.

† If $|a| < b$, then $a \leqslant b$ (but not conversely), and thus inequality (1) implies the inequality given in the text of the problem.

We conduct the proof using the induction principle. For $n = 2$, theorem (1) is true because

$$\sin(x_1+x_2) = \sin x_1 \cos x_2+\cos x_1 \sin x_2,$$

whence

$$|\sin(x_1+x_2)| \leqslant |\sin x_1| \times |\cos x_2|+|\cos x_1| \times |\sin x_2| < \sin x_1+\sin x_2.$$

Suppose that for an integer $k \geqslant 2$

$$|\sin(x_1+\ldots+x_k)| < \sin x_1+\sin x_2+\ldots+\sin x_k$$

and let $0 < x_{k+1} < 180°$. Then:

$$|\sin(x_1+x_2+\ldots+x_k+x_{k+1})| = |\sin(x_1+x_2+\ldots+x_k)\cos x_{k+1}+$$
$$+\cos(x_1+x_2+\ldots+x_k)\sin x_{k+1}|$$
$$\leqslant |\sin(x_1+x_2+\ldots+x_k)\cos x_{k+1}|+|\cos(x_1+x_2+\ldots+x_k)\sin x_{k+1}|$$
$$< |\sin(x_1+x_2+\ldots+x_k)|+|\sin x_{k+1}|$$
$$< \sin x_1+\sin x_2+\ldots+\sin x_k+\sin x_{k+1}.$$

We conclude by induction that theorem (1) is true for every $n \geqslant 2$.

Method II. To begin with, we shall prove the following lemma (auxiliary theorem). If $n \geqslant 2$ and $0° < x_i < 180°$ for $i = 1, 2, \ldots, n$, then there exist numbers $a_1, a_2, \ldots$ such that

$$\sin(x_1+x_2+\ldots+x_n) = a_1 \sin x_1+a_2 \sin x_2+\ldots+a_n \sin x_n,$$

and

$$|a_1| < 1, \quad |a_2| < 1, \quad \ldots, \quad |a_{n-1}| < 1 \quad \text{and} \quad |a_n| \leqslant 1.$$

Proof. We use the method of mathematical induction. If $n = 2$, the lemma is true because

$$\sin(x_1+x_2) = \cos x_2 \sin x_1+\cos x_1 \sin x_2 = a_1 \sin x_1+a_2 \sin x_2,$$

where

$$|a_1| = |\cos x_2| < 1 \quad \text{and} \quad |a_2| = |\cos x_1| < 1.$$

Suppose that for an integer $k \geqslant 2$

$$\sin(x_1+x_2+\ldots+x_k) = a_1 \sin x_1+a_2 \sin x_2+\ldots+a_k \sin x_k,$$

where

$$|a_1| < 1, \quad |a_2| < 1, \quad \ldots, \quad |a_{k-1}| < 1 \quad \text{and} \quad |a_k| \leqslant 1$$

and let $0° < x_{k+1} < 180°$. Then

$$\sin(x_1+x_2+\ldots+x_k+x_{k+1}) = \sin(x_1+x_2+\ldots+x_k)\cos x_{k+1}+ \\ +\cos(x_1+x_2+\ldots+x_k)\sin x_{k+1}$$

$$= (a_1 \sin x_1+a_2 \sin x_2+\ldots+a_k \sin x_k)\cos x_{k+1}+ \\ +\cos(x_1+x_2+\ldots+x_k)\sin x_{k+1}$$

$$= (a_1 \cos x_{k+1})\sin x_1+(a_2 \cos x_{k+1})\sin x_2+ \\ +\ldots+(a_k \cos x_{k+1})\sin x_k+\cos(x_1+x_2+\ldots+x_k)\sin x_{k+1}$$

$$= b_1 \sin x_1+b_2 \sin x_2+\ldots+b_k \sin x_k+b_{k+1} \sin x_{k+1},$$

where

$$|b_1| = |a_1 \cos x_{k+1}| = |a_1|\,|\cos x_{k+1}| < 1;$$

similarly

$$|b_2| < 1, \ \ldots, \ |b_k| < 1, \ |b_{k+1}| = |\cos(x_1+x_2+\ldots+x_k)| \leqslant 1.$$

We conclude by induction that the lemma is true for every $n \geqslant 2$.

We now pass to the proof of the theorem proper; using the lemma, we can put it very briefly:

Since in the above equality

$$\sin(x_1+x_2+\ldots+x_n) = a_1 \sin x_1+a_2 \sin x_2+\ldots+a_n \sin x_n$$

the right-hand side is the sum of the products of the positive numbers $\sin x_1$, $\sin x_2$, ..., $\sin x_n$ by coefficients which, with the exception of the last, are less than 1, the last being at most equal to 1, we have

$$\sin(x_1+x_2+\ldots+x_n) < \sin x_1+\sin x_2+\ldots+\sin x_n.$$